T0136063

About Island Press

Since 1984, the nonprofit Island Press has been stimulating, shaping, and communicating the ideas that are essential for solving environmental problems worldwide. With more than 800 titles in print and some 40 new releases each year, we are the nation's leading publisher on environmental issues. We identify innovative thinkers and emerging trends in the environmental field. We work with world-renowned experts and authors to develop cross-disciplinary solutions to environmental challenges.

Island Press designs and implements coordinated book publication campaigns in order to communicate our critical messages in print, in person, and online using the latest technologies, programs, and the media. Our goal: to reach targeted audiences—scientists, policymakers, environmental advocates, the media, and concerned citizens—who can and will take action to protect the plants and animals that enrich our world, the ecosystems we need to survive, the water we drink, and the air we breathe.

Island Press gratefully acknowledges the support of its work by the Agua Fund, Inc., The Margaret A. Cargill Foundation, Betsy and Jesse Fink Foundation, The William and Flora Hewlett Foundation, The Kresge Foundation, The Forrest and Frances Lattner Foundation, The Andrew W. Mellon Foundation, The Curtis and Edith Munson Foundation, The Overbrook Foundation, The David and Lucile Packard Foundation, The Summit Foundation, Trust for Architectural Easements, The Winslow Foundation, and other generous donors.

The opinions expressed in this book are those of the author(s) and do not necessarily reflect the views of our donors.

Oceans and Marine Resources in a Changing Climate

A Technical Input to the
2013 National Climate Assessment

Note: This technical input document in its current form does not represent a Federal document of any kind and should not be interpreted as the position or policy of any Federal, State, Local, or Tribal Government or Non-Governmental entity

Suggested Citation: Griffis, R. and Howard, J. [Eds.]. 2013. *Oceans and Marine Resources in a Changing Climate: A Technical Input to the 2013 National Climate Assessment.* Washington, DC: Island Press.

Keywords: Climate change, climate variability, oceans, adaptation, extreme events, ocean acidification, loss of sea ice, sea surface temperature, precipitation, ocean circulation patterns, marine ecosystems, regime shifts, tipping points, fisheries, socio-economics, tourism, energy, international implications, decision making, vulnerability

This Technical Input was produced by a team of experts at the request of the NCA Development and Advisory Committee. It will be available for use as reference material by all NCA author teams.

The scientific results and conclusions, as well as any views or opinions expressed herein, are those of the author(s) and do not necessarily reflect the views of NOAA or the Department of Commerce.

Date Submitted to the National Climate Assessment Development Advisory Committee: April 17, 2012

This report is derived from Howard et al. 2013. Oceans and Marine Resources in a Changing Climate. *Oceanography and Marine Biology: An Annual Review,* **2013, 51, 71-192.**

Technical Inputs Lead Author Contact Information:
Roger Griffis: roger.b.griffis@noaa.gov 301-427-8134
Jennifer Howard: jennifer.howard@noaa.gov 301-427-8173

Front Cover Images: Courtesy of the National Oceanic and Atmospheric Administration.

About This Series

This report is published as one of a series of technical inputs to the Third National Climate Assessment (NCA) report. The NCA is being conducted under the auspices of the Global Change Research Act of 1990, which requires a report to the President and Congress every four years on the status of climate change science and impacts. The NCA informs the nation about already observed changes, the current status of the climate, and anticipated trends for the future. The NCA report process integrates scientific information from multiple sources and sectors to highlight key findings and significant gaps in our knowledge. Findings from the NCA provide input to federal science priorities and are used by U.S. citizens, communities and businesses as they create more sustainable and environmentally sound plans for the nation's future.

In fall of 2011, the NCA requested technical input from a broad range of experts in academia, private industry, state and local governments, non-governmental organizations, professional societies, and impacted communities, with the intent of producing a better informed and more useful report. In particular, the eight NCA regions, as well as the Coastal and the Ocean biogeographical regions, were asked to contribute technical input reports highlighting past climate trends, projected climate change, and impacts to specific sectors in their regions. Each region established its own process for developing this technical input. The lead authors for related chapters in the Third NCA report, which will include a much shorter synthesis of climate change for each region, are using these technical input reports as important source material. By publishing this series of regional technical input reports, Island Press hopes to make this rich collection of information more widely available.

This series includes the following reports:

Climate Change and Pacific Islands: Indicators and Impacts
Coastal Impacts, Adaptation, and Vulnerabilities
Great Plains Regional Technical Input Report
Climate Change in the Midwest: A Synthesis Report for the National Climate Assessment
Climate Change in the Northeast: A Sourcebook
Climate Change in the Northwest: Implications for Landscapes, Waters, and Communities
Oceans and Marine Resources in a Changing Climate
Climate of the Southeast United States: Variability, Change, Impacts, and Vulnerability
Assessment of Climate Change in the Southwest United States

Electronic copies of all reports can be accessed on the Climate Adaptation Knowledge Exchange (CAKE) website at *www.cakex.org/NCAreports*. Printed copies are available for sale on the Island Press website at *www.islandpress.org/NCAreports*.

Oceans and Marine Resources in a Changing Climate

A Technical Input to the
2013 National Climate Assessment

LEAD AUTHORS

Roger Griffis
National Oceanic and Atmospheric Administration

Jennifer Howard
National Oceanic and Atmospheric Administration

ISLANDPRESS

Washington | Covelo | London

Authors

SECTION 1: INTRODUCTION AND CONTEXT

Lead Author: Roger Griffis, National Oceanic and Atmospheric Administration

Jennifer Howard, AAAS Science and Technology Policy Fellow at the National Oceanic and Atmospheric Administration

SECTION 2: CLIMATE-DRIVEN PHYSICAL CHANGES IN MARINE ECOSYSTEMS

Lead Author: Jennifer Howard, AAAS Science and Technology Policy Fellow at the National Oceanic and Atmospheric Administration

Carol Auer, National Oceanic and Atmospheric Administration

Russ Beard, National Oceanic and Atmospheric Administration

Nicholas Bond, University of Washington

Tim Boyer, National Oceanic and Atmospheric Administration

David Brown, National Oceanic and Atmospheric Administration

Kathy Crane, National Oceanic and Atmospheric Administration

Scott Cross, National Oceanic and Atmospheric Administration

Bob Diaz, Virginia Institute of Marine Science

Libby Jewett, National Oceanic and Atmospheric Administration

Rick Lumpkin, National Oceanic and Atmospheric Administration

J. Ru Morrison, North East Regional Association of Coastal and Ocean Observing Systems

James O'Donnell, University of Connecticut

James Overland, National Oceanic and Atmospheric Administration

Rost Parsons, National Oceanic and Atmospheric Administration

Neal Pettigrew, University of Maine

Emily Pidgeon, Conservation International

Josie Quintrell, National Federation of Regional Associations for Ocean Observing Systems

Jeffrey Runge, University of Maine and Gulf of Maine Research Institute

Uwe Send, Scripps Institution of Oceanography (SIO)

Diane Stanitski, National Oceanic and Atmospheric Administration

Yan Xue, National Oceanic and Atmospheric Administration

SECTION 3: IMPACTS OF CLIMATE CHANGE ON MARINE ORGANISMS

Lead Authors: Brian Helmuth, University of South Carolina and Laura Petes, National Oceanic and Atmospheric Administration

Eleanora Babij, U.S. Fish and Wildlife Service

Emmett Duffy, Virginia Institute of Marine Science

Deborah Fauquier, National Oceanic and Atmospheric Administration

Michael Graham, Moss Landing Marine Laboratories

Anne Hollowed, National Oceanic and Atmospheric Administration

Jennifer Howard, AAAS Science and Technology Policy Fellow at the National Oceanic and Atmospheric Administration

David Hutchins, University of Southern California

Libby Jewett, National Oceanic and Atmospheric Administration

Nancy Knowlton, Smithsonian Institute

Trond Kristiansen, Institute of Marine Research

Teri Rowles, National Oceanic and Atmospheric Administration

Eric Sanford, Bodega Marine Laboratory, University of California at Davis

Carol Thornber, University of Rhode Island

Cara Wilson, National Oceanic and Atmospheric Administration

SECTION 4: IMPACTS OF CLIMATE CHANGE ON HUMAN USES OF THE OCEAN

Lead Authors: Amber Himes-Cornell, National Oceanic and Atmospheric Administration and Mike Orbach, Duke University

Stewart Allen, National Oceanic and Atmospheric Administration

Guillermo Auad, Bureau of Ocean Energy Management, Regulation

Mary Boatman, Bureau of Ocean Energy Management, Regulation

Patricia M. Clay, National Oceanic and Atmospheric Administration

Sam Herrick, National Oceanic and Atmospheric Administration

Dawn Kotowicz, National Oceanic and Atmospheric Administration

Peter Little, Pacific States Marine Fisheries Commission

Cary Lopez, National Oceanic and Atmospheric Administration

Phil Loring, University of Alaska, Fairbanks

Paul Niemeier, National Oceanic and Atmospheric Administration

Karma Norman, National Oceanic and Atmospheric Administration

Lisa Pfeiffer, National Oceanic and Atmospheric Administration

Mark Plummer, National Oceanic and Atmospheric Administration

Michael Rust, National Oceanic and Atmospheric Administration

Merrill Singer, University of Connecticut

Cameron Speirs, National Oceanic and Atmospheric Administration

SECTION 5: INTERNATIONAL IMPLICATIONS OF CLIMATE CHANGE

Lead Authors: Eleanora Babij, U.S. Fish and Wildlife Service and Paul Niemeier, National Oceanic and Atmospheric Administration

Brian Hayum, U.S. Fish and Wildlife Service

Amber Himes-Cornell, National Oceanic and Atmospheric Administration

Anne Hollowed, National Oceanic and Atmospheric Administration

Peter Little, Pacific States Marine Fisheries Commission

Mike Orbach, Duke University

Emily Pidgeon, Conservation International

SECTION 6: MANAGEMENT CHALLENGES, ADAPTATIONS, APPROACHES, AND OPPORTUNITIES

Lead Authors: Laura Petes, National Oceanic and Atmospheric Administration and Roger Griffis, National Oceanic and Atmospheric Administration

Jordan Diamond, Environmental Law Institute

Bill Fisher, U.S. Environmental Protection Agency

Ben Halpern, National Center for Ecological Analysis and Synthesis

Lara Hansen, EcoAdapt

Amber Mace, California Ocean Protection Council

Katheryn Mengerink, Environmental Law Institute

Josie Quintrell, National Federation of Regional Associations for Ocean Observing Systems

SECTION 7: SUSTAINING THE ASSESSMENT OF CLIMATE IMPACTS ON OCEANS AND MARINE RESOURCES

Lead Author: Roger Griffis, National Oceanic and Atmospheric Administration

Brian Helmuth, University of South Carolina

Jennifer Howard, AAAS Science and Technology Policy Fellow at the National Oceanic and Atmospheric Administration

Laura Petes, National Oceanic and Atmospheric Administration

Acknowledgements

This report was made possible by the generous assistance of many experts from a variety of fields who contributed their time and information. The author team thanks experts from the NOAA Fisheries Science Centers, academia, and other institutions who provided regional assessments, references, and other information. The team also thanks peer reviewers for their time and comments, which significantly improved the document. Fred Lipshultz, Ralph Cantal, and Anne Waple from the National Climate Assessment are greatly appreciated for their vision, leadership, support, and encouragement throughout the development of this report.

Contents

Key Terms

Adaptation – Adjustment in natural or human systems in response to actual or expected climatic stimuli or their effects for the purposes of moderating harm or exploiting beneficial opportunities.

Anthropogenic – Of, relating to, or resulting from the influence of human beings on nature.

Biodiversity – The variability among living organisms from all sources including terrestrial, marine, and other aquatic ecosystems and the ecological complexes of which they are a part; this includes diversity of ecosystems as well as diversity within and among species.

Blue Carbon – The carbon sequestered and stored by marine and coastal organisms including coastal seagrasses, tidal marshes, and mangroves.

Carbon sequestration – A long-term storage plan for carbon dioxide or other forms of carbon that has been proposed as method to slow the atmospheric and marine accumulation of greenhouse gases that are released by burning fossil fuels.

Climate – In a narrow sense, climate is the 'average weather.' More rigorously, climate is the statistical description in terms of the mean and variability of relevant quantities over a period of time ranging from months to thousands or millions of years. These quantities are most often surface variables such as temperature, precipitation, and wind. The classical period of time is 30 years, as defined by the World Meteorological Organization (WMO). In a wider sense, climate is the state, including a statistical description, of the climate system.

Climate Change – Climate change refers to any change in climate over time, whether as a result of natural variability or human activity.

Climate Modeling – Quantitative methods used to simulate the interactions of the atmosphere, oceans, land surface, and ice. They are used for a variety of purposes such as studying the dynamics of the climate system and generating projections of future climate.

Disaster – Severe alterations in the normal functioning of a community or a society due to the interaction of hazardous physical events and vulnerable social conditions, which leads to widespread adverse human, material, economic, or environmental effects that require immediate emergency response to satisfy critical human needs and may require external support for recovery.

Ecosystem – A biological environment consisting of all of the organisms living in a particular area as well as all of the nonliving, or abiotic, physical components of the environment, such as air, soil, water and sunlight, with which the organisms interact.

Ecosystem Services – The benefits people obtain from ecosystems. These include provisioning services such as food, water, timber, and fiber; regulation services such as the regulation of climate, floods, disease, wastes, and water quality; cultural services such as recreation, aesthetic enjoyment, and spiritual fulfillment; and supporting services such as soil formation, photosynthesis, and nutrient cycling.

Eutrophication – The movement of a body of water's trophic status in the direction of increasing biomass through the addition to an aquatic system of artificial or natural substances such as the nitrates and phosphates found in fertilizers or sewage.

Exclusive Economic Zone (EEZ) – A seazone over which a state has special rights related to the exploration and use of marine resources including the production of energy from water and wind. It stretches from the seaward edge of the state's territorial sea out to 200 nautical miles from its coast.

Exposure – The nature and degree to which a system is exposed to significant climatic variations.

Extreme Events – Includes weather phenomena that are at the extremes of the historical distribution, especially severe or unseasonal weather such as heat waves, drought, floods, storms, and wildfires.

Greenhouse Gas – A gas in an atmosphere that absorbs and emits radiation within the thermal infrared range. This process is the fundamental cause of the greenhouse effect. The primary greenhouse gases in the Earth's atmosphere are water vapor, carbon dioxide, methane, nitrous oxide, and ozone.

Hydrology – The movement, distribution, and quality of water on Earth and other planets, including the hydrologic cycle, water resources, and environmental watershed sustainability.

Hypoxia – A phenomenon that occurs in aquatic environments as the concentration of dissolved oxygen becomes reduced to a point that is detrimental to aquatic organisms living in the system.

Invasive Species – Non-indigenous species of plants or animals that adversely affect the habitats and bioregions that they invade economically, environmentally, and/or ecologically. They disrupt by dominating regions, wilderness areas, particular habitats, and/or wildland-urban interface land and causing a loss of natural controls.

Large Marine Ecosystems (LMEs) – Areas of the ocean characterized by distinct depth, hydrology, productivity, and trophic interactions.

Maximum Sustainable Yield (MSY) – The largest long-term average catch or yield that can be taken from a stock or stock complex under prevailing ecological and environmental conditions.

Mitigation – An anthropogenic intervention to reduce anthropogenic influences on the climate system. Mitigation includes strategies to reduce greenhouse gas sources and emissions and enhance greenhouse gas sinks.

Ocean Acidification – The ongoing decrease in pH and increase in acidity of the Earth's oceans, caused by the uptake of anthropogenic carbon dioxide (CO_2) from the atmosphere.

Resilience – The ability of a system and its component parts to anticipate, absorb, accommodate, and recover from the effects of a hazardous event in a timely and efficient manner through ensuring the preservation, restoration, or improvement of its essential basic structures and functions.

Resistance – The capacity of the ecosystem to absorb disturbances and remain largely unchanged.

Risk – The combination of the probability of an event and its consequences.

Risk Assessment – The determination of quantitative or qualitative value of risk related to a concrete situation and a recognized threat such as climate change.

Sensitivity – The degree to which a system is affected, either adversely or beneficially, by climate variability or change. The effect may be direct,such as a change in crop yield in response to a change in the mean, range, or variability of temperature, or indirect, such as damages caused by an increase in the frequency of coastal flooding due to sea-level rise.

Stakeholders – A person, group, organization, or system who affects or can be affected by an organization's actions.

Stratification – Water stratification occurs when water masses with different properties such as salinity, oxygenation, density, or temperature form layers that act as barriers to water mixing. These layers are normally arranged according to density, with the least dense water masses sitting above the more dense masses.

Stock Assessment – The process of collecting and analyzing demographic information about fish populations to describe the condition or status of a fish stock. A stock assessment results in a report that often includes an estimation of the amount or abundance of the resource, the rate at which it is being removed due to harvesting and other causes, and one or more reference levels of harvesting rate and/or abundance at which the stock can maintain itself in the long term.

Transformation – The alteration of fundamental attributes of a system including value systems; regulatory, legislative, or bureaucratic regimes; financial institutions; and technological or biological systems.

Upwelling – The oceanographic phenomenon in which dense, cooler, and usually nutrient-rich water is driven towards the ocean surface by wind, replacing the warmer, usually nutrient-depleted surface water.

Vulnerability – The propensity or predisposition to be adversely affected.

Acronyms

ACAP – Agreement for the Conservation of Albatross and Petrels

ACL – Annual Catch Limits

ADFG – Alaska Department of Fish and Game

AMO – Atlantic Multidecadal Oscillation

AMSA – Arctic Marine Shipping Assessment

BOEM – Bureau of Ocean Energy Management

BSAI – Bering Sea–Aleutian Islands

CAKE – Climate Adaptation Knowledge Exchange

CBD – Convention on Biological Diversity

CCAMLR – Commission for the Conservation of Antarctic Marine Living Resources

CDC – Centers for Disease Control and Prevention

CDM – Clean Development Mechanisms

CFP – Ciguatera Fish Poisoning

CI – Conservation International

CITES – Convention on International Trade in Endangered Species

CMS – Convention on Migratory Species

CMSP – Coastal and Marine Spatial Planning

CO_2 – Carbon Dioxide

CoP – Conference of the Parties, United Nations Framework Convention on Climate Change

CREST – Coral Reef Ecosystem Studies

CWA – Clean Water Act

CyanoHABs – Cyanobacterial Blooms

EBS – Eastern Bering Sea

EEDP – Energy, Environment and Development Program of the United Nations

EEZ – Exclusive Economic Zone

ENSO – El Niño Southern Oscillation

EPA – U.S. Environmental Protection Agency

FDA – U.S. Food and Drug Administration

FFA – South Pacific Forum Fisheries Agency

FMP – Fishery Management Plans

GAO – U.S. Government Accountability Office

GEF – Global Environment Facility

GET – General Excise Tax

GHG – Greenhouse Gas

GOA – Gulf of Alaska

HABs – Harmful Algal Blooms

IAC – Inter-American Convention for the Protection and Conservation of Sea Turtles

ICCATF – Interagency Climate Change Adaptation Task Force

IGO – Inter-Governmental Organizations

IOC – Intergovernmental Oceanic Commission

IPCC – Intergovernmental Panel on Climate Change

IUCN – International Union for Conservation of Nature

IWC – International Whaling Commission

LME – Large Marine Ecosystems

LULUCF – Land-Use, Land-Use Change and Forestry

MPA – Marine Life Protection Act

MPA – Marine Protected Area

MSY – Maximum Sustainable Yield

NAFO – Northwest Atlantic Fisheries Organization

NAMA – National Appropriate Mitigation Actions

NAO – North Atlantic Oscillation

NASCO – North Atlantic Salmon Conservation Organization

NCA – National Climate Assessment

NEUS – Northeast U.S. Shelf Ecosystem

NGO – Non-Governmental Organizations

NMFS – National Marine Fisheries Service (a division within NOAA)

NOAA – National Oceanic and Atmospheric Administration

NPAFC – North Pacific Anadromous Fish Commission

NPCREP – North Pacific Climate Regimes and Ecosystem Productivity

NPFMC – North Pacific Fishery Management Council

NS – National Standards

PDO – Pacific Decadal Oscillation

PEIS – Programmatic Environmental Impact Statement

PFMC – Pacific Fishery Management Council

PICES – North Pacific Marine Science Organization

PICTs – Pacific Island Countries and Territories

PLA – Participatory Learning Assessment

PWS – Prince William Sound

REDD – Reducing Emissions from Deforestation and Forest Degradation

RFMO – Regional Fisheries Management Organizations

SHARC – Subsistence Halibut Registration Certificate

SSB – Spawning Stock Biomass

SST – Sea Surface Temperature

TAC – Total Allowable Catch

TAT – Transient Accommodations Tax

THC – Thermohaline Circulation

UNCLOS – United Nations Convention on the Law of the Sea

UNESCO – United Nations Educational, Scientific and Cultural Organization

UNFCCC – United Nations Framework Convention on Climate Change

USCOP – U.S. Commission on Ocean Policy

USFWS – U.S. Fish and Wildlife Service

USGCRP – U.S. Global Change Research Program

WCPFC – Western and Central Pacific Fisheries Commission

WECAFC – Western Central Atlantic Fishery Commission

WHO – World Health Organization

WIDECAST – Wider Caribbean Sea Turtle Conservation -Network

ZSL – Zoological Society of London

Communicating Uncertainty

Based on the Guidance Note for Lead Authors of the third U.S. National Assessment, this technical input document relies on two metrics to communicate the degree of certainty, based on author teams' evaluations of underlying scientific understanding, in key findings:

- Confidence in the validity of a finding by considering (i) the quality of the evidence and (ii) the level of agreement among experts with relevant knowledge.

Table 1: Communicating Uncertainty

Confidence level	Factors that could contribute to this confidence evaluation
High	Strong evidence (established theory, multiple sources, consistent results, well documented and accepted methods, etc.), high consensus
Moderate	Moderate evidence (several sources, some consistency, methods vary and/or documentation limited, etc.), medium consensus
Fair	Suggestive evidence (a few sources, limited consistency, models incomplete, methods emerging, etc.), competing schools of thought
Low	Inconclusive evidence (limited sources, extrapolations, inconsistent findings, poor documentation and/or methods not tested, etc.), disagreement or lack of opinions among experts

- Probabilistic estimate of uncertainty expressed in simple quantitative expressions or both the quantitative expressions and the calibrated uncertainty terms.

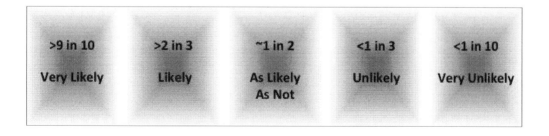

Executive Summary

The United States (U.S.) is an ocean nation; our past, present and future are inextricably connected to and dependent on oceans and marine resources. Marine ecosystems under U.S. jurisdiction extend from the shore to 200 nautical miles seaward, include waters around U.S. Territories, and span 3.4 million square nautical miles of ocean, an area referred to as the U.S. exclusive economic zone (NMFS, 2009a), which is an area 1.7 times the land area of the continental U.S.. This area encompasses an incredible diversity of species and habitats as well as 11 different large marine ecosystems (LMEs) that provide many important services, including jobs, food, transportation routes, recreational opportunities, health benefits, climate regulation, and cultural heritages that affect people, communities, and economies across America and internationally every day.

U.S. marine ecosystems and services are increasingly at risk from climate change and other human pressures. A wealth of information documents strong linkages between the planet's climate and ocean systems as well as changes in the climate system that can produce dramatic changes in the physical, chemical, and biological characteristics of ocean ecosystems on a variety of spatial and temporal scales. Additionally, a growing body of literature provides evidence of the current impacts of increasing atmospheric carbon dioxide and the associated global warming and ocean acidification on physical, chemical, and biological components of ocean ecosystems. Conversely, relatively little information shows how these climate-driven changes in ocean ecosystems may impact ocean services and uses, although it is predicted that the vulnerability of ocean-dependent users, communities, and economies increases in a changing climate. In addition, non-climatic stressors resulting from a variety of human activities, including pollution, fishing impacts, and over-use, can interact with and exacerbate impacts of climate change. Collectively, climatic and non-climatic pressures on marine ecosystems are having profound and diverse impacts that are expected to increase in the future.

Based on current understanding of these linkages between the planet's climate and ocean systems as well as projections of future changes in the global climate system, the marine ecosystems under U.S. jurisdiction and U.S. interest internationally are likely to continue to be affected by anthropogenic-driven climate change and rising levels of atmospheric CO_2. These impacts are set in motion through a collection of changes in the ocean's physical (e.g., temperature, circulation, stratification, upwelling), chemical (e.g., acidification, nutrient input, oxygen content), and biological (e.g., primary production, species distributions, phenology, foodweb structure, community composition and ecosystem functions/services) components and processes. Given the value and U.S. dependence on ocean products and ecosystem services found in U.S. and international ocean ecosystems, these climate-driven changes are likely to have significant implications for U.S ocean uses and international activities as well as the communities and economies that depend on them. Effectively responding to these challenges requires mitigating the pace and scope of climate change through concerted efforts to reduce emissions of greenhouse gases and protect and enhance those natural environments that act as carbon sinks as well as helping species, ecosystems, and the societies that depend on them to adapt to the changes we can no longer avoid.

This report was produced by a team of experts as a contribution to the third National Climate Assessment (NCA), conducted under the auspices of the U.S. Global Change Research Program (USGCRP). The report provides an assessment of scientific knowledge of the current and projected impacts of climate change and ocean acidification on the physical, chemical, and biological components and human uses of marine ecosystems under U.S. jurisdiction. It also provides assessment of the international implications for the U.S. due to climate impacts on ocean ecosystems and of efforts to prepare for and adapt to climate and acidification impacts on ocean ecosystems.

U.S marine ecosystems are inherently connected to U.S. coastal and terrestrial areas through the flow of water from land to sea, the effect of oceans on the physical climate system, the connectivity and movement of species, and the extensive and diverse uses of marine resources and services that occur throughout the Nation. Therefore, climate impacts on oceans and marine resources intersect with many regions and sectors that are also considered in the NCA.

The following is a summary of key findings from each of the six major sections of this report.

Chapter 2: Climate-Driven Physical and Chemical Changes in Marine Ecosystems

1. The Earth's oceans are warming as a result of increasing concentrations of atmospheric carbon dioxide and other greenhouse gases.

   - From 1955 to 2008, approximately 84 percent of the added heat from climate change has been absorbed by the oceans, thereby increasing the average temperature of the upper 700 meters of water by 0.2° Celsius; this trend is likely to continue.

   - Rising global temperatures are likely to cause greater oceanic evaporation of water vapor, which is a potent greenhouse gas, into the atmosphere, thus amplifying warming due to climate change.

2. Arctic ice has been decreasing in extent since the second half of the 20th century as a result of oceanic and atmospheric warming.

   - Arctic ice has been decreasing throughout the early 21th century. The summer of 2012 saw a record low, when sea ice extent shrank to 3.6 million km2, approximately 1 million km2 less than the previous minima of 2007. Arctic sea ice volume has decreased by 75% over the previous decade.

   - The most significant consequence of melting Arctic ice will be sea level rise. A growing body of recently published work suggests that accelerating loss of the great polar ice sheets in Greenland and Antarctica will likely result in global mean sea level rise of more than 1 meter above present day sea level by 2100.

   - Reductions in ice may occur more rapidly than previously suggested by coupled air-sea-ice climate models. These models may have overestimated ice thickness; more recent modeling predicts that a seasonal ice-free state could occur as early as 2030.

3. The oceans play an important role in climate regulation through the uptake and sequestration of anthropogenic CO_2.

   - The annual accumulation of atmospheric CO_2 has increased. In 2010, the overall CO_2 concentration was 39 percent above the concentration at the start of the Industrial Revolution in 1750.
   - Ocean water holds approximately 50 times more CO_2 than the atmosphere and holds the majority of that in the deeper, colder waters; however, the ability of oceans to absorb CO_2 decreases with increasing temperature and decreasing pH.
   - Healthy coastal wetlands are effective at storing and sequestering carbon. This carbon is referred to as "coastal blue carbon." Unfortunately, our nation has lost more than half of its wetlands in the past 200 years.

4. The oceans absorb carbon dioxide, causing a series of chemical reactions that reduce ocean pH, a process known as ocean acidification.

   - Over the past century, the average ocean pH has decreased by 30 percent; however, in coastal areas this number may be much higher following upwelling events.
   - Predictions indicate that ocean acidification will lead to an under-saturation of argonite in ocean surface waters by the year 2050. Argonite is a form of calcium carbonate impacted by the ocean acidification process and a key component of the exoskeleton of corals and some plankton.
   - Polar ecosystems may experience exacerbated ocean acidification because CO_2 is more soluble in colder water and therefore more likely to be absorbed.

5. Ocean temperature and precipitation/evaporation rates have a direct influence on the Earth's weather and precipitation patterns that will likely increase storm intensity but decrease (or have no effect on) storm frequency.

   - Climate change impacts on winds may be moderate and vary regionally but precipitation events are predicted to grow in intensity.
   - Shifts in precipitation patterns from snow to rain are likely, resulting in decreased snow pack and earlier snow melts that may impact water availability.
   - Storm intensity is likely to increase and could have a large impact, especially on coastal communities because of the increases in coastal populations and infrastructure expected over the next century. The impact of storm surges will be greater for the same reasons as well as the expected rise in sea level.

6. Given that climate and oceans interact to produce intra- and inter-decadal variability in ocean currents, upwelling, and basin-scale circulation, climate change will likely influence these important ocean features, although the ability to detect and project such impacts is low.

   - Melting of polar ice will reduce the salinity, and thus the density, of polar waters, which could weaken the rate at which this water sinks, possibly impairing circulation.

- However, uncertainty remains about the quantity of freshwater input necessary to slow ocean currents like the Thermohaline Circulation.

Chapter 3: Impacts of Climate Change on Marine Organisms

1. Throughout U.S. ocean ecosystems, climate change impacts are being observed that are highly likely to continue into the future.

   - Observed impacts include shifts in species distributions and ranges, effects on survival, growth, reproduction, health, and the timing of life-history attributes, and alterations in species interactions, among others.

   - Impacts are occurring across a wide diversity of taxa in all U.S. ocean regions, but high-latitude and tropical areas appear to be particularly vulnerable.

2. The vulnerability and responses of marine organisms to climate change vary widely, leading to species that are positively impacted, or "winners," and species that are negatively impacted, or "losers."

   - Species with more elastic life histories and/or higher physiological tolerance for changes in temperature and other environmental conditions will likely experience fewer climate-related impacts and may therefore out-compete less tolerant species.

   - Species such as corals and other calcifying organisms that are exposed to ocean conditions that inhibit their ability to produce calcium carbonate shells, will likely experience impacts. These impacts may result in cascading effects on marine ecosystems.

3. Climate change interacts with and can exacerbate the impacts on ocean ecosystems of non-climatic stressors such as pollution, overharvesting, disease and invasive species.

   - Climate-related stressors such as changes in temperature can operate as threat multipliers, worsening the impacts of non-climatic stressors.

   - Opportunities exist for ameliorating some of the impacts of climate change through reductions in non-climatic stressors at local-to-regional scales.

   - Predicting the effects of multiple stressors is difficult given the complex physiological effects and interactions between species.

4. Past and current responses of ocean organisms to climate variability and climate change are informative, but extrapolations to future environmental conditions will differ from the range observed in recent history.

   - Observed responses to ongoing environmental change often vary in magnitude across space and time, suggesting that extrapolations of responses from one location to another may be challenging.

   - In addition to gradual declines in ecosystem function, potential threshold effects, or "tipping points," that could result in rapid ecosystem change are a particular area of concern.

Chapter 4: Impacts of Climate Change on Human Uses of the Ocean

1. Significant effects of climate change on all sectors pertaining to human uses of the ocean, including but not limited to fisheries, energy, transportation, security, human health, tourism, and maritime governance, are already being observed and are predicted to continue into the future.

   - Some effects are predicted to be "positive," in that they expand the extent of individual sectors, while others are predicted to be "negative," in that they reduce the ability of humans to use the ocean in a given sector, and virtually all effects will result in some distributional changes in how and where, as well as by whom, marine resources are used.

   - Most of the effects of climate change on U.S. fisheries will stem from the changes to the fish stocks brought about by direct and indirect climate impacts on productivity and location; others stem from impacts that climate has on the fisheries themselves as well as fishing-dependent communities across the country, which could experience changes in distribution and abundance of their available stocks.

   - Aquaculture stocks are expected to be more resilient to climate change than wild stocks due to selective breeding and vaccination.

   - The main climate-related effects on the oil and gas industry are anticipated to be increases in demands on water sources, failure of infrastructure not designed to withstand new climatic conditions, and changes in the extraction of financial assets and resources as energy production moves from the traditional oil and gas industry to renewable sources of energy.

   - In the face of climate change, impacts to marine resource distribution, variable weather conditions, and extreme events such as typhoons and hurricanes are expected to pose the most significant impacts on the tourism industry; these effects will be positive in some regions of the country, negative in others, and mixed in others.

   - The scale and scope of climate impacts such as increased economic access and ecosystem shifts in polar areas (Arctic, Antarctic) are already having significant impacts on ocean uses and users. There is a high likelihood that these impacts will grow and have serious consequences as well as opportunities on human uses of the oceans in the future.

2. As a result of climate change, governance regimes for ocean environments and resources, as well as human health, will be challenged, and will likely have to change significantly in character and configuration.

   - Geographic shifts in the distribution of fishery resources, ocean energy exploitation, and ocean transportation routes may require restructuring policy arenas for those resources and may require amending policy and governance systems with new statutory law.

3. Ocean "health," both in terms of the functioning of biophysical ocean ecosystems and the factors related to oceans and human health will likely be affected by climate change.

  • The distribution and abundance of water-born human disease vectors are expected to change significantly as a result of climate change.

  • The potential for climate change to affect the boundaries and determinants of human health is significant, including increases in extreme weather-related injuries, morbidity and mortality, declines in access to drinkable water, increased food insecurity and malnutrition, rising pollutant-related respiratory problems, and spread of infectious disease.

4. Impacts of climate change on human social and economic systems provide critical insight into societal responses and adaptation options.

  • Additional disciplinary and interdisciplinary research will be necessary to improve understanding of the interactions between physical, biological, economic, and social systems in the future.

Chapter 5: International Implications of Climate Change

1. Many migratory species that span jurisdictional boundaries are exhibiting shifts in distribution and abundance.

  • Current policy and structured agreements assume species stationarity, which is no longer the case.

  • Many species will continue to shift significantly, expanding their ranges in countries where they previously have not lived. As a result, current protected area networks may need to be expanded to match critical sites needed in the future.

  • Several global conventions, as well as regional and bilateral migratory species agreements, are well-placed to support coordinated, multi-country climate change adaptation efforts on behalf of migratory species.

2. International partnerships will be necessary to ensure that monitoring protocols, management plans, and training programs are robust and coordinated for effective long-term implementation on shared marine resources.

  • Multinational large-scale and long-term work is needed to better understand the risks to oceans and marine resources as a result of climate change.

  • Strengthening synergies across existing treaties and conventions would provide better coordination and improved focus and facilitate the development of shared priorities.

3. Although aware of climate change, only half of the existing Regional Fisheries Management Organizations (RFMO) have addressed the issue.

  • Agreements, both multilateral and bilateral, will need flexibility to adapt to changing circumstances, particularly unanticipated, climate-driven changes in stock levels or distribution across Exclusive Economic Zones or high-seas areas.

- The potential for spatial displacement of aquatic resources and people as a result of climate change impacts will require existing regional structures and processes to be strengthened or enhanced.

4. Climate change will affect transportation and security issues in both the short and long term.

- Changes in available shipping lanes in the Arctic created by a loss of sea ice have generated an expanded geopolitical discussion involving the relationship among politics, territory, and state sovereignty on local, national, and international scales.

5. Accounting for the carbon sequestration value of coastal marine systems has the potential to be a transformational tool in the implementation of improved coastal policy and management.

- A number of countries including Indonesia, Costa Rica, and Ecuador have identified "Blue Carbon" as a priority issue and are currently developing strategies and approaches.

Chapter 6: Ocean Management Challenges, Adaptation Approaches, and Opportunities in a Changing Climate

1. Climate change presents both challenges and opportunities for marine resource managers and decision makers.

- In comparison to terrestrial, freshwater, and coastal systems, relatively few adaptation actions have been designed and implemented for marine systems.
- Barriers to ocean adaptation currently exist and include a lack of usable scientific information, awareness, and institutional capacity.
- Despite barriers, creative solutions are emerging for advancing adaptation planning and implementation for ocean systems.

2. Ocean-related climate information, tools, and services are being developed to address the needs of decision makers and managers.

- Long-term observations and monitoring of ocean physical, ecological, social, and economic systems provide essential information on past and current trends as well as insight into future conditions. An impending challenge is ensuring that the provision and accessibility of high-quality information regarding resolutions is commensurate with the scales at which adaptation decisions are made.
- User-friendly tools, guidance, and services are emerging to foster dialogue, grow communities of practice, and inform and support decisions to enhance ocean resilience in the face of climate change.

3. Opportunities for adaptation include incorporation of climate change into existing ocean policies, practices, and management efforts.

- Climate considerations can be integrated into fisheries management and spatial planning, as well as the design of marine protected areas to enhance ocean resilience and adaptive capacity.
- Application of existing legislative and regulatory frameworks can advance climate adaptation efforts in the marine environment.

4. Progress is being made across the U.S. to enhance ocean resilience to climate change through local, state, national, federal, and non-governmental adaptation frameworks and actions; however, much work still remains.

Chapter 7: Sustaining the Assessment of Climate Impacts on Oceans and Marine Resources

1. Sustained assessment of climate impacts on U.S. ocean ecosystems is critical to understanding current impacts, preparing for future impacts, and taking action for effective adaptation to a changing climate.
2. Assessing what is known about past, current, and future impacts is a challenging task for a variety of reasons. Many uncertainties and gaps still exist in understanding about the current and future impacts of climate change and ocean acidification on marine ecosystems.
3. A number of steps could be taken in the near term to address these challenges and advance assessment of impacts of climate change on oceans and marine resources.

- Identify and collect information on a set of core indicators of the condition of marine ecosystems that can specifically be used to track and assess the impacts of climate change and ocean acidification as well as the effectiveness of mitigation and adaptation efforts over time at regional to national scales.

- Assess capacity and effectiveness of existing ocean-observing systems to detect, track and deliver useful information on these indicators as well as other physical, chemical, biological, and social/economic impacts of climate change on oceans and marine resources.

- Increase capacity and coordination of existing observing systems to collect, synthesize, and deliver integrated information on physical, chemical, biological, and social/economic impacts of climate change on U.S. marine ecosystems.

- Conduct regional-scale assessment of current and projected impacts of climate change and ocean acidification on ocean physical, chemical, and biological components and human uses.

- Increase data, information, and capacity needed to assess and project impacts of climate change on marine ecosystems.

- Build and support mechanisms for sustained coordination and communication between decision makers and science providers to ensure the most critical information needs related to impacts, vulnerabilities, mitigation, and adaptation of ocean ecosystems in a changing climate are being met.

- Build and support mechanisms for getting and sharing information and resources on impacts, vulnerabilities, and adaptation of U.S. ocean ecosystems in a changing climate.
- Build and support mechanisms with neighboring countries and other international partners for assessing and addressing impacts of climate change and ocean acidification on marine ecosystems of key interest to the U.S.

Chapter 1

Introduction

The U.S. is an ocean nation—our past, present and future are inextricably connected to and dependent on oceans and marine resources. Marine ecosystems of the U.S. support an incredible diversity of species and habitats (NMFS, 2009a,b) and provide many valuable ecosystem services, including jobs, food, transportation routes, recreational opportunities, health benefits, climate regulation, and cultural heritages, that affect people, communities, and economies across America every day and that affect the nation's international relations in many ways (NOC, 2012; NMFS, 2011; U.S. USCOP, 2004). In 2004, the ocean-dependent economy, which is divided into six industrial sectors, generated $138 billion or 1.2 percent of U.S. Gross Domestic Product (GDP) (Kildow et al., 2009). U.S. ocean areas are also inherently connected with the nation's vital coastal counties, which make up only 18 percent of the U.S. land area but are home to 36 percent of the U.S. population and account for over 40 percent of the national economic output (Kildow et al., 2009).

Marine ecosystems under U.S. sovereignty generally extend from the shore to 203 nautical miles seaward including areas under State (0-3 nautical miles except 0-9 nautical miles off the shores of Texas, the Gulf Coast of Florida, and Puerto Rico) and federal (3-200 nautical miles) jurisdiction. The area under federal jurisdiction spans 3.4 million square nautical miles of ocean, an area referred to as the U.S. exclusive economic zone (EEZ) (National Marine Fisheries Service, 2009a). The U.S. has the largest EEZ in the world, an area 1.7 times the land area of the continental U.S. and encompassing 11 different large marine ecosystems (LMEs) (Figure 1-1).

These valuable marine ecosystems and services are increasingly at risk from a variety of human pressures, including climate change and ocean acidification. Climate change and acidification are affecting oceans in a number of ways over multiple temporal and spatial scales (Figure 1-2a) (Doney et al., 2012; Osgood, 2008). In addition, non-climatic stressors resulting from a variety of human activities, including pollution, fishing impacts, and over-use, can interact with and exacerbate impacts of climate change. Collectively, climatic and non-climatic pressures are having profound and diverse impacts on marine ecosystems (Figure 1-2b). These impacts are expected to increase in the future with continued changes in the global climate system and increases in human population levels.

Climate change is affecting ocean physical, chemical, and biological systems, as well as human uses of these systems. Rising levels of atmospheric CO_2 is one of the most serious problems because its effects are globally pervasive and irreversible on ecological timescales (NRC, 2011). The two primary direct consequences of increased atmospheric CO_2 in marine ecosystems are increasing ocean temperatures (IPCC, 2007a) and acidity (Doney et al., 2009). Increasing temperatures produce a variety of other ocean changes including rising sea level, increasing ocean stratification, decreased extent of sea ice, and

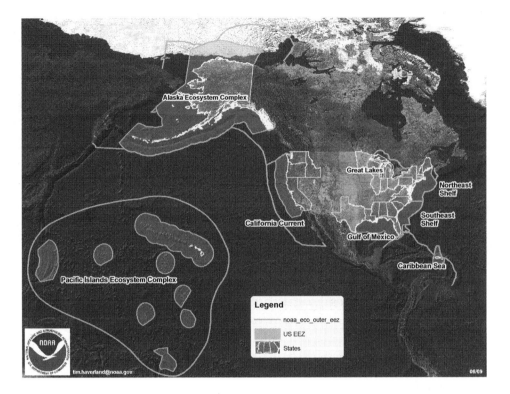

Figure 1-1 Large marine ecosystems within the U.S. Exclusive Economic Zone (NOAA Fisheries 2009c).

altered patterns of ocean circulation, storms, precipitation, and freshwater input (Doney et al., 2012). These and other changes in ocean physical and chemical conditions, such as changes in oxygen concentrations and nutrient availability, are impacting a variety of ocean biological features including primary production, phenology, species distribution, species interactions, and community composition, which in turn can impact vital ocean services across the Nation (Figure 1-3). Projections of future change show that it is likely that marine ecosystems under U.S. jurisdiction and U.S. interest internationally will continue to be affected by anthropogenic-driven climate change and rising levels of atmospheric CO_2. Interactions of climate impacts vary by region and complexity. Figure 1-4 is an illustrative example of this in the California Current.

1.1 Scope and Purpose

This report provides an assessment of current scientific knowledge on the climate impacts, vulnerabilities, and adaptation efforts related to U.S. oceans and marine resources. The report was produced by a team of experts charged with synthesizing and assessing climate-related impacts on U.S. oceans and marine resources as a contribution to the third National Climate Assessment (NCA), which was conducted under the auspices of the U.S. Global Change Research Program (USGCRP). The U.S. Global Change Research Act of 1990 requires that periodic national climate assessments be conducted and submitted to the President and the Congress. Two previous national assessment reports published in 2000 and 2009 contained little information on climate impacts on U.S. oceans and marine resources. This report is intended to increase understanding and emphasis on this topic for the 2013 NCA.

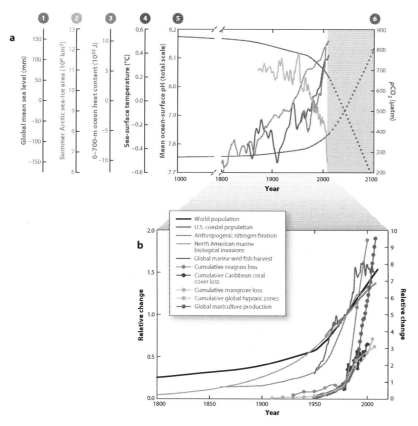

Figure 1-2 (a) Changes in (1) global mean sea level (data starting in 1800 with an upward trend; Jevrejeva et al., 2008), (2) summer Arctic sea-ice area (data starting just prior to 1900 with a downward trend; Walsh and Chapman, 2001),(3) 0-700-m ocean heat content (data starting around the mid 1900's with an upward trend; Levitus et al., 2009),(4) sea-surface temperature (data starting around the mid-1800's with a general upward trend; Rayner et al., 2006), (5) mean ocean surface pH (data starting around 1000 with an downward trend into the future; NRC, 2010b), and (6) pCO_2 (data starting around 1000 with an upward trend tinto the future; Petit et al., 1999). Shaded region denotes projected changes in pH and pCO_2 consistent with the Intergovernmental Panel on Climate Change's 21st-century A2 emissions scenario with rapid population growth. **(b)** Time series (as identified in figure key): trends in world population (solid line, data starting in the 1800s with an upward trend; Goldewijk, 2005), U.S. coastal population (solid line, data staring in the 1950s with a general upward trend; Wilson and Fischetti, 2010), anthropogenic nitrogen fixation (solid line, data starting in the late 1850s with a general upward trend; Davidson, 2009), North American marine biological invasions (solid line, data starting in the 1800s with a general upward trend; Ruiz et al., 2000), global marine wild fish harvest (solid line, data starting in the 1950s with a general upward trend; Food Agricultural Organization [FAO] U.N., 2010), cumulative seagrass loss (dotted line, data starting around the mid 1920's with a general upward trend and a sharp increase after the mid 1970s; Waycott et al., 2009), cumulative Caribbean coral cover loss (dotted line, data starting around the mid 1970s with a general upward trend; Gardner et al., 2003), cumulative mangrove loss (dotted line, data starting around the mid 1920's with a general upward trend and a sharp increase after the mid 1970s; FAO U.N., 2007), cumulative global hypoxic zones (dotted line, data starting in the early 1900's with a general upward trend; Diaz and Rosenberg, 2008), and global mariculture production (dotted line, data starting around 1950 with an upward trend; FAO U.N., 2010). All time series in (b) are normalized to 1980 levels. Trends with <1.5-fold variation are depicted as solid lines (left axis), and trends with >1.5-fold variation are depicted as dotted lines (right axis) (Source: Doney et al., 2012).

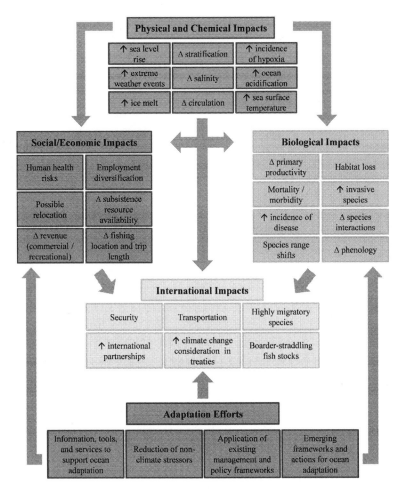

Figure 1-3 Imapcts of Climate Change on Marine Ecosystems. This table is intended to provide illustrative examples of how climate change is currently affecting U.S. ecosystems, the species they support, and the resulting impacts on ocean services. It is not intended to be comprehensive or to provide any ranking or prioritization. Black arrows represent impacts driven by climate change either directly or indirectly. Gray arrows represent countering effects of various adaptation efforts. ↑ indicates where climate change is predicted to increase the incidence or magnitude of that attribute and Δ indicates attributes where the impact of climate change on that attribute is variable.

This assessment is organized into the following major Sections:

- Sections 2-4 assess the state of knowledge on the impacts and vulnerabilities of ocean physical and chemical conditions (Section 2), biological systems (Section 3), and ocean uses and services (Section 4) in a changing climate.

- Section 5 assesses the international implications of these climate impacts and vulnerabilities because U.S. oceans and marine resources are inherently connected to ocean areas beyond U.S. borders.

- Section 6 assesses the status of efforts to prepare for and adapt to the impacts of climate change on U.S. oceans and marine resources.

- Section 7 identifies key steps to sustain and advance assessment of climate impacts on U.S. oceans and marine resources

1.2 Linkages with Other Parts of the National Climate Assessment

U.S marine ecosystems are inherently connected to U.S. coastal and terrestrial areas through many important linkages including the:

- Flow of water, organic matter, and sediments from land to sea;
- Effect of oceans on the physical climate system, including water, wind, and heat energy;
- Connectivity and movement of species; and
- Extensive and diverse uses of marine resources and services that occur throughout the Nation.

This means that climate impacts on ocean ecosystems intersect with, and have major implications for, many regions and sectors across the nation that are also considered in the NCA. As part of larger marine ecosystems and global oceans, U.S. marine ecosystems influence and are strongly influenced by ocean conditions beyond U.S. jurisdiction. This means that changes in these systems can have implications for U.S. efforts internationally.

The following is a brief summary of some of the key intersections with other parts of the NCA:

- **Regional Assessments:** Seven of the eight regions of the NCA include coastal areas and marine ecosystems. Climate change impacts on marine ecosystems may have significant implications in these regions especially for marine-dependent species, habitats, users, and communities;
- **Coastal Areas:** The oceans and marine resources considered in this report are directly tied to species, processes, and services of the coastal zone. Climate change impacts on marine ecosystems have significant implications for coastal areas, especially for marine-dependent species, habitats, users, and communities;
- **Public Health:** Marine ecosystem conditions can directly impact public health through harmful algal blooms, contaminated seafood, the spread of disease, and other mechanisms. Climate change impacts that increase these conditions in marine ecosystems could have significant implications, especially in coastal areas;
- **Transportation:** Marine transportation is critical to the nation's economy, health, and safety, as well as national security. Climate change impacts on marine ecosystems, such as changes in ocean circulation, storms, and other features, could have significant implications for the nation's vital marine transportation system;
- **Energy Supply:** The nation's energy supply from marine-related sources is increasing and may grow as the nation seeks alternative sources of energy.

Climate change impacts on marine ecosystems, such as changes in ocean circulation, storms, and other features, could have significant implications for the ocean
energy sector; and

- **Ecosystems and Biodiversity:** Marine ecosystems are some of the nation's most
complex, biologically rich, and valued ecosystems. Climate change is already
impacting marine ecosystems and biodiversity and these impacts are expected
to increase with significant implications for communities and economies dependent on marine resources.

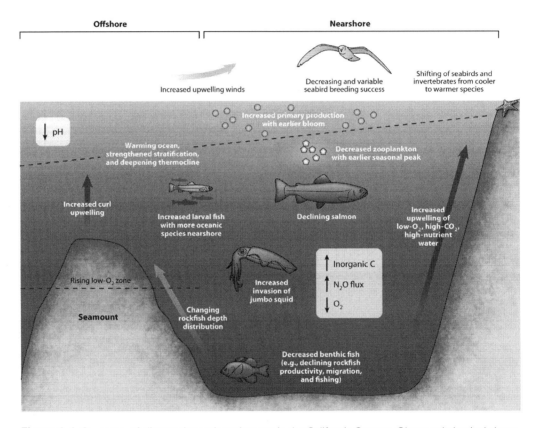

Figure 1-4 Summary of climate-dependent changes in the California Current. Observed physical changes include surface warming, strengthened stratification, and a deepening thermocline that is super-
imposed on strengthened upwelling wind stress and resultant increases in coastal and curl-driven upwelling. Long-term declines in dissolved oxygen have resulted in intensification of shelf hypoxia and
vertical displacement of the hypoxia horizon, which reduces the habitat for certain oxygen-sensitive
demersal fish species and increases the inorganic carbon burden and potential for N_2O flux. The surface
inorganic carbon load has also increased because of anthropogenic CO_2 uptake. Time series from the
past two to three decades indicate increasing trends in phytoplankton biomass; longer time series for
zooplankton indicate decreasing biovolumes over the past six decades as well as shifts toward an earlier
and narrower window of peak abundance. Increases in oceanic larval fish have been observed as have
declines in salmon and rockfish productivity. Seabirds have experienced more variable and, in some cases, declining breeding success. Distributional shifts toward species with subtropical or southern ranges
that are warmer and away from species with subarctic or northern ranges that are cooler have been
evident in intertidal invertebrate, zooplankton, and seabird communities (Source: Doney et al., 2012).

Chapter 2

Climate-Driven Physical and Chemical Changes in Marine Ecosystems

Executive Summary

The Earth's oceans are gradually warming as a direct result of increasing atmospheric carbon dioxide and other greenhouse gasses. The Intergovernmental Panel on Climate Change (IPCC) reported that, between 1961 and 2003, the average temperature of the upper 700 meters of water increased by 0.2° Celsius (IPCC, 2007a). Average global land and sea surface temperatures are continually increasing; 2010's temperatures were the hottest on record (Blunden et al., 2011). Warming of the Earth's oceans has multiple consequences, including sea level rise, changes in global climate patterns, increased stratification of the water column, and changes in ocean circulation and salinity. In addition to warming the oceans, CO_2 is being absorbed by the oceans, causing a series of chemical reactions that lead to a reduction in ocean pH.

Arctic sea ice volume has shrunk by 75% over the last decade (Laxton et al. 2013); incidences of hypoxia (regions where the oxygen concentration has been depleted) within U.S. estuaries have increased thirty-fold since 1960 (Diaz and Rosenberg, 2008); and ocean acidity has increased by 30 percent over the past century (Feely et al., 2004). Warming of ocean waters increases the available energy used to create short-lived storms, and while the frequency of hurricanes and typhoons may not change, a warming ocean will likely result in increased storm intensity (Knutson et al., 2010). As the ocean surface warms, stratification increases, resulting in the warmer water remaining at the surface instead of mixing with the cooler water below. Warm water is not as efficient at absorbing CO_2, and while this might have a slowing effect on ocean acidification, consequences include potential reductions in the uptake of atmospheric CO_2 by the oceans. While there is some variability in salinity levels globally, recent analyses of water density and atmospheric data collected from 1970-2005 suggests that there are overriding changes, including acceleration, in the global hydrological cycle (Helm 2010).

Some evidence suggests that warming oceans affect salinity, circulation patterns, and climate regimes, but uncertainty remains. Many climate regimes follow oscillation patterns that can take place over time periods lasting between several years to decades, which can make it difficult to tease apart natural patterns and variability from anthropogenic climate change. A great deal of uncertainty remains about how rapidly the physical and chemical attributes of the oceans will change in the future, as well as the magnitude of specific impacts and what, if any, potential feedbacks will occur.

The physical and chemical changes taking place in the global oceans set the stage for subsequent effects on marine organisms (Section 3), U.S. communities and economies dependent on marine services (Section 4), U.S. governance and interactions with neighboring countries (Section 5), and potential adaptation strategies (Section 6).

Key Findings

1. The Earth's oceans are warming as a result of increasing concentrations of atmospheric carbon dioxide and other greenhouse gases.

   - From 1955 to 2008, approximately 84 percent of the added heat from climate change has been absorbed by the oceans, thereby increasing the average temperature of the upper 700 meters of water by 0.2° Celsius; this trend is likely to continue.

   - Rising global temperatures are likely to cause greater oceanic evaporation of water vapor, which is a potent greenhouse gas, into the atmosphere, thus amplifying warming due to climate change.

2. Arctic ice has been decreasing in extent since the second half of the 20th century as a result of oceanic and atmospheric warming.

   - Arctic ice has been decreasing throughout the early 20th century. The summer of 2012 saw a record low, when sea ice extent shrank to 3.6 million km2, approximately 1 million km2 less than the previous minima of 2007. Arctic sea ice volume has decreased by 75% over the previous decade.

   - The most significant consequence of melting Arctic ice will be sea level rise. A growing body of recently published work suggests that accelerating loss of the great polar ice sheets in Greenland and Antarctica will likely result in global mean sea level rise of more than 1 meter above present day sea level by 2100.

   - Reductions in ice may occur more rapidly than previously suggested by coupled air-sea-ice climate models. These models may have overestimated ice thickness; more recent modeling predicts that a seasonal ice-free state could occur as early as 2030.

3. The oceans play an important role in climate regulation through the uptake and sequestration of anthropogenic CO_2.

   - The annual accumulation of atmospheric CO_2 has been increasing and in 2010 the overall CO_2 concentration was 39 percent above the concentration at the start of the Industrial Revolution in 1750.

   - Ocean water holds approximately 50 times more CO_2 than the atmosphere and holds the majority of that in the deeper, colder waters; however, the ability of oceans to absorb CO_2 decreases with increasing temperature and decreasing pH.

   - Healthy coastal wetlands are effective at storing and sequestering carbon. This carbon is referred to as "coastal blue carbon." Unfortunately, our nation has lost more than half of its wetlands in the past 200 years.

4. The oceans absorb carbon dioxide, causing a series of chemical reactions that reduce ocean pH, a process known as ocean acidification.

- Over the past century, the average ocean pH has decreased by 30 percent; however, in coastal areas this number may be much higher following upwelling events.

- Predictions indicate that ocean acidification will lead to an under-saturation of argonite in ocean surface waters by the year 2050. Argonite is a form of calcium carbonate impacted by the ocean acidification process and a key component of the exoskeleton of corals and some plankton.

- Polar ecosystems may experience exacerbated ocean acidification because CO_2 is more soluble in colder water and therefore more likely to be absorbed.

5. Ocean temperature and precipitation/evaporation rates have a direct influence on the Earth's weather and precipitation patterns that will likely increase storm intensity but decrease (or have no effect on) storm frequency.

    - Climate change impacts on winds may be moderate and vary regionally but precipitation events are predicted to become more intense.

    - Shifts in precipitation patterns from snow to rain are likely, resulting in decreased snow pack and earlier snow melts that may impact water availability.

    - Storm intensity is likely to increase and could have a large impact, especially on coastal communities because of the increases in coastal populations and infra-structure expected over the next century. The impact of storm surges will be greater for the same reasons as well as the expected rise in sea level.

6. Given that climate and oceans interact to produce intra- and inter-decadal variability in ocean currents, upwelling, and basin-scale circulation, climate change will likely influence these important ocean features, although the ability to detect and project such impacts is low.

    - Melting of polar ice will reduce the salinity and thus density of polar waters, which could weaken the rate at which this water sinks, possibly impairing circulation.

    - However, uncertainty remains about the quantity of freshwater input necessary to slow ocean currents like the Thermohaline Circulation.

Key Science Gaps/Knowledge Needs

Many critical research gaps related to impacts of climate change on physical and chemical ocean systems remain, including:

- Improved understanding of and model projection for thermal expansion of the oceans and associated sea level rise;

- Improved projections for Arctic ice melt and the associated impacts on the oceans' currents and stratification;

- Advanced integration of observations and predictive modeling, particularly at regional scales, in order to gain insight into future impacts of climate change;

- Determination of how changes in upper ocean temperature distributions impact the atmosphere;
- Successful monitoring of tropical cyclone activity globally for emergence of trends, as well as further research concerning earlier detection and/or anticipation of future storms; and
- Improved understanding of the role of "blue carbon" science in ecosystem management issues and what its implications mean for future climate adaptation strategies as well as coastal habitat conservation.

2.1 Introduction

Covering more than two-thirds of the Earth's surface, the oceans are a central component of the global climate system. The oceans help to control the timing and regional distribution of the Earth's response to climate change, primarily through their absorption of carbon dioxide (CO_2) and heat. Changes to the physical and chemical properties of the oceans are already being observed. Sea surface temperatures are warming, sea level rise is accelerating, the oceans are becoming increasingly acidic, and the rate of sea ice melt is steadily increasing. The International Panel on Climate Change (IPCC) assessment released in 2007 projects that, due to the persistence of greenhouse gases in the atmosphere, it is highly likely that the oceans will continue to warm and the impacts will be felt for centuries (IPCC, 2007a). This section focuses on the physical and chemical changes currently being observed in the Earth's oceans, including changes in temperature, stratification, salinity, sea ice, climate regimes, ocean circulation, and ocean acidification. Knowledge gaps and research needs will be discussed throughout. The aim is to assess the current state of knowledge related to how climate change may be interacting with, and in some cases, driving, the observed physical and chemical changes in the Earth's oceans, and what that means for the U.S.

2.2 Ocean Temperature and Heat Trapping

The Earth's oceans are gradually warming as a direct result of increasing concentrations of atmospheric carbon dioxide and other greenhouse gases that increase air temperatures. Air temperature and sea surface temperature are strongly correlated, and the oceans are now experiencing some of the highest temperatures on record. According to the IPCC, "most of the observed increase in globally averaged temperatures since the mid-20th century is very likely due to the observed increase in anthropogenic greenhouse gas concentrations," and current modeling systems predict that it is highly likely that the rate of warming will accelerate over the next few decades. Estimates show that, from 1955 to 2008, approximately 84 percent of the added atmospheric heat has been absorbed by the oceans (Levitus et al., 2009), thereby increasing the average temperature of the upper 700 meters of water by 0.2° Celsius (IPCC, 2007a) (Figure 2-1).

Warming is likely to accelerate over the next few decades, with a predicted increase in global mean surface air temperature between 1.1° Celsius under low CO_2 emission scenario B1 and 6.4° Celsius under high CO_2 emission scenario A1FI by the end of the 21st century (IPCC, 2007a), which will continue to increase ocean temperatures. For example,

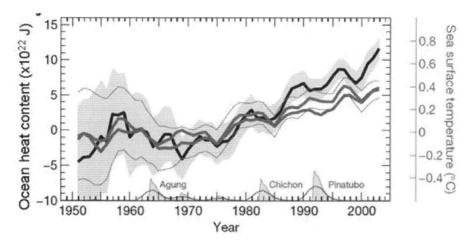

Figure 2-1 Time series of global annual ocean heat content (1022 J for the 0-700 m layer (black) and 0-100 m layer (thick red line; thin red lines indicate estimates of one standard deviation error) and equivalent sea surface temperature (blue; right-hand scale). All time series were smoothed with a three-year running average and are relative to 1961 (Source: Barange and Perry, 2009).

global ocean surface temperatures for January 2010 were the second warmest on record since record keeping began in 1880 and, in 2009, sea surface temperatures from June to August reached 0.58° Celsius above the average global temperature for the 20th century (Hoegh-Guldberg, 2010). Warming of sea surface and atmospheric temperatures is likely to lead to increased evaporation and thus an increase in atmospheric water vapor, which is a potent greenhouse gas, and increased ice melt, which will result in a decrease in Earth's reflection of solar radiation to space. Increased water vapor concentrations are likely to increase the amount of heat that is retained in the atmosphere, causing sea surface temperatures to increase at a faster rate than previously observed (Friedlingstein, 2001). The present CO_2 concentration is the highest on record in at least the last 800,000 years based on ice core data (Lüthi et al., 2008); however, oceanic absorption of CO_2 is dependent on water temperature and pH and these mechanisms are likely to become less efficient as waters warm and pH decreases under future climate scenarios.

Ocean sea surface temperature

The oceans play a dominant role in the Earth's climate system through storage and transport of heat (Levitus et al., 2009). Observed temperature increases are uneven around the globe and both gains and losses in sea surface temperatures are governed by atmospheric and oceanic processes. Dominant atmospheric factors driving ocean temperature include wind speed, air temperature, cloudiness, and humidity; dominant oceanic factors include heat transport by currents and vertical mixing. Fluctuations in sea surface temperatures vary with the seasons; for example, the warming trend seen in the waters off the Northeastern shelf of the U.S. is due primarily to summer water temperatures (Friedland and Hare, 2007), which were the warmest on record in 2010, while winter water temperatures remained near the long-term average (Levy, 2011). Oceans have the ability to release or attain heat from the atmosphere and this exchange of heat is a driving force of atmospheric circulation. Evaporation rates are expected to increase with climate change, resulting in increases in atmospheric water vapor. Water vapor is a much more potent greenhouse gas than carbon dioxide; thus, increased water vapor concentrations contribute to greater downward longwave radiative fluxes that increase the amount of heat retained in the atmosphere (IPCC, 2007a). The potential exists for a cycle of increased water vapor to stimulate increased warming and thus continued production of water vapor, which would lead to a "runaway greenhouse effect" (Trenberth et al., 2009); however, water vapor also stimulates cloud formation as it cools and condenses in the upper atmosphere (Figure 2-2). Clouds can be warming or cooling; high altitude clouds absorb and re-radiate long-wave radiation, which causes warming by augmenting the greenhouse effect, but low altitude clouds reflect incoming solar radiation, which leads to cooling (Stephens et al., 2008). It is uncertain which effect will have the greatest impact, but the preponderance of evidence indicates that it will be an increase in warming (Dessler, 2011), which is likely to cause upper ocean temperatures to increase at a faster rate than observed during the last few decades (Pierre, 2001).

The interplay between the ocean-atmosphere heat exchanges has important implications for global weather and climate patterns; for example, sea surface temperatures in the Pacific Ocean influence winter precipitation in the southwestern U.S. (Wagner, 2010). Observations indicate that changes in sea surface temperature in the tropical North Atlantic affect precipitation in North and Central America, leading to increased incidence of drought throughout the U.S. and Mexico (Kushnir et al., 2010). Tropical storms form over the warm ocean waters that supply the energy for hurricanes and typhoons to grow and move (IPCC, 2007a). Conversely, the oceans have tremendous thermal inertia, which can slow and dampen the rate of climate change (Schewe et al., 2010). The significantly larger heat capacity of the deep ocean is particularly important when looking at time scales of decades to millennia, which are relevant for long-term climate change. Ocean currents and mixing by gyres, winds, and waves can transport and redistribute heat to deeper ocean layers. Heat energy can reside in this deep reservoir for centuries, further stabilizing the Earth's climate and slowing the effects of climate change (Hansen et al., 2007a).

Perhaps one of the most destructive effects of ocean warming comes in the form of sea level rise (discussed in greater detail in the Coastal Impacts, Adaptation and Vulnerabilities: A Technical Input to the 2013 National Climate Assessment, USGCRP, pending 2012, Section 2.2 Overview of Climate Change on Sea Level Rise Effects on Coasts). Most of the sea-level rise observed over the last several centuries can be accounted for by two major variables: the amount of water that is being released by land-locked glaciers and ice sheets and the thermal expansion or contraction of the oceans. In the case of thermal expansion, given an equal mass, the total volume of ocean waters decrease when ocean temperatures drop and expand when temperatures increase. For the period 1961-2003, thermal expansion was responsible for approximately 30 percent of the rise in sea level (Cazenave and Llovel, 2010). As ocean temperatures continue to rise, thermal expansion and associated sea level rise are likely to become more extensive and impact U.S. coastal communities as well as marine species such as seabirds, seals, and turtles that depend on coastal areas for breeding or haul-out areas.

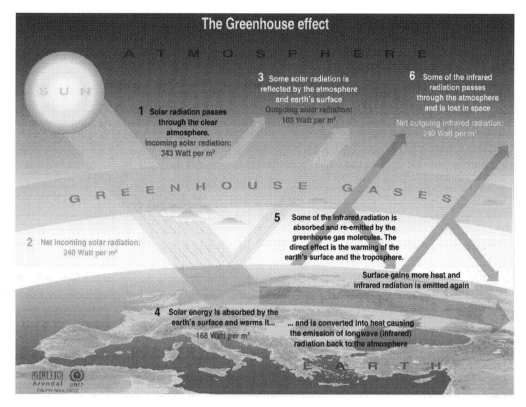

Figure 2-2 Idealized depiction of how solar energy is absorbed by the earth's surface, causing the earth to warm and to emit infrared radiation. The greenhouse gases then trap the infrared radiation, thus warming the atmosphere (Source: Philippe Rekacewicz, UNEP/GRID-Arendal http://www.grida. no/graphicslib/detail/greenhouse-effect_156e).

2.3 Loss of Arctic Ice

As a result of warming temperatures, Arctic sea ice has been decreasing in extent throughout the second half of the 20th century and the early 21st century (Figure 2-3) (AMSA, 2009; Alekseev et al., 2009; Comiso and Nishio, 2008; Deser and Teng, 2008; Maslanik et al., 2007; Nghiem et al., 2007). Arctic sea ice naturally extends surface coverage each winter and recedes each summer, but the rate of overall loss since satellite records began in 1978 has accelerated (IPCC, 2007a). The summer of 2012 saw a record low, when sea ice extent shrank to approximately 3.6 million km2, approximately 1 million km2 less than the previous minima of 2007. Arctic sea ice volume has shrunk by 75% over the last decade (Laxton et al. 2013). Every year since then, September ice extent has been lower than in years prior to 2007, with the 2011 extent being second lowest compared with 2007 (Perovich et al., 2011; Stroeve et al., 2011). Overall, the observed sea ice extent for the years 1979 to 2006 indicates an annual loss of approximately 3.7 percent per decade (Comiso et al., 2008) and an overall 42 percent decrease in thick, multi-year ice between 2004 and 2008 (Giles et al., 2008; Kwok and Untersteiner, 2011). The relative impacts of the loss of thicker, older ice versus the younger, thinner ice are currently not well-understood.

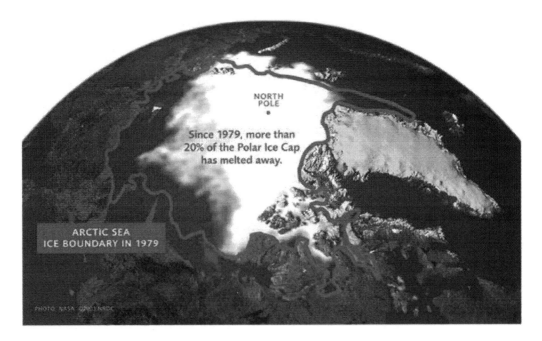

Figure 2-3 Arctic sea ice extent. Satellite imagery of sea ice extent in September 1979 (outlines in red), and at a record low in September 2007 (Source: NASA).

Sea ice plays an important role in reducing the ocean-atmosphere exchanges of heat, moisture, and other gases, with implications for the global climate. The most influential property of sea ice related to climate change is its ability to reflect solar energy. Sea ice is efficient at reflecting solar energy; the much darker ocean surface is more effective at absorbing solar energy. Therefore, as reflective sea ice is lost, further warming of the ocean occurs, leading to increasing ice melt (Serreze and Barry, 2011). Decreases in snow and ice cover may ultimately decrease global albedo, which is the reflection of solar radiation\, therefore amplifying warming, particularly in high latitudes (Soden and Held, 2006 and Bony et al., 2006). These complex interactions and feedback systems cause the Arctic Ocean to be extremely sensitive to warming with consequent changes in atmospheric circulation, vegetation, and the carbon cycle, which impacts both the Arctic and beyond.

In recent decades, the magnitude of warming in Arctic near-surface air temperatures has been approximately twice as large as the global average, a phenomenon known as Arctic amplification (Miller et al., 2010a, Serreze and Francis, 2006). Diminishing sea ice has had a leading role in recent Arctic temperature amplification (Screen and Simmonds, 2010). Enhanced heat storage occurs in the ice-free regions of the Arctic Ocean and heat is returned to the atmosphere in the following autumn (Gascard et al., 2008; Serreze and Barry, 2011; Stroeve et al., 2011). The roles of reductions in snow and changes in atmospheric and oceanic circulation, cloud cover, and water vapor in increasing Arctic amplification are still uncertain. Nevertheless, evidence suggests that strong positive ice-temperature feedbacks have emerged in the Arctic, increasing the chances of further rapid warming and sea ice loss. These changes are likely to affect polar ecosystems and human activities (see Sections 3 and 4). Decreasing Arctic sea ice also affects continental shelves downstream of the Arctic. The Northeast U.S. continental shelf has experienced a freshening over the past 30 years as a result of the advective supply of freshwater from the Arctic and / or Labrador Seas (Greene et al., 2008). The consequence of melting polar ice will be sea level rise. A growing body of recently published work suggests that increasing loss of the great polar ice sheets in Greenland and Antarctica could result in global mean sea level rise of substantially more than 1 meter above present day sea level by 2100 (Grinsted et al., 2010; Jevrejeva et al., 2010; Pfeffer et al., 2008; Rignot et al., 2011).

The Arctic is likely moving toward a new state as more and more old ice is removed and new ice becomes the dominant feature. The IPCC (2007) projections suggest that the Arctic may be virtually ice-free in the summer by the late 21st century; however, others predict that reductions may actually happen more rapidly than previously indicated. The previous projections are from coupled air-sea-ice climate models that tend to overestimate ice thickness, and hence an ice-free Arctic in summer could occur as early as 2030 (Stroeve et al., 2008).

2.4 Salinity

Salinity refers to the salt content of the oceans. Contributors to salinity are terrestrial mineral deposits, precipitation, freshwater input, which results in the dilution of salts already present, and evaporation, which results in increased concentration of salts already present. Ocean salinity changes are an indirect but potentially sensitive indicator for detecting changes in precipitation, evaporation, river runoff, and ice melt. Thus, salinity may function as a proxy for identifying climate-driven changes in the Earth's hydrological cycle (Helm et al., 2010).

The globally-averaged surface salinity change to date is small, whereas basin averages are more noteworthy: increased salinity for the Atlantic, freshening in the Pacific, and a near-neutral result for the Indian Ocean (Wijffels and Durack, 2009). A recent study by Cravatte et al. (2009) uncovered large changes to ocean properties in the tropical western Pacific, indicating that a freshening and density decrease had occurred in the 1955-2003 period. Despite an overall increase in salinity for the Atlantic, studies conducted in the Scotian Shelf and Gulf of Maine waters show a decrease in salinity (Greene et al., 2008). One of the main reasons for this is the melting of Arctic sea ice and the resulting freshwater input into the Labrador Current. Another influence on the decrease in salinity in the Gulf of Maine is increased precipitation. Many climate models predict that all forms of precipitation will increase in the Gulf of Maine area (Wake et al., 2006), which will increase river flow and terrestrial runoff into the Gulf. This input of freshwater will be especially large in the spring when pack ice melts within the watershed and causes typically high river flow (Wake et al., 2006). These inputs of fresh surface water inhibit the vertical mixing of the upper portion of the water column within the Gulf of Maine. This stratification prevents nutrients from being brought into the surface layer that phytoplankton need to grow. Reductions in available nutrients will affect phytoplankton produc¬tivity, which is the base of the food web, meaning that these effects will cascade up to larger organ¬isms like fish and marine mammals (see Section 3).

Changes to the global hydrological cycle are anticipated as a consequence of anthropogenic climate change (IPCC, 2007a; Held and Soden, 2006). Even small variations in ocean salinity can have a dramatic effect on the water cycle and circulation (Holland et al., 2001). Because salt water is denser than fresh water, changes in salinity affect density parameters, which in turn affect stratification (see Section 2.5). Salinity changes can also impact ocean circulation, which is driven by temperature and salinity (thermohaline circulation). As the salty cold water sinks at the poles, warmer, fresher water is pulled from the equator to replace it (see Section 2.7) (Broecker, 1991). If salinity declines due to climate-related increases in precipitation and glacial melt, ocean circulation may begin to slow (IPCC, 2007a), which could result in potentially dramatic changes to the Earth's climate.

2.5 Stratification

Stratification, or layering, of ocean water is a naturally occurring phenomenon that is important to water column structure, circulation, and marine productivity. Water density is strongly influenced by temperature and salinity, with less dense, warmer surface waters floating on top of denser, colder waters. Warming of the upper layers of the oceans enhances the density difference between the surface-mixed layer and the deeper waters beneath. All else being equal, as waters warm, this increased density difference strengthens the vertical stratification of the oceans and suppresses vertical mixing across the density gradient. This results in contrasting effects on nutrient and light availability for phytoplankton growth. On the one hand, stratification reduces nutrient influx from deep, nutrient-rich waters into the surface-mixed layer, thus limiting the availability of nutrients for phytoplankton growth (Behrenfeld et al., 2006; Huisman et al., 2006). On the other hand, stratification keeps phytoplankton in the surface-mixed layer, resulting in improved light conditions for growth (Berger et al., 2007; Huisman et al., 1999). Increased stratification and its effect on phytoplankton productivity can have negative consequences for coastal ecosystems; for example, warming of the surface waters in Southern California between 1951 and 1993 has resulted in an 80 percent decrease in phytoplankton and zooplankton biomass as well as decreased coastal upwelling and nutrient availability (Palacios et al., 2004).

Stratification varies regionally due to mixing, upwelling, and the uneven distribution of sea surface temperatures. Many waters in the tropics and subtropics, such as those surrounding the Pacific Islands, are permanently stratified. Nutrient concentrations in the surface-mixed layer of these waters are strongly depleted and are characterized by extremely low primary production. Climate-ocean models predict that, by the year 2050, the ocean area covered by permanent stratification will have expanded by 4.0 percent and 9.4 percent in the Northern and Southern hemispheres, respectively, due to ocean warming (Sarmiento et al., 2004), thereby reducing overall ocean productivity (Behrenfeld et al., 2006); however, these predictions are surpassed by recent observations that indicate a much faster expansion of the ocean's least productive waters over the past nine years (Polovina et al., 2008). In the temperate zone and at high latitudes, waters are not permanently stratified and deep mixing during the winter and/or spring provides nutrients into the surface layer. In these regions, phytoplankton growth is often light-limited in winter due to short day lengths and deep vertical mixing. Warming temperatures lead to earlier onset of stratification in spring, which retains phytoplankton in the well-lit surface layer where they can take advantage of nutrients that have not yet been depleted, thereby favoring their growth. This leads to an earlier spring bloom and the potential for a substantially longer growing season in the temperate zone (Winder and Schindler, 2004; Peeters et al., 2007).

Stratification can also be affected through changes in precipitation. Climate-induced precipitation increases will introduce fresh water into nearshore environments, either directly through rainfall, or indirectly through increased runoff from the terrestrial environment. This freshening will tend to increase stratification, resulting in decreased diffusion of oxygen and nutrients. Decreasing oxygen concentrations will negatively affect biological communities and secondary production in addition to disrupting biogeochemical cycles (Rabalais, 2009). Thus, climate-related changes in stratification are likely to cause subsequent impacts on the ecosystem and resulting ecosystem services of inshore and nearshore U.S. ecosystems.

2.6 Changes in Precipitation and Extreme Weather Events

Global climate change affects the Earth's weather and precipitation patterns. As mentioned in Section 2.2, warming of the sea surface will very likely lead to increased ice melt and evaporation, and therefore increased atmospheric water vapor, which may ultimately accelerate warming. Increased humidity may lead to more intense storms, which may lead to more extreme precipitation and wind events. Ironically, because of increased evaporation, some areas, such as the subtropics, will experience intense surface drying, increasing the risk of flooding when intense storms occur (Trenberth, 2011). Climate change impacts on winds and precipitation may be moderate but vary regionally, so that some areas become dryer while other areas, particularly at mid to high latitudes, may become wetter (Trenberth, 2011). Warming temperatures will likely lead to some shifts in precipitation from snow to rain, resulting in decreased snow pack and earlier snow melts. Taken together, climate change is likely to influence the Earth's hydrological cycle and the character of weather events.

Winds

Winds can have a major influence on marine ecosystems. In fact, wind changes may be more important than temperature changes for inducing the stratification effects mentioned above but wind predictions have a higher uncertainty. Global wind stress has increased over the past ten years and shows variable impacts regionally (EAP, 2009). These changes in wind stress may be linked to climate regimes such the North Atlantic Oscillation (NAO), Pacific Decadal Oscillation (PDO), and the El Niño Southern Oscillation (ENSO), as well as a northward shift in the location of the jet stream (Archer and Caldeira, 2008). Increased temperatures caused by climate change may impact regional wind patterns, leading to changes in circulation and mixing. If wind patterns or intensities change, currents and their effects on coastal waters might change, potentially impacting oxygen concentrations. For example, off the Oregon and Washington coasts, wind-driven shifts in the California Current region occurred in 2001 and subsequent years, altering upwelling dynamics and resulting in extensive hypoxia along the inner continental shelf (Chan et al., 2008; Grantham et al., 2004). Climate-related changes to wind and ocean currents are also likely to interact with non-climatic stressors such as the addition of excessive nutrients from terrestrial runoff, which would further increase the incidence of hypoxic and anoxic events (see Section 2.11).

Precipitation

Global-mean precipitation is an important part of the Earth's climate system by linking the global water and energy cycles through condensational heating of the atmosphere and by connecting the hydrological cycle to radiative processes such as cloud feedback (see Section 2.2.1). Precipitation forms as water vapor is condensed usually in rising air that expands and cools. Climate change impacts on the frequency of precipitation events will likely be moderate but precipitation events are predicted to become more intense (Trenberth et al., 2003). As the climate warms, the percentage of precipitation that falls as snow is likely to decrease, resulting in earlier snowmelt and diminished snow pack, and thus a temporal change in stratification due to the timing of fresh water input into coastal and marine systems (see Section 2.5).

Storms

Concern is mounting about the potential for climate-related impacts on the frequency of intense storm events. The record-breaking hurricane season in the North Atlantic in 2005 had the largest number of named storms (28), the largest number of hurricanes (15), the most intense hurricane in the Gulf of Mexico (Rita, 897 hPa), and the most damaging hurricane on record (Katrina), which was also the deadliest in the U.S. since 1928 (IPCC, 2007a). Uncertainty remains as to the effects of climate change on storm intensity. Multi-decadal variability and the lack of high-quality tropical cyclone records prior to routine satellite observations that began in approximately 1970 complicate the detection of long-term trends in extreme weather events (Blunden et al., 2011).

Sea surface temperatures are related to the maximum potential intensity of tropical storms, so theoretically the intensity will rise as temperatures increase. This indicates that storm wind speeds, from which category ratings of storms are based, may increase by 3-11 percent by 2100 (Knutson, 2010). Regions that have historically endured two or three land-falling category 4 or 5 storms per century, such as Florida or Louisiana, are likely to see an increase to three or four storms of that magnitude per century (Kunkel et al., 2008). However, because climate models indicate that the environment will be less favorable for development to maximum potential due to more wind shear and less relative humidity (Vecchi and Soden, 2007), there may be fewer weak storms and fewer storms overall (Knutson et al., 2010). This has the potential to affect water supplies because the weaker storms, which are now more frequent, are an important source of rainfall for coastal regions. However, rainfall may increase by as much as 20 percent in an individual storm (Knutson, 2010), leading to larger short-term pulses of freshwater.

During a tropical storm, strong surface winds take heat out of the ocean in addition to mixing the ocean at depths from tens to hundreds of meters, which cools the surface and creates a cold wake (Trenberth and Fasullo, 2007; Walker et al., 2005). Hence, tropical storm activity depends on sea surface temperatures as well as subsurface temperatures, especially for whether the ocean environment is favorable for the next storm and thus an entire active season. Better understanding of and accounting for these feedbacks should improve ocean model predictions of hurricanes; however, the roles of surface fluxes, ocean spray, and ocean mixing are all highly uncertain for strong winds. Increased sea surface temperatures have been suggested as being directly linked to increased hurricane intensity. The warm and deep "Loop Current" in the Gulf of Mexico, which appeared to play a key role in the intensification of Hurricanes Ivan (Walker et al., 2005), Katrina (Figure 2-4), and Rita (IPCC, 2007a) provides naturally-occurring, contemporary evidence of this theory. Moreover, because current climate models do not include these processes, the role of hurricane-induced mixing in the ocean on currents and the thermohaline circulation (Boos et al., 2004; see Section 2.7) are unresolved issues that could alter views of how future climate may change.

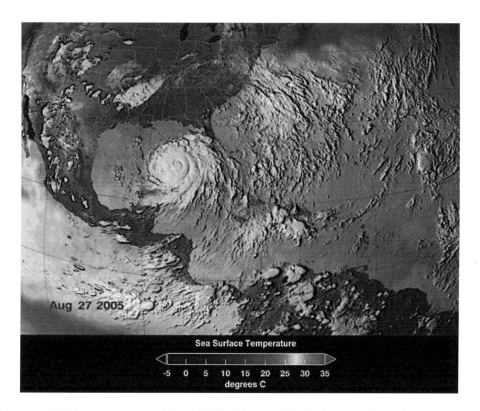

Figure 2-4 This image was created from AMSR-E data on NASA's Aqua satellite and shows a 3-day average of actual sea surface temperatures (SSTs) for the Caribbean Sea and Atlantic Ocean from August 25–27, 2005. 80 degrees Fahrenheit is necessary to maintain hurricanes, during this time period the Gulf of Mexico up to the coast of North Carolina averaged 82 degrees Fahrenheit or above (Source: NASA Goddard's Scientific Visualization Studio).

Although storm events in this area already impact marine fisheries, commerce, and coastal ecosystems, storm activity is expected to increase in the Aleutian Islands region (Ulbrich et al., 2008). In November 2011, thirty-seven coastal communities reported damages (D'Oro, 2011) after the worst storm in 40 years hit the Alaska coast with hurricane-strength winds and large storm surges. The increasing frequency of extreme events may be amplified by the interactions with the loss of sea ice, which typically buffers the effect of winter storm surge on Alaskan coastlines. Added heat to the lower atmosphere can generate storms in newly ice-free areas (Inoue and Hori, 2011) and can impact larger regions (Overland et al., 2011).

Though modeling projections for climate change and storms predict fewer storms overall, a small increase of the most destructive storms will have a large impact because: (1) damage increases exponentially with wind speed, (2) coastal population and infrastructure will increase over the century, which will result in greater vulnerability to strong storms, (3) storm surge has historically been responsible for the greatest loss of life and property as well as contributing to the incidence of marine debris (see Case Study 2-A), and (4) storm surges are greater for stronger storms and because of the expected rise in sea level (for more information please see Coastal Impacts, Adaptation and Vulnerabilities: A Technical Input to the 2013 National Climate Assessment, US-GCRP, pending 2012, Section 2.2 Overview of Climate Change on Sea Level Rise Effects on Coasts).

2.7 Ocean Circulation

Together with tidal and other energy sources, wind, heat, and fresh-water fluxes at the ocean surface are responsible for global ocean circulation, mixing, and the formation of a broad range of water masses. On the global scale, individual shallow and deep-ocean currents form an interconnected pattern known as the thermohaline circulation (THC), sometime referred to as the "Global Conveyor Belt" (Broecker, 1991). The path of the THC is generally described as originating in the Northern Atlantic Ocean, where cold, dense water sinks and travels across ocean basins to the tropics, where it warms and upwells to the surface. The warmer, less dense, tropical waters are then drawn to polar latitudes to replace the cold, sinking water. However, in practice, the THC interacts with other currents and mixing occurs among the waters traveling along these intersecting pathways, creating a more complex situation (Broecker, 1991). The THC plays an important role in transferring heat from the oceans to the atmosphere by causing the water to become colder and denser, thus renewing the cycle in a process that can take hundreds of years to complete (Lumpkin and Speer, 2007; Kanzow et al., 2010). The THC is responsible for much of the distribution of heat energy from the equatorial oceans to the polar regions and has a large influence on the Earth's climate.

It is possible that the THC may weaken by the end of the 21st century as a result of climate change (IPCC, 2007a). Melting of polar ice will reduce the salinity and thus the density of polar waters, which could weaken the rate at which this water sinks and possibly impair circulation (Hu et al., 2011). However, the quantity of freshwater input necessary for such slowing of the THC is uncertain (Kuhlbrodt et al., 2009). The pattern of temperature change as a consequence of THC slowing is complex, with predicted cooling over the North Atlantic (Vellinga and Wu, 2004; Wood et al., 2003, 2006) and warming occurring farther east (Stouffer et al., 2006). Another possible impact of weakening circulation may be an increase in sea level rise as a result of circulation-induced pressure gradients (Sturges, 1974). Weakening of the THC is likely to increase sea level rise by

Case Study 2-A
The "garbage patches"

The "garbage patches" are areas of marine debris concentration in the North Pacific Ocean. Because little scientific research has been conducted in these areas, the exact size and content of these areas are difficult to predict accurately.

What's In a Name?

The name "garbage patch" is a misnomer. No island of trash forms in the middle of the ocean nor can a blanket of trash be seen with satellite or aerial photographs. Instead, much of the debris found here is formed of small bits of floating plastic not seen easily from a boat.

Eastern and Western Patches

- **Eastern garbage patch** – Concentrations of marine debris have been noted in an area midway between Hawaii and California known as the North Pacific Subtropical High or the "eastern garbage patch." The High is not a stationary area, but one that rotates, moves, and changes.

- **Western garbage patch** – Another area of marine debris concentration is located off the coast of Japan, and researchers believe it to be a small recirculation gyre (ocean feature made up of currents that spiral around a central point) likely created by winds and ocean eddies.

Impact of Climate Change

- Floatable marine debris from land- and ocean-based sources (e.g., tiny pieces of plastic) enter the marine environment through several pathways; for example, debris resulting from wind and wave damage during storms may be swept into the ocean. The number of intense storms capable of this action is predicted to increase with climate change (Section 2.6).

- Ocean currents and atmospheric conditions (e.g., waters rotating—large or small area, fast or slow rotation—in cyclone-like fashion) result in the concentration of marine debris; changes in current location and strength may result in more or less concentration of debris. However, the impact of climate change on these processes is uncertain (Section 2.7).

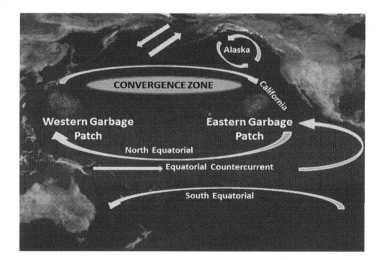

This map shows the locations of the eastern and western "garbage patches." Keep in mind that this is an over simplification of the constantly moving and changing features of the North Pacific Ocean. Source: NOAA

Case Study 2-A (Continued)

Can It Be Cleaned Up?

The answer is not as simple as you might think. Even a cleanup focusing on the "garbage patches" would be a tremendous challenge because:

- Concentration areas move and change throughout the year
- These areas are typically large
- The marine debris is not distributed evenly in these areas
- Modes of transport and clean up will likely require fuel of some sort
- Most of the marine debris found in these areas is small bits of plastic

Image of marine debris on a coral reef. Source: NOAA

This all adds up to a bigger challenge than even sifting beach sand to remove bits of marine debris. In some areas where marine debris concentrates so does marine life. This makes simple skimming of the debris risky; if not conducted carefully, clean-up may result in more harm than good for marine life.

approximately 20 centimeters along the East Coast of the U.S. above and beyond what is predicted as a result of thermal expansion and glacial melt (Yin et al., 2010). Thus, the U.S. East Coast is likely to experience enhanced impacts due to climate-related changes to the THC.

The following is a brief overview of possible impacts of climate change on two ocean current systems: the Califorina Current and the Gulf Stream. These currents were chosen due to the amount information available related to climate change impacts. Additional currents important to the U.S., such as the Alaska Current, may also experience changes in behavior as a result of climate change, but the information on those currents is more limited and will not be mentioned in this report. Hopefully, these will be included in the 2017 report.

California Current

The California Current is a Pacific Ocean current that moves southward along the western coast of North America, beginning off southern British Columbia and ending off southern Baja California in Mexico. The California Current spans strong physical gradients in circulation and water column structure (King et al., 2011). Considerable evidence suggests that physical changes, including significant warming and concomitant changes in water column stratification have occurred across the California Current in the past century (Palacios et al., 2004); for example, declines in oxygen concentrations due to changes in upwelling have already been measured along the coasts of California, Oregon, and Washington (Bograd et al., 2008; Chan et al., 2008; Grantham et al., 2004). Possible climate-related effects on the California Current are highly uncertain. Some observations indicate that upwelling events are becoming less frequent, stronger, and longer in duration in accordance with climate change, which has impacted primary productivity along the U.S. West Coast (King et al., 2011). Others suggest only moderate oceanographic changes such as mild surface warming accompanied by relatively minor increases in upwelling-favorable winds in northern portions of the California Current, suggesting that natural variability may overshadow climate change signals (Wang et al., 2010). Two critical questions for further research concern whether observed changes in wind forcing reflect anthropogenic changes in variability of the California Current climate system, and what possible impacts of climate-driven delayed onset of upwelling will exist for future regional oceanic production in the California Current (Di Lorenzo et al., 2005).

Gulf Stream

The Gulf Stream transitions from the Gulf of Mexico, passes through the Straits of Florida, travels off the U.S. coast until Cape Hatteras, and then moves through the Mid-Atlantic Bight until it reconnects to the coast at the Grand Banks off Newfoundland. Long-term observations of the Gulf Stream exist at only at a few locations along the path. As stated above, numerical climate models have projected a slow-down of the THC in the Atlantic over the next several decades. The Gulf Stream is one component of the THC, so a slowdown could result in a reduction in Gulf Stream strength; however, the Gulf Stream also carries the western boundary flow of the subtropical wind-driven gyre. It cannot yet be excluded that a slowdown of the THC might be accompanied by an increase in the wind-driven gyre, leading to no net impact on the Gulf Stream. This is a question for future research. If the Gulf Stream were to decrease in strength, the sea level along the U.S. eastern seaboard may rise as a result of a relaxation of the pressure gradient created by the Gulf Stream (Kelly et al., 1999). For example, an anomalously high sea level along the U.S. East Coast in the summer of 2009 was related to a short-term weakening of the Florida Current (Sweet et al., 2009). A decrease in the Gulf Stream volume transport might also have significant impacts on precipitation patterns, hurricane tracks, and surface air temperatures throughout the northern hemisphere. The long-term variations of the Gulf Stream are not yet clear and warrant future observation and study.

2.8 Climate Regimes

Regime shifts refer to a broad set of often basin-wide changes in the characteristic behavior of the physical environment, such as persistent increases in ocean and atmospheric temperatures or shorter-term perturbations related to climatic events. These shifts have impacts on climate conditions in oceans and on land. Climate change and the subsequent warming of the atmosphere and oceans will likely impact regional climate regimes such as those encapsulated by the North Atlantic Oscillation (NAO), Pacific Decadal Oscillation (PDO), and the El Niño Southern Oscillation (ENSO); however, the nature of these oscillations over varying time scales creates challenges for climate prediction models, resulting in a moderate level of uncertainty with regards to climate change impacts. Establishing a clear set of considerations for assessing uncertainty in regional climate change impacts assessments is vital for enabling an informed response to potential climate risks to the Earth's oceans on a multi-decadal to century time scale (see Case Study 2-B). One reason for the uncertainty is that even small changes in atmospheric warming can have cumulative and multiplying effects across regions, and even globally, resulting in climate regime shifts.

The following is a brief overview of possible impacts of climate change on three climate regimes: the North Atlantic Oscillation, the Pasific Decadal Oscillation, and the El Niño/Southern Oscillation. These regimes were chosen due to the amount information available related to climate change impacts. Additional regimes important to the U.S., such as the North Pacific Gyre Oscillation, may also experience changes in behavior as a result of climate change; however, the information on those regimes is more limited and will not be mentioned in this report. Hopefully, these will be included in the 2017 report.

North Atlantic Oscillation (NAO)

The climate of the Atlantic region exhibits considerable variability on a wide range of time scales. A substantial portion of that variability is associated with the NAO (Stenseth et al., 2002), a measure of the fluctuation in the sea level pressure difference between the Icelandic low in Stykkisholmur, Iceland and the Azores high in Lisbon, Portugal. The NAO fluctuates between positive and negative states (Arzel et al., 2011; Czaja et al., 2003). When the NAO index is high (positive NAO state), precipitation increases over the eastern seaboard of the U.S. and ocean temperatures are relatively warm. Conversely, when the NAO index is low (negative NAO state), decreased storminess, drier conditions, and relatively cool ocean temperatures occur in the eastern U.S. (Hurrell and Deser, 2010). Although the NAO index varies from year-to-year, it also exhibits a tendency to remain in one phase for intervals lasting more than a decade. An unusually long period of positive phase from 1970-2000 led to the theory that climate change was likely affecting the behavior of the NAO (Goodkin et al., 2008). Uncertainty remains as to whether climate change is influencing the timing of the NAO phases, but evidence indicates that the strength of its variability is increasing as phases are becoming more strongly positive and negative (Rind et al., 2005).

Pacific Decadal Oscillation (PDO)

The North Pacific Ocean has experienced different climatic regimes and shifts in the recent past. The two principal modes of sea surface temperature anomalies acting in the North Pacific Ocean are the longer-term Pacific Decadal Oscillation (PDO) events, which occur on inter-decadal time scale of roughly 25-30 years, and the shorter-term El Niño-Southern Oscillation (ENSO, discussed below) events, which occur at a time scale of approximately 3-7 years (Mantua et al., 1997). The PDO, like the NAO, fluctuates between positive and negative phases. During the positive phase of the PDO,

Case Study 2-B
Assessing confidence in regional climate ecosystem projections

Establishing a clear set of considerations for assessing uncertainty in regional climate change impacts assessments is vital for enabling an informed response to potential climate risks to the Earth's oceans. Global climate model simulations are a primary tool used in analyzing climate dynamics and making projections of future climate change on multi-decadal to century time scales. Climate model outputs are routinely used in a variety of climate change impact studies and assessment products, including the reports of the IPCC. However, a number of global climate model limitations must be carefully considered when interpreting and assessing uncertainty in regional climate-marine ecosystem impacts studies (Stock et al., 2011).

1. The spatial resolution of most global climate models is coarse. At regional scales, model projections reflect the influence of broad-scale radiative and circulation changes and may not capture the impact of unresolved regional processes. Such processes may significantly alter broad-scale trends.

2. Changes in climate are driven by changes in radiative forcing arising from greenhouse gases, aerosols, and other factors (often referred to as the "forced change") and internal climate variability. Climate variability is often prominent at regional scales and makes attribution of projected regional-scale impacts to climate change difficult on time-scales of a few decades or less.

3. Although model projections often agree on the direction and magnitude of multi-decadal changes in many climate variables at

continental scales and above, disagreements are common at smaller spatial and temporal scales. Consideration of "inter-model spread" is essential for establishing uncertainty in projected impacts.

4. Global climate models can have significant regional biases. Although progress can be made using simple bias corrections, these should be applied with caution and require long observational records to establish mean climate conditions.

Uncertainties in the physical climate system are only part of the challenge. To assess the ecosystem response to climate changes, projected changes in physical climate variables must be integrated with ecological models. Confidence in ecological models for climate projections must be built in a manner similar to climate models, according to their reliance on fundamental ecological and physiological principles expected to be robust as climate changes, and according to their ability to match past observed ecological changes.

Continuing developments in climate and ecological models will ameliorate the limitations described above and improve our understanding of climate impacts on ocean ecosystems. However, existing tools and understanding have progressed to a point where substantial progress can be made, particularly in establishing first estimates of the direction and magnitude of potential impacts. Efforts to improve models should be conducted with and informed by applications of present tools.

Provided by Dr. Charles Stock, NOAA

stronger-than-normal downwelling winds along the northern California and Alaska regions generate a local convergence of water masses at the coast that result in higher sea surface height, warmer sea surface temperature, and anomalous poleward circulation near the coast (King et al., 2011). The opposite occurs during the negative phase. Overland and Wang (2007) suggest that under the A1B (middle range) CO_2 emission scenario, the change in winter sea surface temperatures due to anthropogenic influences will surpass the natural variability in most of the North Pacific in less than 50 years.

El Niño/Southern Oscillation (ENSO)

ENSO events exhibit a see-saw pattern of reversing surface air pressure and surface ocean temperature between the eastern and western tropical Pacific. The ENSO cycle is comprised of two interacting climate regimes: El Niño, which produces warmer conditions sea surface temperatures in the eastern Pacific, and La Niña, which produces cooler temperatures (Lumpkin et al., 2011). A balance of amplifying and dampening feedbacks in ocean-atmosphere exchange controls year-to-year ENSO variability and one or more of the physical processes responsible for determining the characteristics and global impacts of ENSO will likely be modified by climate change; for example, climate-driven changes in the upwelling cycle, with delayed and weak seasonal upwelling, have been documented in the central California Current region during El Niño years (Bograd et al., 2009). Higher-than-normal uplifting of the thermocline and strengthened poleward circulation in the proximity of the southern California coast has occurred during the latest (2010-2011) La Niña events (Nam et al., 2011). These events influence ocean productivity, as well as atmospheric circulation, and, consequently, regional rainfall rates and extreme weather events at interannual time scales. Despite considerable progress in our understanding of the impact of climate change on many of the processes that contribute to ENSO variability, it is not yet possible to say whether ENSO activity will be enhanced or dampened or if the frequency of events will be altered under climate change (Collins et al., 2010).

2.9 Carbon Dioxide Absorption by the Oceans

The annual accumulation of atmospheric CO_2 has been increasing and, in 2010, the overall CO_2 concentration was 39 percent above the concentration at the start of the Industrial Revolution in 1750 (IPCC, 2007a). The IPCC indicates that reducing the CO_2 deposited into the atmosphere by 85 percent by the year 2050 would prevent a global mean temperature increase in excess of 2.0° Celsius, a temperature increase that could result in extreme global changes (IPCC, 2007a). CO_2 reductions can be achieved both through reducing anthropogenic sources of CO_2 and supporting CO_2 uptake and storage through the conservation of natural ecosystems with high carbon sequestration rates and capacity (Canadell and Raupach, 2008). The Earth's oceans and coastal ecosystems, including mangrove forests, seagrass beds, and salt marshes, are proportionately more effective on a per acre basis at sequestering carbon dioxide than are terrestrial ecosystems.

The oceans play an important role in climate regulation through the uptake and sequestration of anthropogenic CO_2 (Sabine et al., 2004). Ocean water holds approximately 50 times more CO_2 than the atmosphere, with the majority being held in the deeper, colder waters, but the ability of oceans to absorb CO_2 is finite. Currently the oceans absorb more CO_2 than they release but CO_2 is less soluble in warmer waters, so as the sea surface temperature increases, a decrease in oceanic uptake of CO_2 from the air is likely (IPCC, 2007a). Second, the ocean carbon sink is finite because CO_2 uptake also depends on the pH of ocean water, which decreases as more CO_2 is absorbed (see Section 2.10), decreasing the buffering capacity, or the ability of the ocean to continue to take up CO_2 (Andersen and Malahoff, 1977; Broecker and Takahashi, 1966; Revelle and Seuss, 1957; Skirrow and Whitfield, 1975; Stumm and Morgan, 1970). However, future projections of oceanic sink strength and regional distribution are highly uncertain (Doney, 2010).

"Blue carbon" is a term used to describe the biological carbon sequestered and stored by marine and coastal organisms. A significant fraction of blue carbon is stored in sediments, coastal seagrasses, tidal marshes, and mangroves. When degraded or disturbed, these systems release carbon dioxide into the atmosphere or ocean. Currently, carbon-rich coastal ecosystems are being degraded and destroyed at a global average 2 percent annually, resulting in significant emissions of carbon dioxide and contributions to climate change. Mangrove areas alone lost 20 percent of global cover between 1980 and 2005 (Giri, 2010; Spaulding et al., 2010). Carbon continues to be lost from the most organic soils in coastal areas. For instance, analysis of the agricultural soils of Sacramento's San Joaquin Delta, a diked and drained former tidal wetland, documents emissions of CO_2 at rates of 5 to 7.5 million tCO_2 each year, or 1 percent of California's total GHG emissions. Each year, an inch of organic soil evaporates from these drained wetlands, leading to releases of approximately 1 billion tCO_2 over the past 150 years (Crooks et al., 2009; Deverel and Leighton, 2010, Hatala et al., 2012). Similar emissions are likely occurring from other converted wetlands along the East and Gulf Coasts of the U.S. Conservation and improved management of these systems brings climate change mitigation benefits in addition to increasing their significant adaptation value (Crooks et al., 2011; McLeod et al., 2011). Developing a better understanding of blue carbon science and ecosystem management issues has implications for future climate adaptation strategies as well as coastal habitat conservation.

2.10 Ocean Acidification

Ocean chemistry is changing in response to the absorption of CO_2 from the atmosphere at a rate unprecedented for perhaps more than 50 million years (Hönisch et al., 2012). Ocean acidification refers to the decrease in the pH of the Earth's oceans associated with the uptake of atmospheric CO_2 and subsequent chemical reactions. Ocean acidification is related to, but distinct from, climate change; however, it is important to include in this report because both climate change and ocean acidification share a common cause: increasing carbon dioxide concentration in the atmosphere. Acidification is not a climate process, but instead a direct impact of rising CO_2 absorption on ocean chemistry. The distinction between climate change and acidification is important because discussions regarding both solutions and adaptation to climate change often ignore ocean acidification.

It is estimated that the surface waters of the ocean have absorbed approximately 25 percent of all anthropogenically-generated carbon since 1800 (Sabine et al., 2004). In the past, it was believed that the oceans would offset the effects of greenhouse gas emissions, but it is now understood that although absorption of CO_2 by the ocean slows the atmospheric greenhouse effect, CO_2 reacts with seawater to fundamentally change the chemical environment (Feely et al., 2010). Carbon occurs naturally and in abundance in seawater, simultaneously occurring as a suite of multiple compounds or ions, including dissolved carbon dioxide ($CO_{2(aq)}$), carbonic acid (H_2CO_3), bicarbonate ions (HCO_3), and carbonate ions (CO_3^{2-}). The relative proportion of these compounds and ions adjusts to maintain the ionic charge balance in the ocean. The addition of CO_2 to seawater alters the carbonate equilibrium, with an associated decline in pH due to the production of excess hydrogen ions, and changes in the relative concentrations of bicarbonate and carbonate ions (Andersen and Malahoff, 1977; Broecker and Takahashi, 1966; Revelle and Seuss, 1957; Skirrow and Whitfield, 1975; Stumm and Morgan, 1970; Figure 2-6).

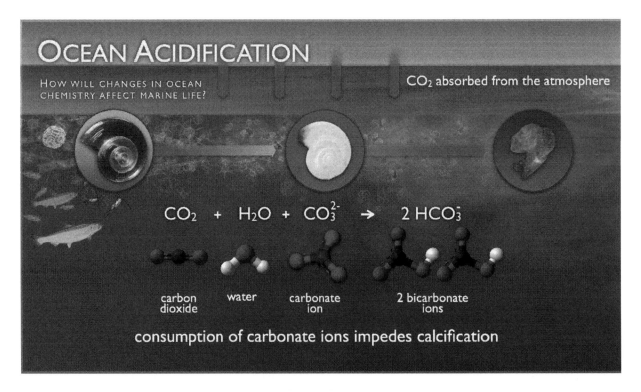

Figure 2-6 As CO_2 is absorbed by the atmosphere it bonds with sea-water to form carbonic acid. This acid then releases a bicarbonate ion and a hydrogen ion. The hydrogen ion bonds with free carbonate ions in the water to form another bicarbonate ion. This free carbonate would otherwise be available to marine animals for making calcium carbonate shells and skeletons. (Source: NOAA).

The average pH of the upper layers of the world's oceans has already declined from an average value of 8.2 to 8.1 over the past century. Given that pH is measured on a logarithmic scale, this represents a 30 percent increase in ocean acidity (Caldeira and Wickett, 2003, 2005; Feely et al., 2004). Under current CO_2 emission rates, a further decline in pH of 0.3 to 0.4 units could occur by the year 2100 (Orr et al., 2005). Ocean acidification has recently caught much attention because it affects the health of calcifying organisms and the rates of biogeochemical processes in the oceans (see reviews by Doney et al., 2009 and Riebesell et al., 2007).

Different regions will respond variably to the rapidly changing carbon chemistry (Figure 2-7). Regions with strong upwelling regimes, such as the U.S. West Coast, could be more vulnerable because upwelled waters are naturally higher in CO_2 and thus lower in pH. These waters are now also showing signs of anthropogenically-augmented CO_2 levels (Feely et al., 2008). Coastal regions with high freshwater input, such as areas of the Chesapeake Bay that receive input from the James River, may also be more vulnerable to acidification because fresh water has a decreased ability, in general, to neutralize acids and may carry other acidifying solutes (Kelly et al., 2011; Salisbury et al., 2008). Ocean acidification will also be exacerbated in polar ecosystems such as the Arctic Ocean because CO_2 is more soluble in colder water and the loss of sea ice in summer results in greater exposure of seawater to the atmosphere, which allows more exchange of CO_2 across the ocean-atmosphere interface. Fresh water dilution from ice melt at high latitudes also exacerbates acidification and results in undersaturation of carbonate minerals (Denman et al., 2011). Finally, predictions indicate that surface waters in the Arctic will be undersaturated with respect to aragonite, the more soluble form of calcium carbonate, as soon as within the next decade, or even sooner if other climate-related factors such as shrinking sea ice and increased fresh water are taken into account (Orr et al., 2005; Steinacher et al., 2009). Similarly, projections for the Southern Antarctic Ocean surface waters suggests that it too is likely to become undersaturated with respect to aragonite by the year 2050 (Orr et al., 2005). However, it should be noted that low latitude systems are not immune to such changes. In fact, the greatest rate of change in carbonate mineral saturation state has unfolded within the Atlantic tropical waters. Because the effect on marine organisms is not necessarily a threshold effect (e.g., some coral growth rates decline in proportion to a changing carbonate mineral saturation state) but a continuous effect, then these ecosystems could be at considerable risk even though they will remain in waters supersaturated with aragonite for the foreseeable future.

Acidification can also be exacerbated by changes in oceanic circulation. Deep waters that are upwelled are relatively acidic, which can negatively impact marine organisms. Acidified waters resulting from upwelling have already been shown to negatively affect oyster hatcheries in the U.S. Pacific Northwest (see Section 3). The California Current has been identified as an area of particular concern because coastal upwelling brings to the sea surface "old waters" that are naturally low in pH. These deep waters carry the cumulative signature of decomposition of organic matter through respiration processes that have taken place over hundreds of years. When this old water upwells in the California Current, it is naturally rich in CO_2 and is high concentration of nutrients, a low concentration of oxygen, and a low pH. Few measurements have been made of CO_2 and pH in upwelled waters of the California Current, but available data indicate that upwelled waters are undersaturated with respect to aragonite and have a pH of 7.6 to 7.7 (Feely et al., 2008), which are values already lower than those expected for global oceans by 2100. This also makes the coastal marine ecosystem on the U.S. West Coast particularly vulnerable to the ocean acidification impacts described above, as well as the impacts of ocean acidification on biological resources and ocean services (see Sections 3–4).

One of the more remarkable effects of the ocean's rapidly changing pH is the impact on low-frequency sound absorption (Hester et al., 2008). Sound is produced as a by-product of many anthropogenic activities such as shipping, oil and gas exploration, etc., as well as natural sources of noise in the ocean such as marine mammals, wind, earthquakes, etc. A decline in pH of approximately 0.3 causes a 40 percent decrease in the intrinsic sound absorption properties of surface seawater (Hester et al., 2008). Because acoustic properties are measured on a logarithmic scale, and neglecting other losses, low frequency sounds at frequencies important for marine mammals and for naval and industrial interests may travel some 70 percent farther with the reduction in ocean pH expected from a doubling of CO_2 (Brewer and Hester, 2009). Conversely, More recent modeling suggests that, due to the complexities of sound traveling through the ocean, actual increases in background noise are likely to range from being negligible (Joseph and Chiu, 2010; Reeder and Chiu, 2010) to a few decibels within the next 100 years (Udovydchenkov et al., 2010). With the magnitude of potential impact uncertain and generating debate among researchers, the effect of ocean acidification on background sound levels in the ocean is an area that deserves further study.

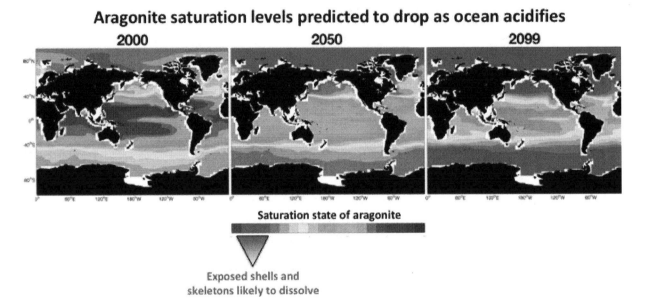

Figure 2-7 Calculated saturation states of aragonite, a form of calcium carbonate often used by calcifying organisms. By the end of this century, polar and temperate oceans may no longer be conducive for the growth of calcifying organisms such as some mollusks, crustaceans, and corals. (Source: Feely, Doney, and Cooley, 2009).

2.11 Hypoxia

Hypoxia has been recognized as one of the most important water quality problems worldwide. The term "hypoxia" refers to a condition where an aquatic environment has oxygen concentrations that are insufficient to sustain most animal life (Diaz and Rosenberg, 1995; Vaquer-Sunyer and Duarte, 2008) and can lead to dead zones. The number of water bodies with recorded and published accounts of hypoxia from around the globe has increased from less than 50 in the 1960s to approximately 400 by 2008 (Diaz and Rosenberg, 2008). The number of water bodies in the U.S. with documented hypoxia follows the same trend, increasing from 12 prior to 1960 (Bricker et al., 2007) to over 300 by 2008 (CERN, 2010).

Hypoxia naturally develops when the water column becomes stratified, isolating an oxygen-depleted layer of bottom water and sediments that accumulate through the decomposition of organic matter from a usually well-oxygenated surface layer created by interactions with the atmosphere. Stratification resulting in hypoxia occurs in Long Island Sound and the New York Bight (Boesch and Rabalais, 1991; Lee and Lwiza, 2007). Development and maintenance of hypoxia are strongly affected by the mixing of water columns. In the fall, cooling temperatures and decreasing freshwater inputs can destabilize summer stratification, leading to relatively abrupt mixing, or "turnover", of the water column that eliminates seasonal hypoxia. Strong tidal mixing prevents water column stratification almost entirely in some systems. Estuaries in the northeastern Gulf of Mexico, such as Mobile Bay, Alabama, and Pensacola Bay, Florida, have low amplitude tides (Hagy and Murrell, 2007), and others, such as the Albermarle-Pamlico Estuarine System on the North Carolina coast, are virtually tideless (Luettich et al., 2002). These estuaries, which are also in a warmer climate resulting in warmer waters, are particularly susceptible to stratification and hypoxia (Hagy and Murrell, 2007; Park et al., 2007; Reynolds-Fleming and Luettich, 2004).

Hypoxia can be a natural occurrence, but human activities are increasing the frequency, duration, and intensity of naturally-occurring hypoxia (Cooper and Brush, 1991; Diaz and Rosenberg, 2008; Helly and Levin, 2004). Eutrophication, "an increase in the rate of supply of organic matter to an ecosystem," is most often associated with anthropogenic nutrient enrichment of coastal and ocean waters from urban and agricultural land run-off, wastewater treatment plant discharges, and air deposition of nutrients (Figure 2-8) (Bricker et al., 2007; Galloway et al., 2008). Eutrophication encourages algal blooms and, subsequently, the population growth of organisms that feed on algae. As organic matter sinks into the deeper waters, respiration by senescing algae as well as animals and bacteria that consume the material may utilize more oxygen than is being mixed into these deep waters. This net loss of oxygen may result in the development of hypoxic waters. The second-largest eutrophication-related hypoxic area in the world, after the Baltic Sea, which is approximately 80,000 km^2 (Hansen et al., 2007b; Karlson et al., 2002), occurs in the U.S., and is associated with the discharge from the Mississippi/Atchafalaya Rivers in the northern Gulf of Mexico (Alexander et al., 2008). The northern Gulf of Mexico hypoxic area has increased substantially in size since the mid-1980s when it was first measured at about 4,000 km^2 (Rabalais et al., 2007). In 2008, it encompassed 20,719 km^2, the second largest area on record (http://www.gulfhypoxia.net). The most commonly reported eutrophication-related problems included hypoxia, losses of submerged grasses, numerous occurrences of nuisance and toxic harmful algal blooms, and excessive algal blooms, which is the most commonly reported problem (see Case Study 2-C, Section 3, and Section 4). Upwelling of nutrient-rich deep ocean water into shallow areas also can support large blooms of phytoplankton (Chan et al., 2008) and may result in hypoxia. Climate-related changes in regional wind patterns might already be enabling the extent and severity of hypoxia off the coast of Oregon, a system dominated by upwelling (Chan et al., 2008). These changes led to the development of severe widespread hypoxia and, in 2006, for the first time, anoxia on Oregon's inner continental shelf (Chan et al., 2008; Grantham et al., 2004). Transport of hypoxic water from the continental shelf into coastal embayments has also been documented along the Oregon coast (Brown et al., 2007). These developments seem to be linked to impacts of climate variation on ocean processes, such as intensity of upwelling winds, oxygen concentration in upwelled water, and water column respiration (Chan et al., 2008; Grantham et al., 2004). Subtler upwelling has also been observed along the New Jersey coast and has been implicated in the development of nearshore hypoxia (Glenn et al., 1996, 2004).

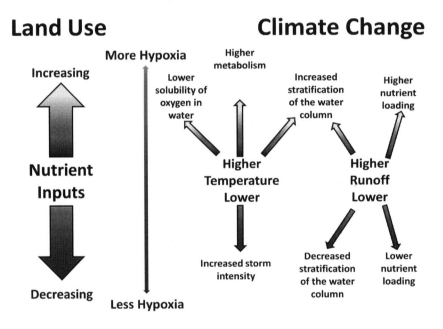

Figure 2-8 Relative magnitude and contribution (the larger the arrow, the larger the contribution) of land management practices versus climate change factors to the expansion or contraction of low-dissolved oxygen. (Source: modified from Diaz and Breitberg, 2009).

Climate change will almost certainly exacerbate both naturally occurring and eutrophication- related hypoxia as well as the incidence of harmful algal blooms. In general, the expected long-term ecological changes favor progressively earlier onset of hypoxia each year and, possibly, longer overall duration (Boesch et al., 2007). Increasing average water temperature is one mechanism by which climate change may increase susceptibility of systems to hypoxia. Higher water temperatures promote increased water column stratification, decreased solubility of oxygen, and increased metabolic rates, including oxygen consumption and nutrient recycling.

Climate predictions also suggest large changes in precipitation patterns, but with significant uncertainty regarding what changes will occur in any given watershed (IPCC,

2007a). Increased precipitation can be expected to promote increased runoff of nutrients to estuarine and coastal ecosystems and increased water column stratification within these systems, contributing to more severe oxygen depletion (Karl et al., 2009; Justić et al., 2007). Climate predictions for the Mississippi River basin suggest a 20 percent increase in river discharge (Miller and Russell, 1992), which is expected to increase the average extent of hypoxia on the northern Gulf of Mexico shelf (Greene et al., 2009). Climate-related changes in regional wind patterns may already be enhancing the extent and severity of hypoxia off the coast of Oregon, a system dominated by upwelling (Chan et al., 2008).

Case Study 2-C
Hypoxia in the Gulf of Mexico

In August 1972, scientists found severe hypoxia in bottom waters of the southeastern Louisiana shelf. In 1985, annual surveys began to map the extent of this seasonally hypoxic region in the Gulf of Mexico and now we know that it is the second-largest eutrophication-related hypoxic area in the world after the Baltic Sea. It encompasses 20,719 km2 in 2008 and is associated with the discharge from the Mississippi/Atchafalaya Rivers. Because dissolved oxygen levels below 2 parts per million cause suffocation in many fish species, this region has come to be known as the "Gulf of Mexico Dead Zone."

Not surprisingly, discovery of the Dead Zone immediately led scientists to seek the cause of hypoxic conditions. Investigators soon agreed that the most probable cause of these conditions was the influx of fertilizer and animal waste from the Mississippi River watershed, coupled with seasonal stratification of Gulf waters and subsequent upwelling events.

Fertilizer and animal waste include nutrients such as nitrogen and phosphorus that are essential for the growth of algae normally found in healthy marine and freshwater ecosystems. An overabundance of nutrients can cause excessive algal growth, so that more algae grow than can be consumed by other organisms in the system. Excess algae can reduce sunlight, crowd out other organisms, and lead to oxygen depletion when the algae die and begin to decompose.

Upwelling in the Gulf of Mexico is induced by dynamic uplift of deeper waters onto shelf regions and by prevailing winds favorable to pushing overlying waters offshore. The most consistent upwelling occurs on the western and southern boundaries of the Gulf of Mexico, along the Texas-Louisiana shelf, western Mexican states, and the Campeche Bank in the south. During the 1998 El Niño event, upwelling favorable winds

Algal blooms in Gulf of Mexico. Dead zones are areas of oxygen-depleted waters that form when nutrients stimulate the growth of algae blooms (Source: NASA).

Case Study 2-C (Continued)

in the northeast Gulf, coupled with other favorable conditions for phytoplankton growth, caused large blooms. High stratification in the water column suppressed aeration, created hypoxic conditions, and resulted in mass fish mortality.

Climate change will almost certainly exacerbate both naturally occurring and eutrophication related hypoxia, as well as the incidence of harmful algal blooms. Changes in temperature, precipitation, and winds will likely lead to long-term ecological changes that favor progressively earlier onset and duration of hypoxia each year. Climate predictions for the Mississippi River basin suggest a 20 percent increase in river discharge, which is expected to increase the average extent of hypoxia on the northern Gulf of Mexico and result in the continued expansion of the Gulf of Mexico Dead Zone.

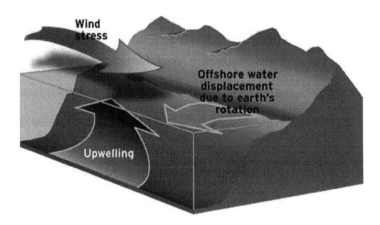

These displaced surface waters are replaced by cold, nutrient-rich water that wells up from below (Source: NOAA).

Chapter 3

Impacts of Climate Change on Marine Organisms

Executive Summary

Climate change effects on marine organisms and ecosystems are occurring throughout the U.S. These effects are occurring across scales, from changes in physiology including growth, reproduction, and survival, to alterations in interactions between species, to changes in the timing of life-history events, and ecosystem regime shifts. Climate-related impacts on ocean systems include shifts in species' phenology and ranges, increases in species' invasions and disease, and changes in the abundance and diversity of marine plants and animals, among others. Observations and research have demonstrated high variability in the vulnerability and responses of organisms to changes in climate, resulting in species that are positively impacted, or "winners", and species that are negatively impacted, or "losers."

These climate change effects on marine organisms and ecosystems are very likely to increase based on projected changes in the magnitude and variability of temperature, ocean pH, and other environmental parameters described in Section 2. These impacts result from changes in average conditions as well as the occurrence of rare but extreme events. In addition, unprecedented effects may be seen due to the complexity of marine ecosystems and the nonlinear interactions of multiple stressors on organisms and ecosystems. Progress is being made in forecasting the ecological responses of ocean systems to climate change, but the development of robust methods for projections into the future, including potentially novel environments, remains a challenge.

Climate change is often a threat multiplier, meaning that it impacts marine organisms by interacting with non-climatic stressors such as nutrient pollution and fishing pressure. Although in many cases these "multiple stressors" are simply additive in their impacts, interactions that are synergistic, or more than the sum of the individual effects, and antagonistic, or less than the sum of the individual effects, are also common. Scientific studies that address both climatic and non-climatic stressors can provide critical insight into these interactions and feedbacks. From a policy and management perspective, reducing non-climatic stressors at local-to-regional scales can provide an opportunity to enhance the resilience of marine ecosystems to climate change.

Evidence suggests that stressors can have both direct physiological effects on organisms as well as indirect effects through impacts on the species with which they interact. In addition, these responses are often non-linear, making threshold effects, or "tipping points", an area of concern. The sustained, long-term monitoring of ecological responses, with concomitant measurements of physical drivers and associated socio-economic impacts, will be key to advancing knowledge and improving projections of change.

Research is also needed to improve understanding of the processes and mechanisms by which changing conditions affect organisms and ecosystems. Past and current responses of marine organisms and ecosystems to climate change can provide important insight into patterns, trends, and trajectories of change; however, extrapolations to future responses must be made with caution, given that novel environmental conditions will occur. Additionally, local-to-regional context can have a major influence on organismal responses to climate change, meaning that conclusions regarding responses to environmental change at one location cannot necessarily be used to predict responses in another location without a more mechanistic understanding of underlying drivers and processes. For example, the degree to which organisms can acclimate and populations can genetically adapt to climate-driven environmental change is highly uncertain.

Key Findings

1. Throughout U.S. ocean ecosystems, climate change impacts are being observed that are highly likely to continue into the future.

   • Observed impacts include shifts in species distributions and ranges, effects on survival, growth, reproduction, health, and the timing of life-history attributes, and alterations in species interactions, among others.

   • Impacts are occurring across a wide diversity of taxa in all U.S. ocean regions, but high-latitude and tropical areas appear to be particularly vulnerable.

2. The vulnerability and responses of marine organisms to climate change vary widely, leading to species that are positively impacted, or "winners", and species that are negatively impacted, or "losers".

   • Species with more elastic life histories and/or higher physiological tolerance for changes in temperature and other environmental conditions will likely experience fewer climate-related impacts and may therefore out-compete less tolerant species.

   • Species such as corals and other calcifying organisms that are exposed to ocean conditions that inhibit their ability to produce calcium carbonate shells, will likely experience impacts. These impacts may result in cascading effects on marine ecosystems.

3. Climate change interacts with and can exacerbate the impacts on ocean ecosystems of non-climatic stressors such as pollution, overharvesting, disease and invasive species.

   • Climate-related stressors such as changes in temperature can operate as threat multipliers, worsening the impacts of non-climatic stressors.

   • Opportunities exist for ameliorating some of the impacts of climate change through reductions in non-climatic stressors at local-to-regional scales.

   • Predicting the effects of multiple stressors is difficult given the complex physiological effects and interactions between species.

4. Past and current responses of ocean organisms to climate variability and climate change are informative, but extrapolations to future environmental conditions will differ from the range observed in recent history.

- Observed responses to ongoing environmental change often vary in magnitude across space and time, suggesting that extrapolations of responses from one location to another may be challenging.

- In addition to gradual declines in ecosystem function, potential threshold effects, or "tipping points", that could result in rapid ecosystem change are a particular area of concern.

Key Science Gaps/Knowledge Needs

Improving understanding and projections of climate change impacts on natural and managed ecosystems remains a critical challenge. How species, habitats, and ecosystems will respond when confronted by environmental conditions that are often outside the range of conditions experienced today is difficult to forecast. In addition, the interactions between climatic and non-climatic stressors can result in "surprises", including the potential for tipping points. Therefore, there is a growing need to:

- Investigate the degree to which marine organisms can acclimate and populations can genetically adapt to rapid environmental change.

- Determine the cumulative impacts of multiple climatic and non-climatic stressors on marine organisms within ecologically-relevant contexts.

- Improve understanding and prediction of environmental and ecological conditions that lead to non-linearities and tipping points in coastal and marine ecosystems.

- Enhance the development of spatially-explicit predictions of ecosystem responses to climate change, particularly for local-to-regional scales, including estimates of uncertainty.

- Further the ability to measure and forecast physical variables at scales and resolutions that are relevant to ecological responses.

- Improve understanding and valuation of climate-related impacts on ocean ecosystem services.

3.1 Physiological Responses

Considerable progress has been made in understanding physiological responses of marine organisms to climate change (Pörtner and Farrell, 2008; Somero, 2011) and in projecting future responses of individual species (Chown and Gaston, 2008; Helmuth, 2009). Key to this understanding are findings that indicate that multiple climate-related and non-climate-related stressors interact in their impacts on marine organisms, and that physiological responses to these stressors can be highly variable across species and life-history stages.

The ability of marine organisms to grow, reproduce, and survive is affected by their environment as well as the other organisms and species with which they interact. Virtually all physiological processes are affected by an organism's body temperature (Somero, 2011) and recent studies have also begun to explore the important impacts of the ocean's changing pH (ocean acidification; Hoegh-Guldberg et al., 2007; Hofmann et al., 2010; Widdicombe and Spicer, 2008), alterations in salinity (Gedan and Bertness, 2010; Lockwood and Somero, 2011a) and changes in food supply (Lesser et al., 2010). Physiological responses of marine organisms to environmental change include effects on survival (Jones et al., 2009), growth (Menge et al., 2008), changes in the timing (Carson et al., 2010) and magnitude (Petes et al., 2008) of reproduction, and metabolism (Jansen et al., 2007), as well as other consequences, such as increased susceptibility to disease (Anestis et al., 2010a; Mikulski et al., 2000). Importantly, climatic and non-climatic stressors such as pollution (Sokolova and Lannig, 2008), physical disturbance (Bussell et al., 2008), and overharvesting (Hsieh et al., 2008, Sumaila et al., 2011), interact with one another, both in their physiological effects on individuals (Hofmann and Todgham, 2010; Hutchins et al., 2007) and in their cumulative impacts on ecological communities (Crain et al., 2008, Dijkstra et al., 2011, Pandolfi et al., 2011). Thus, although considerable progress has been made in understanding physiological and ecological responses to climate change, and in making predictions about the likelihood of future physiological and ecological responses (Helmuth, 2009; Mislan and Wethey, 2011; Nye et al., 2011), additional work is needed to more clearly understand the impacts of the temporally- and spatially-complex changing environment on marine organisms and ecosystems. Advances in molecular technology for detecting physiological responses of organisms to stress and the genetic underpinnings of stress responses offer a promising approach for furthering understanding of the cellular and genetic bases of stress responses (Dahlhoff, 2004; Hofmann and Place, 2007; Place et al., 2012; Pörtner, 2010; Somero, 2011; Tomanek, 2011; Tomanek and Zuzow, 2010; Trussell and Etter, 2001).

One area of uncertainty is the ability of marine organisms to acclimatize, or populations to locally adapt, to new and rapidly changing environmental conditions (Bell and Collins, 2008; Hofmann and Todgham, 2010; Sorte et al., 2011; Trussell and Etter, 2001). Some evidence exists for local adaptation of marine organisms to high-stress environments. For example, marine snails on the Oregon coast experience higher levels of aerial thermal stress due to local environmental conditions experienced at low tide than do southern populations of the same species in California (Kuo and Sanford, 2009). Adult snails in Oregon have higher thermal tolerance than do their counterparts at cooler California sites and they transmit this high tolerance to their offspring, suggesting that local populations have genetically adapted to the more extreme conditions they experience (Kuo and Sanford, 2009). Some researchers have suggested that such acclimatization and adaptation to thermal stress by corals and their symbionts may help reefs to maintain their structure in the future (Baskett et al., 2009; Pandolfi et al., 2011), but this prediction is still debated (Hoegh-Guldberg et al., 2011). Most research to date on local adaptation in marine systems has been conducted on organisms with fast reproductive cycles, such as microorganisms (Bell and Collins, 2008; Collins and Bell, 2004) and copepods (Kelly et al., 2012). Additional research is needed to better elucidate the potential roles of acclimatization and local adaptation to current and future change in marine environments (Kelly et al., 2012; Schmidt et al., 2008).

Effects of temperature change

Ocean temperature change is affecting a diversity of physiological and ecological processes in marine organisms, which is projected to increase in the future. Temperature has diverse effects on physiological processes in marine organisms (Somero, 2011), including changes in metabolic rate (Jansen et al., 2007) as well as other cellular functions such as the performance of critical enzymes (Somero, 2011). Thermal stress can lead to an increase in the metabolic oxygen organisms need, and ultimately to oxygen deficiency at the cellular level (Pörtner, 2010; Pörtner and Farrell, 2008). Metabolic oxygen deficiency has already been documented in commercially-important species such as Atlantic cod (Pörtner et al., 2008; Sartoris et al., 2003) and may become increasingly problematic in the future as increasing ocean temperatures are expected to exacerbate ambient low-oxygen concentrations through decreased oxygen solubility and increased oxygen demand by algae (Hofmann et al., 2011).

All organisms display tolerance limits that, when exceeded, lead to impacts on metabolism, growth, reproduction, and/or survival. Endothermic, or "warm-blooded," organisms such as mammals and birds must maintain a relatively constant body temperature; therefore, changes in the ambient temperature outside of their preferred range require additional expenditures of energy. If temperatures become either too warm or too cold to maintain body temperature within tolerable limits, sub-lethal and lethal effects can occur. For example, manatees living in Florida experience a cold stress syndrome when water temperatures fall below 20°C for several days; consequences can include emaciation, immunosuppression, and increased mortality (Bossart et al., 2002). A recent unusual morality event occurred from January to April 2010, when a total of 480 manatees, 70 percent of which were juveniles, were found dead, with greater than 50 percent of the mortality attributed to cold stress (Barlas et al., 2011) associated with a record-setting negative phase of the Arctic Oscillation (NCDC, 2010). Increasing episodes of extreme weather events or alterations in average winter or summer temperatures can have negative impacts on endothermic marine species and repeated mortality events resulting from thermal stress can lead to population decreases.

The vast majority of marine organisms, other than birds and mammals, are ectothermic, or "cold-blooded," which means that their internal temperatures are driven by ambient environmental conditions. Although some marine organisms exhibit broad tolerances, others already live close to or at their thermal tolerance limits (Hochachka and Somero, 2002; Somero, 2011). Extreme or prolonged high or low temperature events can lead to sub-lethal effects such as reduced growth and changes in the timing and magnitude of reproductive output (Anestis et al., 2010b; Dijkstra et al., 2011; Petes et al., 2008) as well as mortality (Harley, 2003; Harley and Paine, 2009; Petes et al., 2007). Elevated temperature has been linked to risk of dislodgement by waves in mussels, potentially because byssal threads, the animals' primary anchoring system, decay more quickly in warmer water (Moeser et al., 2006). Loss of mussels would translate into a decline in local diversity because hundreds of invertebrate species rely on mussel beds for habitat (Smith et al., 2006a). Declines in such physiologically vulnerable "foundation species" may multiply the detrimental impacts of climate change on ocean ecosystems (Gedan and Bertness, 2010). In contrast, warm-adapted species may increase in abundance and range as they are able to invade new territory due to warming temperatures (Urian et al., 2011).

Reef-building corals around the world, including those in U.S. states and territories, have been negatively impacted by increasing water temperatures. Corals are among the most vulnerable organisms to even slight changes in temperature. When water temperatures exceed "normal" summer extremes by as little as 1-2° Celsius for 3-4 weeks (Gleeson and Strong, 1995), corals eject their symbiotic dinoflagellates, known as zooxanthellae, which most coral species depend on to meet their metabolic requirements and to form skeletons (Hoegh-Guldberg et al., 2007). This phenomenon, known as "coral bleaching," is not necessarily immediately fatal, but it can lead to severe reductions in reef health and resilience (Hoegh-Guldberg et al., 2007). If populations of zooxanthellae are unable to reestablish themselves within host coral tissue, corals often suffer mortality in the mid- to long-term (Pandolfi et al., 2011). Corals that have recently bleached may also be more susceptible to disease outbreaks (Miller et al., 2009). Although debate on the ability of corals to adapt to environmental stress still exists (Hoegh-Guldberg et al., 2011; Pandolfi et al., 2011), a recent report by the World Resources Institute (Burke et al., 2011) indicated that 75 percent of the world's coral reefs, including almost all of the reefs in Florida and Puerto Rico, are threatened by the interactive effects of climate change and local sources of stress such as nutrient pollution. The same report projected that roughly 50 percent of the world's reefs will experience severe bleaching due to thermal stress by the 2030's, and more than 95 percent by the 2050's based on current trajectories of greenhouse gas emissions (Burke et al., 2011). Carpenter et al. (2008) suggest that 1/3 of all reef-building coral species are at risk of extinction due to the combined effects of climate change and local stressors. Loss of coral cover and reef three-dimensional complexity leads to losses of the many species of associated fishes and invertebrates that depend directly and indirectly on corals (Alvarez-Filip et al., 2009; Graham et al., 2006; Idjadi and Edmunds, 2006). Therefore, continued loss of coral reefs is highly likely to have cascading effects on diversity, structure, function, and the valuable ecosystem services on which humans depend (Mumby and Steneck, 2011).

Interactions between thermal stress and food availability can also affect physiology. For some marine animal species, increasing food supply can result in higher levels of thermal tolerance (Schneider et al., 2010). Climate change is expected to impact individual nutrition status as prey species shift geographic and depth ranges, altering the dynamics of food webs (Bluhm and Gradinger, 2008). These impacts could be either negative, for organisms that depend on specific prey items, or neutral or positive, for species that have generalized diets. Food depletion and resultant nutritional stress are well-known causes of immune suppression across several taxa (Burek et al., 2008). Increased exposure to unfavorable environmental conditions may exacerbate nutritional deficiencies, leaving individuals weakened and more susceptible to stress.

Ocean acidification impacts

As the oceans absorb increasing levels of atmospheric carbon dioxide, chemical reactions occur that lead to decreases in ocean pH, a phenomenon known as ocean acidification (see Section 2), and a concomitant reduction in the solubility of calcium carbonate, which many organisms use to create shells. Biological processes known to be affected by ocean acidification include calcification, photosynthesis, nitrogen fixation and nitrification, ion transport, enzyme activity, and protein function (Gattuso and Hansson, 2011; Hutchins et al., 2009; Hofmann et al., 2010).

A growing number of laboratory and field studies have documented the negative impacts of ocean acidification on calcifying organisms (Doney et al., 2009; Fabry, 2008; Gattuso and Hansson, 2011). A recent meta-analysis (Kroeker et al., 2010) found an overall negative effect for the many types of organisms studied; however, the responses were variable when taxonomic groups were separated. Physiological studies have likewise shown that organisms can vary greatly in their responses to decreasing pH, even among closely-related species (Byrne, 2011; Doney et al., 2009; Fabry et al., 2008; Hofmann et al., 2010; Ries et al., 2009). One challenge to enhanced understanding is that few natural modern analogs exist to a world under increased acidification (Hönisch et al., 2012); thus, recent studies have examined natural pH gradients surrounding underwater volcanic vents (Fabricius et al., 2011; Hall-Spencer et al., 2008; Porzio et al., 2011) and acidification in the fossil record (Crook et al., 2011; Hönisch et al., 2012; Pandolfi et al., 2011; Ries, 2010) to gain insight into acidified conditions.

As with warming, ocean acidification is unevenly distributed with latitude (Hoegh-Guldberg and Bruno, 2010). The higher solubility of CO_2 in colder water leads to disproportionately large pH decreases and potentially greater impacts in the polar oceans (McNeil and Matear, 2008; Orr et al., 2005). Although the biological implications of acidification for open ocean ecosystems remain an active area of research, some general trends are emerging. Calcification by the planktonic algal group coccolithophores, which forms massive open ocean blooms, will almost certainly be greatly reduced (Beaufort et al., 2011; Feng et al., 2008; Hutchins, 2011; Riebesell et al., 2000). Zooplankton groups such as foraminifera and pteropods that produce calcium carbonate shells will also be adversely affected (Lombard et al., 2010; Moy et al., 2009; Orr et al., 2005). Pteropods are an especially critical link in high-latitude food webs because commercially-important species such as salmon depend heavily on them as prey (Fabry et al., 2009).

In addition to calcification, another critical biogeochemical process that appears to be strongly inhibited by ocean acidification is nitrification. This process is a key link in the ocean's nitrogen cycle by which certain prokaryotes convert ammonia to nitrate, thereby making oxidized nitrogen forms available to marine biota (Beman et al., 2011). Other ecologically important plankton groups may benefit from increased seawater CO_2 levels. In particular, nitrogen fixation rates of some dominant marine cyanobacteria increase substantially at low pH (Fu et al., 2008; Hutchins et al., 2007), as does cellular toxin production by some harmful algal bloom species (Sun et al., 2011; Tatters et al., 2012).

Ocean acidification can interact with other climatic and non-climatic stressors; for example, increased acidity can decrease thermal tolerance of some marine animals due to oxygen limitation (Pörtner, 2010). Not all interactions lead to negative responses; one study found that fertilization success of many species was not affected by the interaction of increased temperature and ocean acidification (Byrne et al., 2010). Such complexities emphasize the importance of examining interacting environmental factors, variability in the responses of ecologically and economically important species, and long-term impacts (Doney, 2010; Hofmann et al., 2010).

Certain species and ecosystems are particularly vulnerable to ocean acidification. Because coral skeletons are formed of calcium carbonate, the future of coral reefs under ocean acidification has received considerable attention. Studies have suggested that reef accretion stops at atmospheric CO_2 concentrations of 480 ppm (Kleypas and Langdon, 2006), and climate models predict that the worldwide dissolution of coral reefs is possible at atmospheric CO_2 levels of 550 ppm (Silverman et al., 2009). A study based on cores from the Great Barrier Reef in Australia showed that coral calcification rates declined 21 percent between 1988 and 2003 (Cooper et al., 2008). The observed decrease exceeds that predicted by changes in pH alone, suggesting potential effects of multiple and interacting stressors, such as ocean acidification, temperature, and nutrients (Cooper et al., 2008; Doney et al., 2009). Studies of reef communities at varying distances from natural CO_2 seeps have documented large declines in both coral colony size, coral species richness, and coralline algae, which is the preferred settlement substrate of coral larvae, at low pH (Fabricius et al., 2011). Research has also demonstrated the negative effects of ocean acidification on calcification by shellfish (Gazeau et al., 2007), particularly in areas such as the U.S. Pacific Northwest (see Case Study 3-A), where acidified waters have been documented (Feely et al., 2008).

The geological past shows that the abundance and diversity of calcifying organisms is reduced when large amounts of CO_2 are rapidly released into the atmosphere (Zachos et al., 2005); however, this evidence is only partially helpful in understanding potential impacts of ocean acidification in the future because even relatively abrupt geological changes in CO_2 levels in the past likely occurred over thousands of years, allowing the ocean enough time to chemically buffer these increases (Gattuso and Hansson, 2011). The rate of acidification is much more rapid today (Gattuso and Hansson, 2011), creating a magnitude of ocean change that is potentially unparalleled in at least the past ~300 million years of the Earth's history (Hönisch et al., 2012).

Notably, most studies to date of the physiological impacts of ocean acidification have been based on short-term experiments that range in duration from hours to weeks (Doney et al., 2009). Therefore, determining what the biological effects of chronic exposure to decreased pH might be and how ocean acidification interacts with other stressors in intact ecosystems remain critically necessary topics for future research.

Case Study 3-A
Ocean acidification impacts on the oyster industry

Commercial bivalve production on the West Coast of North America is a $273M venture, with oyster hatcheries providing most of the seed used by growers. In recent years, the primary hatcheries (Taylor's and Whiskey Creek Shellfish Hatcheries) supplying the seed for U.S. West Coast oyster growers suffered persistent production failures. Non-hatchery stocks of these oysters also showed non-sustaining recruitment, putting additional strain on the limited seed supply. Potential causes, including low oxygen and pathogenic bacteria, were explored and discounted. The Whiskey Creek Shellfish Hatchery in Oregon began to suspect elevated CO_2 as the culprit and sent samples to Oregon State University and the NOAA Pacific Marine Environmental Laboratories for analyses.

Case Study 3-A (Continued)

Findings showed that Netarts Bay, on which the Hatchery is located, experiences a wide range of carbonate chemistry fluctuations in response to upwelling events and the metabolic state of the Bay. In addition, larval oysters are strongly sensitive to the carbonate chemistry of the water in which spawning takes place. Most important was the relative favorability of the water for the formation of aragonite (ΩA), the mineral of which larval shell material consists.

Two important findings result from this study. First, the production break-even point corresponds to atmospheric CO_2 levels predicted within the next 2-3 decades. West Coast oysters thus serve as a "canary in the coal mine" for other locations where elevated CO_2 levels have not yet become persistent. Second, larval oysters show delayed response to carbonate chemistry during spawning and initial shell growth conditions.

Why are oyster larvae so dependent on the chemistry of ambient waters and what is the mechanism for that dependence? New work is beginning to unravel these questions. Within 48 hours of fertilization, larval oysters transition from having no shell at all to having 70 percent of their mass consist of shell mineral material. During this period, they depend on carbonate drawn from seawater rather than from internal reserves for shell carbon. In addition, lipid (i.e., fat) reserves are severely depleted, highlighting the need for additional energy during this critical bottleneck in development.

Together, these results paint a picture of larval development that depends on favorable ambient conditions during critical and energetically expensive early-growth bottlenecks, with results that do not express themselves clearly until later in the organisms' lives. Hatchery managers who embrace quality measurement technology can optimize operations for favorable conditions and even control conditions with active manipulation of inflow water chemistry, but natural populations will be subject to stress from additional acidification as CO_2 levels rise.

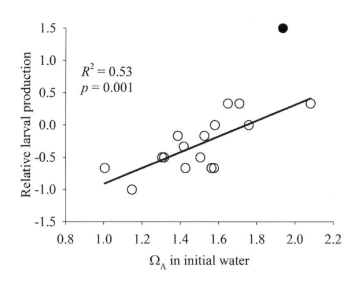

$R^2 = 0.53$
$p = 0.001$

Relative larval production at Whiskey Creek Shellfish Hatchery in response to the favorability of ambient waters with respect to aragonite (Ω_A). Relative production is positive when growth exceeds mortality; in the waters adjacent to the Hatchery, this condition is met at Ω_A ~1.7, corresponding to a pCO_2 of ~450 µatm. Reproduced from Barton et al., 2012

Exposure to toxicants

Toxicants are poisonous substances that can be produced by organisms (i.e., biotoxins), released from geologic stores such as heavy metals and some hydrocarbons, or result from a variety of anthropogenic sources such as persistent organic pollutants, petroleum hydrocarbons, heavy metals, and radioactive substances (Burek et al., 2008). Climate change can alter toxicant exposure levels for marine organisms through changes in the distribution, frequency, and toxicity of harmful algal blooms (see Sections 3.2, 4.6) and other toxins. Climate-related changes can also occur through alterations in ocean currents, which can carry both toxicants and organisms into novel environments, and rates of precipitation, which, when increased, can lead to additional runoff of toxicants into estuaries. Additionally, changes in feeding ecology can propogate toxicants throughout the food web (Macdonald et al., 2005) such that toxicant exposure levels change as animals alter their diets based on shifts in their prey items (Burek et al., 2008). Moreover, in an example of interactions between climatic and non-climatic stressors, organisms experiencing thermal stress may be more susceptible to the effects of contaminants (Schiedek et al., 2007). Responses of marine organisms and ecosystems to changes in toxicant exposure are difficult to predict and will largely depend on the chemical features of the specific toxicants to which they are exposed (Segner, 2011).

Effects on life history tradeoffs and larval dispersal

Physiological tradeoffs occur when resources are limited because each organism has a certain amount of energy available to maintain physiological processes such as growth, reproduction, metabolism, and immune functioning (Roff, 1992; Stearns, 1992). The production of gametes and offspring, which are energetically costly, may therefore be compromised under stressful conditions as energy is redirected toward defense and survival mechanisms (Wingfield and Sapolsky, 2003). Evidence of these life-history tradeoffs has been documented for U.S. West Coast intertidal mussels, which exhibited reduced relative allocation of energy toward growth and reproduction and increased energy toward physiological defenses under high-stress conditions (Petes et al., 2008). In the western North Pacific, differences in optimal temperatures for growth of larvae of Japanese anchovy (22°C) and sardine (16°C) lead to contrasting fluctuations in larval growth rates of these two species based on ambient temperature (Takasuka et al., 2007). Such preferences for thermal regimes create out-of-phase stock oscillations for anchovies and sardines off California; in other words, when conditions are optimal for sardine growth, they are sub-optimal for anchovies, and vice versa (Chavez et al., 2003). Changes in water temperature may alter these oscillations and therefore the relative abundance of these species in the future.

Phenology, the timing of annual life-history events such as migration and breeding can provide valuable insight into the impacts of climate change. Thermal stress has been found to alter the timing of spawning events in marine organisms, leading to mismatches of larval production with the peak in phytoplankton that serves as their food supply (Durant et al., 2007; Edwards and Richardson, 2004; Philippart et al., 2003). These mismatches can lead to starvation, lower growth and development rates, reduced survival probabilities, and decreased recruitment (Cushing, 1996). Impaired reproduction can have large, negative consequences for population dynamics and, in the most extreme cases, can lead to species collapse (Beaugrand et al., 2003). Differential changes in the timing of reproduction across biogeographic gradients can have subsequent effects on patterns of larval dispersal (Carson et al., 2010).

Exchange rates of adults, juveniles, larvae, and gametes determine levels of connectivity between populations, and can drive both local processes and meta-population dynamics (Erlandsson and McQuaid, 2004; Gouhier et al., 2010). Over large spatial scales, larval supply can determine biogeographic range boundaries (Herbert et al., 2009) as well as the colonization and spread of invasive species (EPA, 2008; Zardi et al., 2011). Thus, changing patterns of ocean circulation (see Section 2) are likely to have a significant influence on the ecology and population genetics of marine organisms. Temperature-dependent metabolism leads to an inverse relationship between temperature and the duration of planktonic larvae in the water column, suggesting that some species may develop more quickly under elevated temperatures, spending less time as larvae in the water column, and therefore potentially dispersing shorter distances (O'Connor et al., 2007). On the other hand, faster growth and developmental rates under increased temperatures may increase survival probabilities through the larval stage (Hare and Cowen, 1997; Kristiansen et al., 2011). Larval stages of some marine organisms are more vulnerable to stress than are their corresponding juvenile (Talmage and Gobler, 2011; Zippay and Hofmann, 2010) or adult (Matson and Edwards, 2007) stages. These findings emphasize the importance of considering the relative vulnerabilities of different life-history stages to climate change in order to better understand and predict changes in future population sizes (Russell et al., 2011).

3.2 Population and Community Responses

There is strong evidence that climate-driven changes in environmental conditions are affecting the survival, growth, and reproduction of a diversity of marine species, resulting in alterations in population sizes and subsequent effects on marine communities. Shifts in the distribution of many marine species that are consistent with changes in climate have been observed in coastal waters of all U.S. regions. In general, warm-adapted species are moving poleward (Parmesan and Yohe, 2003), although these responses are highly variable due to the impacts of local environmental conditions including non-climatic drivers (Helmuth et al., 2006a). Population size and distribution are also indirectly affected by climate-related changes in species interactions, such as competition and predation. In addition, strong evidence indicates that many marine species appear to be more vulnerable to disease when exposed to climate-related environmental stress. Collectively, these impacts are leading to observed changes in community composition and ecosystem processes. Exploring the relative sensitivity of marine species and their interactions to changing environmental conditions within an ecological context is key to advancing understanding and projections of future change.

Primary productivity

Marine primary productivity by both microscopic and macroscopic photosynthetic organisms forms the base of most of the ocean's food webs. The majority of marine primary producers are phytoplankton, a diverse suite of microscopic photosynthetic organisms.

Macroalgae, which are seaweeds, and seagrasses are important primary producers that also provide nearshore habitat and food sources to a diversity of marine organisms. Shifts in primary productivity are frequently linked with patterns of oceanic circulation. Due to the complex linkages between air and water temperature, oceanic circulation, and atmospheric conditions (see Section 2), the consequences of climate change on primary productivity are frequently non-intuitive in both coastal and open-ocean systems. In the open ocean, primary productivity can be impacted in numerous ways by increases in water temperature through increased metabolic rates (increased productivity; Doney et al., 2012) or increased stratification (decreased productivity; Behrenfeld et al., 2006). Warming also can lead to dominance by small-celled phytoplankton, or picophytoplankton, which reduces energy flow to higher trophic levels (Hare et al., 2007; Morán et al., 2010). One climate model found that when comparing projected changes in phytoplankton primary production, total density, and size structure for the North Pacific over the 21st century, the dominant response was a shift in size structure, typically towards smaller-sized phytoplankton, which alters food chain length (Polovina et al., 2011).

It is uncertain whether marine primary productivity will increase or decrease under future climate change scenarios. On a global scale, a recent study suggested that the past several decades have shown an overall increase in marine primary productivity (Chavez et al., 2011). Primary productivity in the central and southern California Current System has increased over the past three decades (Chavez et al., 2011). This is correlated with increases in the intensity and duration of wind-driven coastal upwelling events, which drive cold, nutrient-rich water from deep water to the sea surface (Garcia-Reyes and Largier, 2010). In contrast, satellite-derived time series of chlorophyll have shown significant changes in phytoplankton, notably that the most chlorophyll-poor areas have been expanding (Behrenfeld et al., 2006; , Irwin and Oliver 2009; McClain et al., 2004; Polovina et al., 2008), which indicates a reduction in primary productivity. Comparing the output from different earth system models, Steinacher et al. (2010) predict reductions in global primary production of 2-20 percent by 2100, with declines in mid-to-low latitudes due to reduced nutrient input into the euphotic zone and gains in polar regions due to warmer temperatures and less sea ice. Additional observations, research, and modeling efforts will be necessary for improving understanding of the complex relationships between climate change and marine primary productivity.

Harmful algal blooms (HABs) of macroscopic and microscopic, or single-celled, species have been recorded on nearly all of the world's coastlines. Over the past ten years, HABs have been reported in all major U.S. coastal locales (Anderson, 2012). HABs have increased in duration, number, and species diversity over the past three decades (Anderson, 1989, 2009; Hallegraeff, 1993) and can have large negative ecological and economic consequences. Some microalgal HABs cause direct impacts through their production of toxins that are harmful or lethal to consumers, including shellfish, seabirds, and humans (See Section 4.6). Levels of cellular toxin production by some HAB species increase dramatically under high CO_2 conditions, especially when growth is limited by nutrients (Fu et al., 2010; Sun et al., 2011; Tatters et al., 2012). These findings have implications for increased HAB impacts in a more acidified, stratified future ocean. HABs involving the red tide organism *Karenia brevis*, a dinoflagellate that produces potent neurotoxins called brevetoxins, occur frequently along Florida's southwest coast, causing episodes of high mortality in fish, sea turtles, birds, bottlenose dolphins, and manatees (Bossart et al., 1998; Flewelling et al., 2005; Gunter et al., 1948; Kreuder et al., 2002; Landsberg et al., 2009). Although brevetoxin exposure increases during *K. brevis* blooms, the persistence of the toxin in the food web and the long-term effects of exposure on marine mammals are unclear (Fire et al., 2007). Domoic acid, a potent neurotoxin produced by the diatom *Pseudo-nitzschia spp.*, has been linked to mortalities in California sea lions, sea otters, common dolphins, and sea birds along the Pacific coast (Kreuder et al., 2005; Scholin et al., 2000; Torres de la Riva et al., 2009). As well as causing acute mortality, domoic acid can have chronic effects that lead to epilepsy, memory loss, and decreased foraging success in California sea lions (Goldstein et al., 2008; Thomas et al., 2010). Little is currently known about the effects of low-dose chronic or repeat exposures of HAB-associated biotoxins on marine organisms and humans.

Blooms of "nuisance" macroalgae may shade out other benthic primary producers, either seagrasses or perennial macroalgae, and negatively impact coral reefs through competitive interactions and reductions in coral larval settlement (Diaz-Pulido et al., 2011; Hughes et al., 2007; Lapointe et al., 2005; Taylor et al., 1995; Case Study 3-B). Additionally, when blooms of micro- or macroalgae senesce, their decomposition may cause large-scale mortalities of benthic and pelagic organisms due to lowered water-column oxygen levels (Deacutis et al., 2006; Lopez et al., 2008).

Many mechanisms are potentially responsible for the expansion of algal blooms into new areas and their extended duration in pre-existing areas. Increased nutrient inputs, such as those resulting from sewage treatment plants and fertilizer runoff, may be partially responsible for increases (Teichberg et al., 2008; Thornber et al., 2008; Valiela et al., 1997), whereas reductions in nutrient inputs may result in decreases in bloom density (Johansson, 2002). Although nutrient increases, or eutrophication, are not responsible for all HAB events (Anderson, 2009, 2012), separating out the relative and interacting effects of eutrophication and climate change on HABs can be difficult (Heisler et al., 2008; Rabalais et al., 2009). In addition, new biotoxins are still being discovered, providing further insight into the mechanisms and impacts of HABs on marine organisms.

Macroalgae and seagrasses can be important primary producers in shallow coastal waters. Kelp systems are among the largest and most conspicuous macroalgae, supporting some of the most diverse and productive ecosystems along the U.S. coast (Dayton, 1985; Graham et al., 2008; Mann, 1973), primarily due to the provision of energy and complex habitat by the kelps themselves (Graham et al., 2007). The effect of climate change on the productivity of kelp forests across a variety of temporal and spatial scales also provides insight into macroalgal responses. Climate-induced variability in kelp distribution and abundance can affect the distribution of associated kelp forest fauna (Harley et al., 2006; Holbrook et al., 1997) as well as the productivity and diversity of kelp-associated communities (Dayton, 1985; Graham, 2004).

Most studies on the effects of annual-to-decadal variability of environmental factors on kelp systems have focused on the impacts of rising water temperature and the generally concomitant decrease in coastal upwelling (see Section 2) on their growth and survival (Broitman and Kinlan, 2006; Dayton et al., 1999; Reed et al., 2008). Periods of exposure as short as one month to anomalously warm, nutrient-poor ocean conditions can cause deterioration of kelp biomass, whereas prolonged exposure of years or decades to such conditions can lead to high mortality and distributional shifts of kelp taxa (Schiel et al., 2004). In some kelp systems, a simple shift may occur in the identity of the dominant kelp taxa according to species-specific environmental tolerances (Schiel et al., 2004), whereas in other systems, kelps and their associated communities may disappear altogether, resulting in an alternative habitat state (e.g., the formation of sea urchin barrens; Ebeling et al., 1985; Harrold and Reed, 1985; Ling et al., 2009). Although the global response of kelp systems to rising temperatures and decreasing nutrients may appear ubiquitous, the specific response in any given region will depend on the biogeography and environmental tolerances of the local kelp taxa (Martinez et al., 2003; Merzouk and Johnson 2011; Wernberg et al., 2010, 2011a). Monthly-to-decadal climate-related changes in wave disturbance can also have dramatic negative impacts on kelp forest distribution, abundance, and structure (Byrnes et al., 2011; Dayton et al., 1999; Reed et al., 2008,). Furthermore, kelp systems are predicted to be similarly impacted by changes in ocean conditions over millennial time scales, with kelp forest optima occurring during cool, nutrient-rich, and well-illuminated periods (Graham et al., 2010).

Local-to-regional scale variability in sea level, ocean pH, and atmospheric and oceanic carbon dioxide concentrations are likely also to affect macroalgae and their associated communities; however, direct studies of these environmental impacts are generally lacking. The survivorship of calcified macroalgae, which are present in temperate and tropical habitats throughout U.S. coastal waters, is greatly reduced under ocean acidification scenarios (Anthony et al., 2008). The combined impacts of ocean acidification and warming can increase skeletal dissolution rates (Diaz-Pulido et al., 2012) or lead to necrosis (Martin and Gattuso, 2009) for calcified macroalgae. In contrast, non-calcified macroalgae may have higher tolerance to ocean acidification (Diaz-Pulido et al., 2011). Although one study suggests that certain kelp life-history stages may be sensitive to ocean acidification (Roleda et al., 2012), others predict enhanced seaweed performance with rising CO_2 concentrations (Connell and Russell, 2010; Harley et al., 2006). Canopy-forming kelp systems that can directly access atmospheric CO_2 may be particularly sensitive to increases in CO_2 concentrations, but the degree of carbon limitation in kelp systems is relatively unstudied (Harley et al., 2006). The impacts of ocean acidification and other stressors on calcified and non-calcified macroalgae is an important opportunity for future research.

Shifts in species distribution

Climate change-induced shifts have been shown to influence the local and geographic ranges of many marine species (Burrows et al., 2011; Hoegh-Guldberg and Bruno, 2010). Analyses of shifts in species distributions have demonstrated that marine systems appear to be changing substantially faster than terrestrial ecosystems (Burrows et al., 2011; Helmuth et al., 2006b, Sorte et al., 2010b). Studies have shown range shifts in response to both gradual changes in the environment (Findlay et al., 2010, Lockwood and Somero, 2011b) and lasting, sometimes multi-decadal, changes following rare but extreme events (Denny et al., 2009; Firth et al., 2011; Harley and Paine, 2009; Wethey et al., 2011). Climate-related shifts often occur at range boundaries, but, due to the importance of local environmental factors (Burrows et al., 2011; Helmuth et al., 2006a), responses such as decreased growth and increased mortality can also occur well within a species' range

Case Study 3-B
Shifting interactions between corals and macroalgae

Corals and macroalgae are the dominant competitors for primary space on many coral reefs. Adult and juvenile corals can be shaded, abraded, or smothered by macroalgae (McCook et al., 2001), which typically grow much more quickly than corals. Corals can also be killed directly or indirectly by substances released by macroalgae (Smith et al., 2006b, Rasher et al., 2011). In addition, macroalgae can inhibit coral larval settlement through abrasion, called "whiplashing," and serving as unfavorable settlement surface.

When reef herbivores such as fish and urchins that graze on macroalgae are abundant and nutrient concentrations are low, corals are the competitive dominants on tropical hard bottoms. Local human activities tend to shift the competitive balance in favor of macroalgae at the expense of corals by removing herbivores that would normally keep macroalgal abundance relatively low. Increased nutrient levels in coastal waters can also favor macroalgal growth and lead to stress in corals. On many reefs, these processes have already resulted in a phase shift, which may be sudden or gradual, from coral-dominated reefs to algal-dominated reefs (Hughes et al., 2010).

Ocean warming and acidification amplify the impacts of these non-climatic stressors on corals. The dinoflagellate symbionts (zooxanthellae) that power coral growth are more sensitive to slight increases in temperature than are the macroalgae that are commonly found on reefs. Coral bleaching occurs with temperatures only 1-2°C above normal seasonal maxima, and, when severe, leads to coral death (Hoegh-Guldberg et al., 2007). Even if corals do survive, they may succumb to disease after experiencing thermal stress; for example, in 2005 in the U.S. Virgin Islands, severe bleaching, in which 90 percent of coral cover is pale or white, was followed by even more severe mortality, in which over 50 percent of the living coral cover died from diseases (Miller et al., 2009). Ocean acidification harms corals directly, by making it more difficult for them to secrete their skeletons, and indirectly, via the negative effects that acidification has on crustose coralline algae (Fabricius et al., 2011; Hoegh-Guldberg et al., 2007), a preferred settlement surface for coral larvae. Acidification may also increase coral sensitivity to high temperatures (Anthony et al., 2008).

Reducing local stressors such as overfishing and pollution is the best short-term strategy for increasing coral resilience to climate change (Hughes et al., 2010). This is supported by the fact that reefs remote from local human impacts, such as the uninhabited Northern Line Islands that are part of the U.S. Pacific Remote Islands Marine National Monument, still remain relatively healthy (Dinsdale et al., 2008; Sandin et al., 2008). However, over the longer term, climate change will severely compromise the ability of corals to outcompete macroalgae, even when herbivores and low nutrients favor coral dominance.

Overgrowth of coral by the alga Boodlea in Hawaii (Photo Credit: NOAA).

boundaries (Beukema et al., 2009; Harley, 2008; Place et al., 2008). As temperatures warm, current range limits at poleward range boundaries may shift and warm-adapted species, including certain invasive species, would become able to invade new territory (Urian et al., 2011). Forecasts of future responses to climate change based on observations of present-day changes and knowledge of physiological responses strongly suggest that changes in species distribution will continue (Nye et al., 2011; Runge et al., 2010). As indicated above (see Section 3.1), the pace and precise location of these changes remain uncertain because of the interactive effects of multiple stressors, the species-specific effects of these changes on interacting organisms, the spatial and temporal heterogeneity of environmental drivers, and the ability of organisms to acclimatize or adapt to changing conditions (Denny et al., 2009; Nye et al., 2011; Sagarin and Gaines, 2002; Sanford and Kelly, 2011).

Temperature increases are already thought to be altering the distributions of major phytoplankton groups because of numerous observations of poleward range extensions for temperate species (Cubillos et al., 2007; Hallegraeff, 2010; Hays et al., 2005; Merico et al., 2004; Peperzak, 2003). The responses of phytoplankton communities to experimentally-induced warming include major shifts in dominant species, often away from groups like diatoms that support "healthy" food webs and abundant fisheries and towards others like flagellates that tend to shunt marine productivity towards microbially-dominated systems that are less desirable to humans (Feng et al., 2009; Hare et al., 2007; Lewandowska and Sommer, 2010).

Temperature change is also affecting the abundance and distribution of marine invertebrates. Jones et al. (2009) reported high mortality of U.S. East Coast intertidal mussels at their southern range boundary in North Carolina as a result of warming temperatures between 1956 and 2007. A study of the marine snail *Kelletia kelletii* in California demonstrated that the northern range boundary had extended northward by over 400 km between the late 1970s and early 1980s, which was the first recorded extension north of Point Conception (Zacherl et al., 2003). These distributional shifts were consistent with an observed gradient in seawater temperature and the confluence of two major ocean currents. Dijkstra et al. (2011) compared present-day subtidal species composition at a site in New Hampshire with data collected in 1979, and noted a significant shift from perennial species, such as barnacles and mussels, to annual native and invasive species. A study using trace-elemental fingerprinting of larval mussel shells demonstrated autumn poleward movement and spring equatorward movement in the larvae of two species of mussels on the coast of California (Carson et al., 2010). These results suggest that effects of climate change on larval dispersal due to changes in currents and alterations in the timing of reproduction may lead to shifts in species distributions.

Climate change affects the distribution and abundance of marine fish species through a diversity of physical and biological processes and mechanisms whose relative importance varies across space and time (Ottersen et al., 2010; Overland et al., 2010). For example, fish stocks may shift geographically in response to changes in ocean temperature or the distribution of their prey (Barange and Perry, 2009; Cheung et al., 2009; Hsieh et al., 2008; Humston et al., 2004). Climate-related shifts in the abundance and distribution of commercially important species can have consequences for associated fisheries (Cheung et al., 2010; see Section 4). Evidence of climate-induced shifts in the distribution of marine fish has been recorded in several regions of the U.S. EEZ (Beaugrand et al., 2003). Fodrie et al. (2010) documented changes in assemblages of fish within seagrass beds in the northern Gulf of Mexico between the 1970s and 2007, reporting the addition of numerous fish species that had previously not been observed. Nye et al. (2009) examined the spatial distribution of 36 species of marine fish in from bottom trawl surveys along the continental shelf off the Northeastern coast of the U.S. from 1968-2007 and compared shifts to increases in bottom temperature. A significant poleward shift was found in 17 stocks, a southern shift in 4 stocks, and significant range expansion in 10 stocks (Nye et al., 2009). Shifts in the strength of the poleward undercurrent and ocean temperatures influence the spatial distribution of Pacific hake along the West Coast of North America, with a greater proportion of fish feeding off the coast of Canada in warm conditions (Agostini et al., 2006, 2007). Whether these shifts are responses to changes in oceanographic conditions or to changes in zonal and seasonal variations in the zooplankton community is unclear (Hooff and Peterson, 2006; Keister and Peterson, 2003,). Shifts in thermal habitats and prey fields are expected to influence the distribution of Pacific tuna and other fish populations (Lehodey et al., 2003, 2010; Polovina, 2007; Su et al., 2011). For the Arctic, where the rate and relative magnitude of change in ocean conditions is accelerated, differences in topography and currents suggest a higher likelihood of range expansions from the Atlantic side than from the Pacific side (Sigler et al., 2011). These expansions are likely to occur in response to shifts in population density and productivity (Wassmann, 2011). The ability of subarctic species to compete with species that are uniquely adapted to survive in the conditions of the Arctic is unknown.

Seabirds are also exhibiting shifts in species ranges as well as changes in foraging behavior resulting from shifting prey distribution and abundance. Arctic regions are particularly susceptible to range shifts as warm-adapted species move northward and rapid temperature increases affect the physiology and ecology of cold-adapted species. For example, prey shortages for least auklets in Alaska's Pribilof Islands have occurred as a result of high ocean temperatures (Springer et al., 2007). In the same region, reduced sea ice and increasing temperatures have led to breeding phenology shifts in kittiwakes over a 32-year period (Byrd et al., 2008). Temperature-related shifts in foraging distributions have been documented for little auks across the Greenland Sea related to sea temperature (Karnovsky et al., 2010). In addition, evidence indicates a possible range expansion of the razorbill in the Canadian Arctic in response to range expansion of favored prey (Gaston and Woo, 2008) as well as climate-related mismatches of prey availability and timing of breeding in thick-billed murres (Gaston et al., 2009).

As described above, species range shifts are occurring throughout U.S. ocean waters with cascading effects on ecological systems in many cases. Additional work is needed to determine the ecosystem-level consequences of climate-related shifts in species distribution for coastal and marine systems (Sorte et al., 2010b).

Marine diseases

Over recent decades, a significant increase in disease outbreaks has been reported in corals, urchins, mollusks, marine mammals, and turtles (Ward and Lafferty, 2004). Several disease-causing pathogens that were once thought to only occur on land are now known to have marine counterparts. The impacts of climate change on disease emergence and transmission are likely to act through a combination of several mechanisms, including host and pathogen range shifts, changes in contact frequency, changes in the proportion of individuals carrying disease vectors, introductions from terrestrial systems into marine environments, impacts to pathogen ability to reproduce, and increased environmental stress that leads to increased susceptibility of hosts to infection (Mills et al., 2010).

Evolved balances between disease agents, vectors, and hosts will likely be disturbed by climate change. In some cases, these changes could limit disease; in other cases, diseases could increase, particularly in stressed populations (Altizer et al., 2003; Harvell et al., 1999). Pathogens, including macro- and micro-parasites, are in a constant state of change and pathogen selection or alterations may alter the course of an outbreak or alter host species. Trends in infectious disease correlate with host-pathogen-environmental interactions as either the host becomes more susceptible to disease or the pathogen's virulence increases. Variations in species' ranges may alter pathogen distribution and warmer winters due to climate change can increase pathogen overwinter survivorship (Harvell et al., 2009). For example, the protistan parasite Perkinsus marinus, which causes Dermo disease in oysters, proliferates at high water temperatures and high salinities. Additionally, in oyster populations within Delaware Bay, epidemics followed extended periods of warm winter weather; these trends in time are mirrored by the northward spread of Dermo up the eastern seaboard as water temperatures have warmed (Cook et al., 1998; Ford, 1996). Evidence also suggests that increased water temperature is responsible for the enhanced survival of certain marine *Vibrio* bacteria, which can cause seafood-borne illness in humans (Martinez-Urtaza et al., 2010; see Section 4). Similarly, a survey of American lobsters' shell disease conducted in Massachusetts suggested that higher-than-average water temperatures between 1993 and 2003 led to increased disease prevalence (Glenn and Pugh, 2006).

Pathogens novel to marine organisms can enter coastal and oceans systems as terrestrial species expand their range and run-off from land increases because of higher rates of precipitation,. For example, fecal waste from the invasive Virginia opossum on the U.S. West Coast has resulted in an increase in the spread of *Sarcocystis neurona*, a protozoan parasite that infects and kills marine mammals, including sea otters (Miller et al., 2010b). In addition, the emergence and pathogenesis of the disease leptospirosis has been associated with environmental variability (Case Study 3-C). Leptospirosis causes mortality of California sea lions (Gulland et al., 1996; Lloyd-Smith et al., 2007) as well as effects in harbor seals and northern elephant seals (Colegrove et al., 2005; Kik et al., 2006).

The impacts of climate change on future rates of marine disease are uncertain. Changes in environmental conditions may lead to range shifts of macro- and micro-parasites, but those shifts do not necessarily result in increased disease spread. New habitats may contain barriers to the spread of disease through ecological interactions such as competition, physical barriers, or predation (Slenning, 2010). As with their hosts, pathogens and vectors are susceptible to climate-related stressors (Lafferty, 2009). Parasites that release gametes or larvae into the open marine environment or utilize intermediaries to complete various stages of their life cycle are particularly sensitive to climate change because their success is dependent on environmental conditions as well as the availability and responses of their intermediate host species (Burek et al., 2008; Macey et al., 2008). Rising sea levels, warming ocean temperatures, and changes in ocean circulation and estuarine salinity may alter fish parasite composition and biogeography (Palm, 2011). Reductions in non-climatic stressors provided by protected areas may potentially reduce disease prevalence; for example, a survey of 94 oyster reefs found significantly higher densities of oysters and significantly lower disease prevalence and severity inside of sanctuaries (Powers et al., 2009).

In many cases, the lack of integrated, long-term data on marine diseases limits the ability to predict future climate-related changes in infection prevalence and intensity, emphasizing the need for enhanced and sustained surveillance. Pathogen discovery and identification is in a relatively nascent stage for marine systems. As molecular techniques become more accessible, source tracking is allowing scientists to better understand the connections between marine, terrestrial, and freshwater systems as well as the evolution of marine pathogens. Improved understanding of these relationships will provide insight into current and future impacts of climate change on marine diseases.

Case Study 3-C
Leptospirosis disease in California sea lions

The emergence and pathogenesis of a serious emerging human and animal disease, leptospirosis, has been linked to environmental change. Many different mammalian species, including rats, dogs, and people, become infected with *Leptospira* bacteria through contact with contaminated urine, water, or soil (Guerra, 2009). In recent years, this disease has re-emerged in humans and dogs within the United States (Meites et al., 2004; Moore et al., 2006) and continued, periodic outbreaks in California sea lions along the U.S. West Coast have been noted. Increased leptospirosis has been associated with increases in precipitation and flooding during El Niño events (Levett, 2001; Storck et al., 2008). In Hawaii, increases in the number of reported cases of leptospirosis have been linked to flooding and have also shifted to wetter months of the year (Gaynor et al., 2007; Katz et al., 2011).

Leptospirosis is endemic in California sea lions but also causes cyclical outbreaks of mortality (Gulland et al., 1996; Lloyd-Smith et al., 2007; Vedros et al., 1971). During an outbreak in 2004, over 300 sea lions died along the central California coast, with additional animals dying along the coasts of Oregon, Washington, and British

California sea lion being treated for leptospirosis at The Marine Mammal Center, Sausalito, California (Photo: The Marine Mammal Center).

Case Study 3-A (Continued)

Columbia (Cameron et al., 2008). Sick or injured sea lions either die at sea or are stranded on beaches, which leads to situations in which increased disease transmission to terrestrial animals can occur. Leptospira bacteria has been found to survive longer in freshwater and many of the sea lions stranded themselves near freshwater estuaries, thereby increasing the possibility of transmission of the bacteria to domestic animals, terrestrial wildlife, or humans (Meites et al., 2004; Monahan et al., 2009; Zuerner et al., 2009). Recent information has shown that California sea lions may be chronic, or "inapparent," carriers of Leptospirosis and that the seasonal movements of these animals may contribute to the geographical spread of the disease (Zuerner et al., 2009). In the future, the exposure to and incidence of leptospirosis in both humans and marine mammals may increase in response to the combination of human population increase and urbanization as well as increasing populations and expansions of the ranges of marine mammals and changes in environmental conditions such as extreme weather events, increased flooding, and increased temperatures (Lau et al., 2010).

Invasive species

Invasions by non-native species are widely recognized as significant threats to native biodiversity (Carlton, 1996; Rahel and Olden, 2008; Ruiz et al., 2000; Stachowicz et al., 2002a). The frequency of introductions has increased dramatically over the past two centuries and species introductions have been documented in most marine habitats worldwide (Ruiz et al., 1999, 2000). San Francisco Bay is one of the most heavily impacted coastal systems, with over 230 non-indigenous species (Cohen and Carlton, 1998). Although only a fraction of invasive species have had significant impacts on established food webs and trophic linkages, their ecological and economic impacts can be profound.

Through ballast water and hull fouling, shipping and fisheries cause the majority of marine species introductions (Ruiz et al., 2000), but invasions consistent with climatic drivers have also been reported (Firth et al., 2011; Reid et al., 2007). In addition, climate change-induced shifts in invasive species' distributions are predicted to have significant future impacts (Doney et al., 2012). Ubiquitous invasive species are predicted to thrive in as-of-yet uninvaded habitats (de Rivera et al., 2011). For example, climate change is predicted to result in the movement of many planktonically-dispersing, fast-growing Pacific species such as mollusks into Atlantic waters via the Bering Strait and Arctic Ocean (Vermeij and Roopnarine, 2008), although existing physical and physiological barriers to movement, including seasonal ice cover and cold bottom waters, indicate that the number of species capable of invading the polar region may be limited (Sigler et al., 2011). On the Florida coast, evidence suggests that the ability of northward range movement for the introduced Asian green mussel, *Perna viridis*, may be currently limited by cold temperatures; therefore, increasing temperatures may allow for range expansion of this invasive species (Urian et al., 2011).

Once species become established in new areas, climate change may facilitate their subsequent success (Hellmann et al., 2008) because many invasive species have wider temperature tolerance ranges than their native counterparts (Abreu et al., 2011; Braby and Somero, 2006; , Lockwood and Somero 2011b; Sorte et al., 2010a; Stachowicz et al., 2002a). Climate-mediated invasions and range shifts may also alter species interactions as superior competitors (Stachowicz et al., 2002a) and predators (Smith et al., 2011) move into temperate and polar latitudes. In addition, introduced species may affect the distribution of diversity among trophic levels; in several coastal food webs, introduced species have been at least partially responsible for community-wide shifts towards lower trophic levels (Byrnes et al., 2007). These consequences of invasions present a major threat to the persistence and interactions of native marine species in a changing climate.

Protected species

Strong evidence shows that climate change is already affecting a variety of protected species such as marine mammals, sea turtles (Case Study 3-D), and sea birds, and these impacts are very likely to increase in the future. The effects on these species are expected to be primarily due to shifts in productivity and prey availability, changes in critical habitats such as sea ice due to climate warming and nesting and rearing beaches due to sea-level rise, and increases in diseases and biotoxins due to warming temperatures and shifts in coastal currents (see Section 2). Climate change is a challenge for the sustainable management of protected species (also see Section 4). Predicting the consequences of climate change on marine protected species is difficult because of the relative paucity of data and uncertainties as to how these species will respond if reduced in numbers and densities (Hoegh-Guldberg and Bruno, 2010; Kaschner et al., 2011; Simmonds and Isaac 2007; Wassmann et al., 2011).

Many marine protected species are highly mobile or migratory, occupying and utilizing a wide range of habitats and resources throughout their life history. Animal migration is closely connected to climatic factors; as a result, these species are, in many ways, more vulnerable because of the differential impacts they may experience at various life stages. For example, marine turtles may cross entire ocean basins throughout their lifetimes and can occupy diverse habitats such as sandy beaches, mangroves, and seagrass beds (Hawkes et al., 2006; Musick and Limpus, 1997; Polovina et al., 2006; Shillinger et al., 2008; Case Study 3-D). Projected increases in sea level and extreme events coupled with fortification of coastal areas could erode shorelines and compromise the availability of suitable nesting beaches (Hawkes et al., 2009). In addition, increased beach temperatures have led to altered sex ratios with higher female to male ratios reported in hatchlings of marine turtles for which nest temperature determines the sex of offspring. In parts of the Southern U.S., hatchlings of loggerhead sea turtles are currently female-biased and even moderate further increases in temperature could lead to a severe lack of males (Hawkes et al., 2007), which would reduce population viability (Poloczanska et al., 2009). Anomalously cold temperatures can also affect turtles. Sea turtles along the Atlantic Coast and Gulf of Mexico experience episodic, cold-stunning events when water temperatures drop below 10°C (Foley et al., 2007; Morreale et al., 1992; Witherington and Ehrhart, 1989). During these events, hundreds of cold-stunned turtles float listlessly on the water or are washed onto shore. An open question remains as to how temperature fluctuations, which are expected to increase under climate change, are likely to affect turtle populations (Neuwald and Valenzuela, 2011), including sea turtles.

Changes in thermal regimes are affecting the abundance, distribution, feeding, and phenology of protected seabirds (Bertram and Kaiser, 1993; Chastel et al., 1993; , Grémillet and Boulinier, 2009; Montevecchi and Myers 1997). For example, declines in oceanic productivity around the Northwestern Hawaiian Islands in the 1980s led to a 50 percent reduction in the survival of red-footed booby and red-tailed tropicbird eggs and chicks (Polovina and Haight, 1999). Reduced productivity and warmer temperature in the Southeast Farallon Islands (Sydeman et al., 2009) as well as low prey abundance (Sydeman and Thompson, 2010) have led to delayed breeding and reduced offspring numbers in Cassin's auklet (Wolf et al., 2009). Wolf et al. (2010) have projected additional climate-related population declines of 11-45 percent by the end of the century. The common murre has also exhibited a declining trend in reproductive success in the Southeast Farallon Islands, reflecting reduced availability of rockfish, which is their preferred prey item. In 2009, reproductive success of common murres was among the lowest observed in the last 38 years and the lowest ever recorded during a non-El Niño year (Warzybok and Bradley, 2010).

Warming water temperatures and loss of sea ice are fundamentally changing the behavior, condition, survival, and interactions of Arctic marine mammals (Heide-Jorgenson et al., 2011; Kovacs et al., 2010; Thomas and Laidre, 2011; Wassman et al., 2011), and these changes are expected to continue. Cetaceans, including gray whales (Moore, 2008; Moore et al., 2003; Stafford et al., 2007), killer whales (Higdon and Ferguson, 2009), and sei, fin, and Minke whales (Norwegian Polar Institute Marine Mammal Sighting Data Base: http://www.npolar.no), have been sighted further north and/or at higher northern densities than normal. Similar impacts are occurring for pinnipeds; harbor porpoises are appearing in northern areas and harp seals are being sighted in northern locations during abnormal times of the year (Norwegian Polar Institute Marine Mammal Sighting Data Base: http://www.npolar.no). Major declines in pup production and abundance have been documented for hooded seals in the Northeast Atlantic, ringed seals in Hudson Bay, and harp seals in the White Sea (Chernook and Boltnev, 2008; Ferguson et al., 2005). Polar bears are spending more time on land, resulting in declines in survival, condition, body size, and reproductive rates (Stirling et al., 1999; Stirling and Parkinson, 2006). Landward shifts of polar bear dens (Fischbach et al., 2007) and declines in the condition and survival of polar bear cubs have occurred (Regehr et al., 2006, 2010). Pacific walrus females and pups are also being forced to spend more time resting on land, and abandoned calves at sea suggest nutritional stress (Cooper et al., 2006; Garlich-Miller et al., 2011; Kavry et al., 2008) due to separation from feeding areas and the loss of sea-ice resting platforms (Kovacs et al., 2010). These examples illustrate some of the challenges facing marine-protected-species managers in a changing climate.

3.3 Ecosystem Structure and Function

Climate change affects individual populations and species as well as interactions among species. These changes can have cascading effects on food webs and ecosystem structure and function (Gedan and Bertness, 2010; Harley, 2011; Large et al., 2011). Shifts in

Case Study 3-D
Loggerhead turtles and climate change

A great deal of uncertainty remains regarding how climate change effects will influence future populations of marine turtles. As with many migratory species, marine turtles utilize different geographic regions during their various life stages. Juvenile turtles have a limited capacity to exploit their environment for food as compared to adults in their species, and as such, are considered more susceptible to oceanographic variability (Drinkwater et al., 2003; Lasker, 1981; Ottersen et al., 2001). In a recent study, Van Houtan and Halley (2011) developed a model that examined oceanographic influences on juvenile recruitment and breeding remigration. The findings suggest that the number of turtles that reach sexual maturity is strongly correlated to ocean conditions in the Northwest Atlantic and North Pacific. In particular, the Atlantic Multidecadal Oscillation (AMO) affects loggerhead juveniles in the North Atlantic and the Pacific Decadal Oscillation (PDO; see Section 2) affects juveniles in the North Pacific. In both regions, satellite-tagged juveniles were found to forage in oceanographic areas where prey items are abundant (Van Houtan and Halley, 2011). When prey items are scarce, mortality increases, affecting the future breeding population numbers. The model also showed that changes in loggerhead nesting counts over the past several decades are strongly correlated with the PDO in the Pacific Ocean and the AMO in the Atlantic Ocean (Van Houtan and Halley, 2011).

In addition, Van Houtan and Halley (2011) examined potential impacts of future climate change on the loggerheads in these two areas. In the Atlantic Ocean, warmer years would mean a stronger than average Gulf Stream current, helping juvenile turtles get to the North Atlantic Gyre and leading to increased productivity and population size; however, in the Pacific Ocean, loggerheads perform best under anomalously cold conditions. Therefore, available climate data indicate the potential for significant population declines of the Pacific population by 2040 due to warming temperatures (Van Houtan and Halley, 2011)..

Adult loggerhead turtle (Photo: Sarah Dawsey) and juvenile loggerhead turtle (Photo: Steve Hillebrand).

species distributions and interactions are also beginning to create novel, "no-analog" ecosystems consisting of species with little or no shared evolutionary history (Hobbs et al., 2006; Williams and Jackson, 2007), and this is likely to continue with unprecedented environmental change in the future. Although progress is being made in forecasting future responses, complex, nonlinear effects of changing environmental conditions on marine communities present additional uncertainty and challenges for natural resource managers (Crain et al., 2008).

Particularly problematic is obtaining an understanding of how complex feedback interactions between changes in the physical environment will affect ecological processes. Warming, ocean acidification, stratification, and other climate-related parameters can have both synergistic and antagonistic effects on marine organisms, making whole-ecosystem predictions difficult with the current state of knowledge (Boyd et al., 2008; Gao et al., 2012; Hofmann et al., 2010; Hutchins et al., 2009; Pörtner, 2008). Ongoing research efforts are targeting these interactive, multi-stressor effects, but accurate forecasting of ecosystem-level responses is a challenging undertaking because many environmental factors are simultaneously in flux.

Species interactions and trophic relationships

Marine ecosystems are influenced by the direct effects of climate change on individuals and populations and the indirect effects of environmental change that alters the strength of species interactions, including competition, predation, parasitism, and mutualism (reviewed by Kordas et al., 2011). These indirect effects can arise via different mechanisms.

Environmental change can alter an organism's physiology and behavior and, therefore, its per capita effect on the species with which it interacts. Changing ocean temperature and chemistry can alter the per capita feeding rate of an individual consumer (Gooding et al., 2009; O'Connor 2009; Pincebourde et al., 2008; Sanford, 1999) or modify an individual's competitive ability through effects on its growth rate (Sorte et al., 2010a; Stachowicz et al., 2002b; Wethey, 2002). Additionally, differential impacts of thermal stress on predators and their prey can lead to altered species interactions (Yamane and Gilman, 2009); for example, due to temperature-related changes in metabolism, exposure to warm water can increase the feeding rates of U.S. West Coast sea stars on their mussel prey (Sanford. 1999) until temperatures exceed thermal optima and feeding rates are reduced due to stress (Pincebourde et al., 2008). Changes in hydrodynamic conditions can affect the ability of prey to detect predators, as shown by experiments examining the behavioral responses of whelks to predatory crabs under different flow conditions (Large et al., 2011).

Climate change alters species interactions via changes in the population density of interacting species. Environmental changes affect species differentially and the resulting increases or decreases in population abundance can trigger chains of indirect effects (O'Connor et al., 2009; Poloczanska et al., 2008); for example, changes in ocean temperature and carbonate chemistry frequently alter the relative abundance of macroalgal species, which can in turn affect the abundance of herbivores that feed on them (Kroeker et al., 2011; Schiel et al., 2004). Often, a few key species interactions contribute disproportionately to maintaining community structure and ecosystem function (Paine, 1992); for example, the salt marsh grass *Spartina patens* reduces salinity stresses acting on species living within the plant canopy, and thus the removal of this structural species can have cascading effects on marsh communities (Gedan and Bertness, 2010). If these interactions are sensitive to environmental conditions, they may act as "leverage points," through which small changes in climate are amplified to produce large changes at the community and ecosystem levels (Kordas et al., 2011; Monaco and Helmuth, 2011; Sanford, 1999). Similarly, when the direct effects of climate change negatively impact the abundance of habitat-forming species such as coral, kelp systems, and mussels, cascading effects on ecosystem function can be seen due to loss of the services that these foundation species provide (Pratchett et al., 2008; Schiel et al., 2004; Wernberg et al., 2011b; Wootton et al., 2008). Such rippling effects are often unpredictable because of the complexity of food webs (Doney et al., 2012; Schiel et al., 2004).

Climate-related shifts in the geographic distribution of marine species are altering biogeographic patterns of co-occurrence and interaction (Kordas et al., 2011; Sorte et al., 2010b). Analogous shifts in the vertical distribution of sessile intertidal species have increased their overlap with and vulnerability to predatory sea stars (Harley, 2011). Similarly, as described above, ocean warming can alter the timing of life-history events such as spawning, leading to temporal mismatches or increased overlap between consumers and their food sources (Edwards and Richardson, 2004; Kristiansen et al., 2011; Philippart et al., 2003). Climate-related shifts in species dominance have also been observed throughout the U.S., including the California Current, the Gulf of Alaska, and the Bering Sea (Hare and Mantua, 2000) as well as the North Atlantic (Auster and Link, 2009), but these shifts do not always result in changes in ecological processes. For example, Auster and Link (2009) found that although climate-induced shifts in species dominance were observed in the Georges Bank ecosystem, these often involved switching between species that occupied redundant roles within trophic guilds. This redundancy may buffer against ecosystem reorganization under climate change, but the extent of buffering is unknown in most systems.

Currently, the ability to forecast the impacts of climate change on trophic linkages within marine ecosystems is limited primarily to conceptual models of marine ecosystem organization (Hunt et al., 2011; King et al., 2011). As understanding of the range of complex responses of marine species to environmental disturbance improves and coupled biophysical models of marine ecosystems become available, the ability to predict the likely implications of climate change on marine ecosystems will be enhanced. In the near-term, observations and monitoring systems provide the best method of detection and attribution of changes in the trophic structure of marine ecosystems, as well as validation of models of population- and ecosystem-level responses (Helmuth, 2009; Wethey et al., 2011; Wethey and Woodin, 2008). Integrated, sustained observations of the abundance, diets, distribution, and physiological condition of marine species, coupled with field and laboratory studies that identify species responses to ecosystem change, will be critical. A need to understand patterns of genetic variance (Schmidt et al., 2008; Trussell and Etter, 2001), particularly that which underlies traits that influence susceptibility to environmental stress (Place et al., 2008), is also necessary. Over time, insight gained through these efforts will provide the data and understanding needed to more reliably model complex ecosystem responses to climate change.

Biodiversity

The accelerating changes in species distributions resulting from climate change and associated impacts on oceanic circulation have been shown to cause widespread changes in the composition and diversity of marine communities, leading to the potential for ecosystem reorganizations (Fodrie et al., 2010; Sagarin et al., 1999). Many efforts have been implemented to document, quantify, and assess biodiversity of U.S. marine ecosystems through initiatives such as the Census of Marine Life (Fautin et al., 2010) in order to gain baseline understanding and monitor changes through time. Climate-related distribution shifts have already altered community composition and biodiversity of many systems and taxa, including phytoplankton (Hallegraeff 2010; Merico et al., 2004; Peperzak, 2003), pelagic copepods (Beaugrand et al., 2002), rocky intertidal invertebrates (Barry et al., 1995; Helmuth et al., 2006b; Southward et al., 2005; Wethey et al., 2011), fishes (Fodrie et al., 2010; Last et al., 2011; Nye et al., 2009; Perry et al., 2005), and seabirds (Hyrenbach and Veit, 2003). These changes in community composition are a function of both local extinction of species and invasions from elsewhere.

Ocean acidification is also affecting biodiversity and community structure in marine ecosystems. In a northeast Pacific rocky shore community, declining pH over eight years corresponded to gradual shifts from the mussel-dominated communities typical of such temperate shores to communities more dominated by fleshy algae and barnacles (Wootton et al., 2008). Similarly, in shallow benthic communities near natural CO_2 seeps, calcareous corals and algae are replaced by non-calcareous algae, and juvenile mollusks are sharply reduced in number or absent altogether (Hall-Spencer et al., 2008).

Importantly, biodiversity is not only affected by climate change because levels of biodiversity can also influence the resilience of marine ecosystems to climate change. Experiments show that more diverse communities tend to be more stable and less susceptible to disturbance and variability through time (Jiang and Pu, 2009). Recovery of eelgrass after the 2003 heat wave in Europe was enhanced by greater genetic diversity (Reusch et al., 2005) and the important process of algal grazing was more stable in the face of experimental warming and salinity fluctuations when multiple species of grazers were present (Blake and Duffy, 2010). Analyses of global fisheries time series data also support the hypothesis that marine biodiversity increases ecosystem stability and resilience to perturbations (Worm et al., 2006). Biocomplexity and diversity of fishes has been shown to decrease variability in stock productivity (Hilborn et al., 2003) and increase profitability to resource users (Schindler et al., 2010). Further research is needed to understand the causes and consequences of biodiversity change in marine ecosystems, including impacts on resilience, stability, and the provisioning of the ecosystem services on which humans depend (see Section 4). Nevertheless, sufficient evidence now exists from a wide range of ecosystems to conclude with confidence that, on average, loss of biodiversity reduces ecosystem productivity and stability (Cardinale et al., 2011; Stachowicz et al., 2007).

If species cannot migrate or adapt to a changing environment, they face local or even global extinction. Humans have directly caused the global extinction of more than 20 described marine species, including seabirds, marine mammals, fishes, invertebrates, and algae, and many others have likely disappeared unnoticed (Sala and Knowlton, 2006). Under projected climate change, marine species extinctions are expected to be most frequent in sub-polar regions, the tropics, and semi-enclosed seas (Cheung et al., 2009). Cold- and ice-adapted species are especially vulnerable because ocean warming degrades their preferred habitats and invasions occur from temperate regions. For example, warming waters have recently allowed large lithodid "king" crabs to invade the Antarctic shelf for the first time in 14 million years, where they have reduced benthic diversity and appear to have driven certain species locally extinct (Smith et al., 2011).

Quantitative estimates of species losses based on historical comparisons, measured and projected habitat loss, and demographic trajectories of wild populations suggest that Earth is now approaching, if not already in the midst of, the sixth mass extinction in its history, with rates of species loss 2-5 orders of magnitude above the average over geologic time (Barnosky et al., 2011; Butchart et al., 2010; Dirzo and Raven, 2003; Pereira et al., 2010; Pimm, 2008; Pimm et al., 1995). The principal drivers of the current extinction wave are habitat loss, overexploitation, pollution, and impacts of invasive species (Purvis et al., 2000), but changing climate has contributed to several mass extinction events in the past (Barnosky et al., 2011), and today's much more rapidly changing climate is expected to exacerbate the impacts of these other drivers in the coming century (Brook et al., 2008).

3.4 Regime Shifts and Tipping Points

As a result of environmental and ecological complexity in response to climatic and non-climate stressors, rapid changes in ecosystem structure and function are a particular area of concern. Evidence of rapid phase, or regime, shifts is emerging across diverse U.S. geographic locations and ocean ecosystems (Hoegh-Guldberg and Bruno, 2010). Regime shifts occur when dominant populations of an ecological community respond gradually and continuously to changes in environmental conditions until a particular threshold or "tipping point" is reached, beyond which the community rapidly shifts to a new dominant species or suite of species (Scheffer and Carpenter, 2003; Scheffer et al., 2001, 2009). In many instances, these "replacement" assemblages are less desirable from a human standpoint, such as when coral reefs are replaced by fast-growing macroalgae (Dudgeon et al., 2010, Case Study 3-B). Thus, regime shifts have significant implications for ecosystem functioning and services, with consequences for associated ecological, economic, and human social systems (Mumby et al., 2011b). Systems that are already degraded and depleted by non-climatic stressors often have lower resilience and are therefore more susceptible to climate-related regime shifts and tipping points (Folke et al., 2004). However, as physiological and ecological thresholds, tipping points can be difficult to predict and significant declines in ecosystem services can theoretically be reached prior to any associated large-scale changes in the environment (Harley and Paine, 2009; Monaco and Helmuth, 2011; Mumby et al., 2011b).

Certain U.S. marine systems are on a trajectory for rapid change and others have already crossed a tipping point (Hoegh-Guldberg and Bruno, 2010). In the Chesapeake Bay, eelgrass (Zostera marina) died out almost completely during the record-hot summers of 2005 and 2010, evidently because too many days exceeded the species' tolerance threshold of 30° C (Moore and Jarvis, 2008). Summer Arctic sea ice has dramatically declined (see Section 2), thinning by 43 percent from the late 1980s to the early 2000s

(Lindsay and Zhang, 2005). Coral reefs are undergoing rapid phase shifts from coral-dominated to macroalgal-dominated systems (Case Study 3-B) due to a combination of stressors such as nutrient pollution, overharvesting of herbivorous fishes, disease, and thermal stress (Dudgeon et al., 2010; Hoegh-Guldberg et al., 2007; Hughes et al., 2010). Gardner et al. (2003) reported an 80 percent reduction in Caribbean coral cover in fewer than three decades from 50 percent cover to only 10 percent cover. In the northern California Current Large Marine Ecosystem, severe low-oxygen, or hypoxic, events have recently emerged as a novel phenomenon due to changes in the timing and duration of coastal upwelling (Barth et al., 2007; Chan et al., 2008). These events have led to high mortality of benthic invertebrates and Dungeness crabs (Grantham et al., 2004) as well as the loss of rockfish from low-oxygen areas (Chan et al., 2008), which affected local fisheries. In many instances, it is unknown whether reversing these rapid, climate-related trajectories of ocean disruption and decline will be possible. However, evidence indicates that reducing non-climatic stressors such as overharvesting and pollution can potentially prevent tipping points from occurring (Diaz-Pulido et al., 2009; Hsieh et al., 2008; Sumaila et al., 2011).

Case Study 3-E
Cumulative impacts assessment

Developing climate adaptation strategies requires knowing the relative contribution of each stressor to the overall condition of a marine system in order to inform if and to what extent any action can actually affect meaningful change. Cumulative Impact Assessments are designed to provide this type of information (Halpern et al., 2008a). Although climate change consistently ranks as a top pressure to marine ecosystems, at global (Halpern et al., 2008a) and regional areas within the United States (e.g., California Current: Halpern et al., 2009b; Northwestern Hawaiian Islands: Selkoe et al., 2008, 2009), many other stressors play significant roles in impacting overall condition. In some locations, non-climatic stressors represent the dominant impact (e.g., land-based stressors: Halpern et al., 2009a; fishing: Halpern et al., 2010).

In each of the regional assessments that have been conducted within the United States, commercial shipping, commercial fishing, and land-based pollution have all played significant roles in driving overall impacts to marine systems. For specific climate-related stressors, local sources can dominate impacts over global sources; for example, ocean acidification in coastal waters can be due to land-based runoff of sediments and pollutants as well as global increases in carbon dioxide and can therefore be addressed in part through reducing local drivers of pH change (Kelly et al., 2011). Climate-related increases in precipitation and surface water temperature are predicted to exacerbate factors that cause low oxygen in coastal waters; therefore, reducing the flow of nutrients and sediments into coastal waters could lessen the probability that hypoxia develops in certain water bodies whose climate-related susceptibility to hypoxia is increasing (CENR 2010, Doney et al., 2012).

Case Study 3-E (Continued)

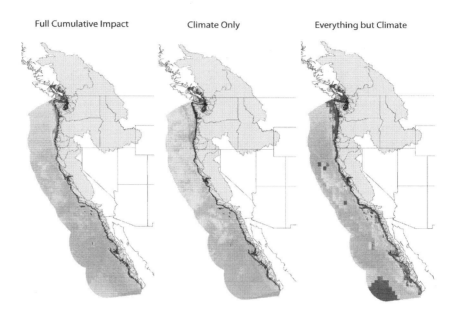

Full Cumulative Impact   Climate Only   Everything but Climate

Cumulative impact map of 25 different human activities on 19 different marine ecosystems within the California Current and impact partitioned into climate change impacts alone (n = 3 layers) and other stressors (land-based sources of stress (n = 9 layers), all types of fishing (n = 6 layers), and other ocean- based commercial activities (n = 7 layers). Figure adopted from Halpern et al., 2009b.

Chapter 4

Impacts of Climate Change on Human Uses of the Ocean and Ocean Services

Executive Summary

The impacts of climate change on oceans include effects on humans and human systems. In addition, climate change is interacting with other anthropogenic impacts such as pollution, habitat destruction, and over-fishing that are currently negatively affecting the marine environment. Although not well-documented across all marine regions of the U.S., evidence to date suggests that substantial socio-economic impacts to marine resource-dependent communities and economies worldwide are likely to result from climate change.

Extensive efforts are underway to understand the socio-economic drivers of and effects from climate change. To date, case studies in which the effects of climate change on ocean services have been documented are few, but data are available regarding the extent of human uses of marine resources as well as the biophysical effects of climate change on the marine resources upon which those uses depend. Using these data and available case studies, this section provides greater understanding and assesses the likelihood and potential consequences of impacts that may occur given certain climate-related changes in specific marine resources and environments for the following sectors: commercial, recreational and subsistence fisheries, offshore energy development, tourism, human health, maritime security, transportation, and governance.

U.S. marine ecosystems are highly valuable and provide a wide variety of resources and services that support a wide array of activities, businesses, communities, and economies across the nation. Commercial and recreational fisheries in the U.S. represent an annual multi-billion dollar industry (NOAA Fisheries, 2010). Subsistence fishing, defined as fishing for direct consumption or barter without the product entering a market, also contributes significantly to the health and well-being of fishing-dependent communities and local economies across the U.S. Most of the effects of climate change on U.S. fisheries will stem from the changes to the fish stocks brought about by direct and indirect climate impacts on productivity and location; others stem from impacts that climate has on the fisheries themselves as well as fishing-dependent communities across the country. Already, case studies have documented climate change effects on fisheries and their stocks, including the Pacific salmon, Bering Sea pollock, Pacific cod, Pacific sardines, Atlantic croaker, and Atlantic cod fisheries. For some stocks, geographic distributions are shifting; for others, climate-induced shifts in marine ecosystems have brought about fluctuations in overall abundance or population characteristics.

The U.S. is a major consumer, but only a minor producer, of aquaculture products. The U.S. imports 86 percent of its seafood and approximately half of that is from aquaculture production. A rise in sea-surface temperatures may trigger the growth of harmful algal blooms that can extend the spatial or temporal scope of a bloom or release toxins into the water and kill cultured fish and shellfish; fish in cage-based aquaculture systems and shellfish beds are particularly at risk. Higher water temperatures may also result in increased disease incidence and parasites, which may develop more rapidly in warmer waters and higher salinities (see Section 3) and threaten the aquaculture sector. Species cultured in temperate regions, predominantly salmon and cod species have a relatively narrow range of optimal temperatures for growth. Temperatures over 17ºC, at which feed intake drops and feed utilization efficacy is reduced, would be detrimental to the salmon farming sector (DeSilva et al., 2009). On the other hand, certain aspects of aquaculture may benefit from climate-related changes. Higher water temperatures may also increase the availability of new culture sites, especially in areas previously too cold to support aquaculture. An increase in water temperature may also have a positive effect on metabolism and stimulate growth of cultured species, provided that the change is gradual and stays within the thermal tolerance of the species.

Offshore energy development from oil, gas, and renewable energy sources has been increasing in recent years. Many factors will impact these industries as a result of global climate change and industry reaction to these changes involves an even larger number of factors. For example, industry response depends largely on whether it involves an international company, a small independent company, or a state-owned enterprise. Other critical factors include, but are not limited to, geographic location; local, regional, and international policies and regulations; the industry's standards of ethical practice; and the industry's influence on national and international markets. The main climate-related effects on these industries are increased pressures on water resources, failure of infrastructure that was not designed to withstand new climatic conditions, and changes in financial assets and resources extracted as energy production moves from the traditional oil and gas industry to renewable sources of energy.

In the U.S., the tourism industry also plays a large economic role in ocean services. Nationally, 2.8 percent of gross domestic product, 7.52 million jobs, and $1.11 trillion in travel and recreational total sales are supported by tourism (OTTI, 2011a, b). In addition, in 2009-2010, nine of the top ten states and U.S. territories and seven of the top ten cities visited by overseas travelers were coastal, including the Great Lakes (OTTI 2011a, b). In the face of climate change, impacts to weather conditions and marine resource distribution, variable weather conditions, and extreme events such as typhoons and hurricanes are expected to pose the most significant impacts on the industry: positively in some regions of the country, negatively in others, and with mixed effects in yet others.

The potential for climate change to affect the boundaries and determinants of human health is equally significant. A number of primary areas exist in which climate change is expected to impact human health (Baer and Singer, 2009) and will likely exacerbate human vulnerability and sensitivity in the future (McGeehin, 2007). These include, but are not limited to, respiratory illness from harmful algal blooms, illness from shellfish contamination; extreme weather-related injuries; morbidity and mortality of marine species;

declines in access to drinkable water; increased food insecurity and malnutrition; spread of infectious disease; and growing violence from climate-triggered civil conflict.

Finally, security, transportation, and governance issues are also at play in terms of expected climate change impacts to ocean services in the U.S. Most notably, climate shifts in the Arctic, especially in regards to sea ice coverage, are provoking discussion on the future of ocean governance, including marine resource and ecosystem-based management. Perhaps the most noteworthy issue in this arena is the increase in shipping accessibility in the Arctic. National security concerns and threats to national sovereignty have also been a recent focus of attention (Borgerson, 2008; Campbell et al., 2007; Lackenbauer 2011). Ocean change will lead to an expanded geopolitical discussion involving the relationships among politics, territory, and state sovereignty struggles as well as opportunities for partnership and cooperation on local, national, and international scales (Nuttall and Callaghan, 2000).

Discussion of these observed and predicted changes to ocean services in the U.S. is raising awareness of the potential socio-economic effects of climate change and highlighting some of the areas where significant changes have not already been observed but are highly likely to be observed over the short- and long-term periods. Information needs are growing with regard to the socio-economic and biophysical effects of climate change and how the U.S. may be able to prepare for and respond to these changes. This section is intended to provide a summary of impacts that are currently known and a framework within which inferences of future impacts can be identified.

Key Findings

1. Significant effects of climate change on all sectors pertaining to human uses of the ocean, including but not limited to fisheries, energy, transportation, security, human health, tourism, and maritime governance, are already being observed and are predicted to continue into the future.

   • Some effects are predicted to be "positive," in that they expand the extent of individual sectors, while others are predicted to be "negative," in that they reduce the ability of humans to use the ocean in a given sector, and virtually all effects will result in some distributional changes in how and where, as well as by whom, marine resources are used.

   • Most of the effects of climate change on U.S. fisheries will stem from the changes to the fish stocks brought about by direct and indirect climate impacts on productivity and location; others stem from impacts that climate has on the fisheries themselves as well as fishing-dependent communities across the country, which could experience changes in distribution and abundance of their available stocks,.

   • Aquaculture stocks are expected to be more resilient to climate change than wild stocks due to selective breeding and vaccination.

   • The main climate-related effects on the oil and gas industry are anticipated to be increases in demands on water sources, failure of infrastructure not designed to withstand new climatic conditions, and changes in the extraction of financial assets and resources as energy production moves from the traditional oil and gas industry to renewable sources of energy.

- In the face of climate change, impacts to marine resource distribution, variable weather conditions, and extreme events such as typhoons and hurricanes are expected to pose the most significant impacts on the tourism industry; these effects will be positive in some regions of the country, negative in others, and mixed in others.

- The scale and scope of climate impacts such as increased economic access and ecosystem shifts in polar areas (Arctic, Antarctic) are already having significant impacts on ocean uses and users. There is a high likelihood that these impacts will grow and have serious consequences as well as opportunities on human uses of the oceans in the future.

2. As a result of climate change, governance regimes for ocean environments and resources, as well as human health, will be challenged, and will likely have to change significantly in character and configuration.

- Geographic shifts in the distribution of fishery resources, ocean energy exploitation, and ocean transportation routes may require restructuring policy arenas for those resources and may require amending policy and governance systems with new statutory law.

3. Ocean "health," both in terms of the functioning of biophysical ocean ecosystems and the factors related to oceans and human health will likely be affected by climate change.

- The distribution and abundance of water-born human disease vectors are expected to change significantly as a result of climate change.

- The potential for climate change to affect the boundaries and determinants of human health is significant, including increases in extreme weather-related injuries, morbidity and mortality, declines in access to drinkable water, increased food insecurity and malnutrition, rising pollutant-related respiratory problems, and spread of infectious disease.

4. Impacts of climate change on human social and economic systems provide critical insight into societal responses and adaptation options.

- Additional disciplinary and interdisciplinary research will be necessary to improve understanding of the interactions between physical, biological, economic, and social systems in the future.

4.1 Introduction

The biophysical impacts of climate change on oceans described in Sections 2 and 3 also affect humans and human systems that interact with the ocean. For example, fishing-dependent communities and the national economy are affected by climate-related impacts on populations of marine resources and understanding climate impacts to fish and shellfish stocks enables improved assessment of the impacts of those changes on fishing behaviors, industries, infrastructure, and communities. This leads to one of the limitations in our current ability to assess these socio-economic impacts: uncertainty regarding the rate and magnitude of change in biophysical aspects of marine resources attributable to climate change. The direction of these changes may be clear but the rate and extent, as well as synergistic, antagonistic, or cumulative impacts that result, are less clear.

Climate change-independent stressors also affect coastal and marine resources as well as the socio-economic factors related to the use of those resources; for example, the demographic phenomenon of human migration towards coasts is not generally driven by climate change, but clearly affects coastal and marine resources. Similarly, human population growth and accompanying growth in food production, infrastructure, and transportation systems clearly affect resources and socio-economic systems. This section, however, will focus on those socio-economic effects of changes in marine resources that are clearly related to climate change, based on our current understanding of the biophysical impacts of climate change (see Sections 2 and 3 for more detail).

The term "ocean services" refers both to the quantifiable monetary and non-market value that use of the ocean provides to humans and to the currently unquantifiable but identifiable benefits that the ocean provides to humans. The U.S. Commission on Ocean Policy recently characterized the "value of the ocean sector to the U.S." in the following way:

> *The ocean economy, the portion of the economy that relies directly on ocean attributes, in 2000 contributed more than $117 billion to American prosperity and supported well over two million jobs. Roughly three quarters of the jobs and half the economic value were produced by ocean-related tourism and recreation (Figure 4-1). For comparison, ocean-related employment was almost 1½ times larger than agricultural employment in 2000, and total economic output was 2 ½ times larger than that of the farm sector.* (USCOP, 2004, page 31)

The report also notes, "Standard government data are not designed to measure the complex ocean economy. They also ignore the intangible values associated with healthy ecosystems, including clean water, safe seafood, healthy habitats, and desirable living and recreational environments. This lack of basic information has prevented Americans from fully understanding and appreciating the economic importance of our oceans and coasts" (USCOP, 2004, page 31).

As described in earlier sections, the physical changes in marine environments expected from climate change include increased ocean temperatures and acidification in addition to changes in salinity, among others (see Section 2). These will in turn lead to impacts on ocean organisms, such as shifts in the distribution of species, populations, and habitats and changes in life histories, growth rates, and survival (see Section 3; MacNeil et al., 2010). In addition, these changes will also lead to new patterns of the human use of and relationships with the ocean, such as possible displacement of fishing fleets from their traditional fishing grounds, increased access to polar region environments for navigation and mineral exploration, and expanded growth and distribution of water-borne pathogens. Information about biophysical change is critical for enhanced understanding of the socio-economic impacts of these changes

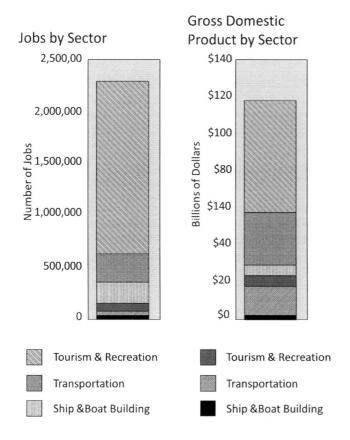

Figure 4-1 Value of the oceans. The ocean economy includes activities that rely directly on ocean attributed or take place on or under the ocean. In 2000, Tourism and Recreation was the largest sector in the ocean economy, providing approximately 1.6 million jobs (Source: USCOP, 2004).

in human use. An interdisciplinary perspective is crucial for analyzing the interwoven impacts of climate change on the socio-economic uses of marine resources.

For example, to the extent that human fishing-dependent communities rely on the exploitation of particular species in particular geographic environments, human populations will likely have to shift their exploitation to other species, particularly those in geographic proximity to the human communities, as the distributions of those species and environments shift. Thus, communities will likely have to adopt new technologies for harvesting marine resources that are new to them, or, if humans desire to continue to exploit the same species, which may have moved to another location, they will have to adapt either through increased vessel transit time, the migration of shore-side human communities, or both to stay in proximity to their "traditional" species of exploitation.

Similarly, if ocean acidification results in certain species, such as corals, shellfish and crustaceans, being no longer viable in their present locations, or perhaps not being viable at all due to diminishing ability to construct shells or skeletons, any human exploitation of such species will be affected. Responses will vary based on the specific marine resources and involved human communities (Hamilton et al., 2003).

The key conclusion of this section is that substantial socio-economic effects in specific areas, some positive and some negative, will unquestionably result from changes in marine resources due to climate change. This section identifies many of the important areas and directions of potential socio-economic effects, although, for the moment, not the details of those effects. The focus is on three approaches to exploring the impacts of climate change on human uses of the ocean. The first approach is to combine a baseline of current human uses of marine resources with case studies of currently document-able changes occurring in specific marine resources and their associated socio-economic impacts. Few of these case studies currently exist, but certain conclusions can be drawn from available data and information. The second approach is the construction of gen-erally expected impacts given certain changes in specific marine resources and envi-ronments. Finally, the implications of these changes for marine resource governance systems will be explored. Will these changes necessitate new or different governance regimes or will they simply require adaptation of existing regimes, and, if so, what kinds of adaptation? Unfortunately, the research and literature on human uses of the ocean and ocean services is not uniform by region of the U.S.; however, an attempt was made to be comprehensive in our coverage of all areas of the U.S. EEZ, and substantial mate-rial and case studies for most regions has been included.

4.2 Climate Effects on Capture Fisheries

> *Fishing is a production activity that takes place under uncommonly uncontrolled conditions. It follows that the output of fisheries, as well as their costs and net benefits, are directly and strongly affected by variations in natural conditions. We can do very little about these conditions but have to adjust to them as best we can. We are not in a position to warm up the ocean or change its salinity as would fit our interests. Neither can we change the ocean currents that affect the migrations of fish. Basically, in fisher-ies, unlike most other types of economic production, we … have to make do with what nature decides to make available..* (Hannesson, 2007, page 157)

Commercial and recreational fisheries in the U.S. represent an annual multi-billion dollar industry (Tables 4-1 and 4-2). Subsistence fishing also contributes significantly to the health and well-being of fishing-dependent communities and local economies across the U.S. In the U.S., fisheries managed by the federal government are generally defined as fishing activities that take place between 3 and 200 nautical miles from the coast-line. Nationwide, National Oceanic and Atmospheric Administration (NOAA) Fisheries oversees the management of 230 major fish stocks or stock complexes that comprise 90 percent of the nation's commercial harvest. In addition, individual states retain manage-ment authority over fishing activities within 3, or in some cases up to 9, nautical miles of their coasts or in their inland waters such as Puget Sound or Chesapeake Bay. For a better understanding of regional fishery characteristics, refer to Appendix A.

Climate change will affect U.S. fisheries through multiple pathways, including the biophysical systems upon which fisheries depend, the methods and locations in which harvests occur, community reliance on fisheries resources, and fisheries' governance systems. Most of these effects will stem from the changes to the fish stocks brought about by direct and indirect climate impacts on stock productivity and location; oth-ers will stem from impacts that climate has on fisheries and fishing-dependent com-munities. Extreme weather events may also disrupt fishing operations and land-based

Table 4-1: 2009 economic impacts of the United States seafood industry

	Jobs	Sales ($1,000s)	Income ($1,000s)	Value added ($1,000s)
Total impacts	1,029,542	116,224,548	31,556,643	48,282,319
Commercial harvesters	135,466	10,349,446	3,435,027	5,340,116
Seafood processors & dealers	183,895	25,240,441	7,965,719	11,073,240
Importers	178,387	49,070,476	7,864,480	14,958,830
Seafood wholesalers & distributors	47,405	6,505,383	2,137,714	3,058,777
Retail	484,389	25,058,802	10,153,704	13,851,356

Source: National Marine Fisheries Service. 2010. Fisheries Economics of the United States, 2009. U.S. Dept. Commerce, NOAA Tech. Memo. NMFS-F/SPO-118, 172p.

Table 4-2: 2009 economic impacts of recreational fishing expenditures

	Jobs	Sales ($1,000s)	Income ($1,000s)	Value added ($1,000s)
Total impacts	327,123	49,811,961	14,574,464	23,196,422
For hire	17,217	1,915,452	606,983	1,039,705
Private boat	31,176	4,243,541	1,253,804	2,158,414
Shore	35,293	4,312,850	1,319,865	2,243,036
Durable equipment	243,438	39,340,118	11,393,812	17,755,268
Retail	484,389	25,058,802	10,153,704	13,851,356

Source: National Marine Fisheries Service. 2010. Fisheries Economics of the United States, 2009. U.S. Dept. Commerce, NOAA Tech. Memo. NMFS-F/SPO-118, 172p.

infrastructure; however, because the management of fisheries in the U.S. is partly based on metrics, such as maximum sustainable yield (MSY), that depend on productivity, the effects of climate change on fisheries will substantially depend on how fisheries' managers respond to those changes.

Changes in the biophysical characteristics of fish stocks can sometimes mean changing gear, which is often expensive, and/or learning new fishing grounds and species' habits. Fishermen often rely on social networks for information-sharing while fishing (Holland et al., 2010; Kitts et al., 2007; Palmer 1990, 1991; St. Martin and Hall-Arber, 2008); thus,changing species may mean needing to cultivate new networks. If fishermen switch to species whose range extends further south, they may need to make longer trips or relocate their home base, either of which affects their families and communities. Fishermen often choose day versus trip fishing based on family consider-ations (Maurstad, 2000). When trips are longer, household dynamics change, affecting time with spouses and children and the ability to participate in community and school events. When households relocate, social networks are lost and part of a community's economic base disappears, although the communities of destination gain. See Fowler and Etchegary (2008) for a more thorough discussion of the social and economic impacts on communities of out-migration, especially in rural areas (Lal et al., 2011). On the other side, gentrification is creating pressure on small fishing-dependent communities (Col-burn and Jepson, 2012; Clay and Olson, 2008; NOAA Fisheries, 2009a), making coastal property less affordable. Any climate-change-related loss of fishing households could exacerbate this trend, but the exact degree or even direction of any of these economic impacts for commercial fishermen depends on: which specific climate impacts occur; factors affecting prices and other market dynamics at that point in time (Markowski et al., 1999); and choices based on social and cultural factors.

Recreational fishermen will largely change target species with unclear economic im-pacts because many aspects of the recreational fishing experience are unrelated to spe-cific species (Fedler and Ditton, 1994). In their study of freshwater sportfishing in the northeastern U.S., Pendleton and Mendelsohn (1998) found that a doubling of atmo-spheric carbon dioxide could lead to anywhere between a $4.6 million loss and a $20.5 million net benefit for the region. Subsistence fishermen generally fish a wider range of species than those fishing for pure recreation, so they would likely be able to adapt provided enough species remain accessible (Steinback et al., 2007); however, to the ex-tent that multiple species become unavailable, these fishermen may experience negative nutritional consequences, especially because they are also more likely to collect non-finfish marine resources (Steinback et al., 2009), such as shellfish, which are expected to be heavily impacted by increasing ocean acidification (see Cooley and Doney, 2009), squid, seaweed, or kelp (see Ling et al., 2009, for potential interactions of fishing, kelp, and climate change).

Effects on the productivity and location of fish stocks

The most direct potential effects of climate change on fisheries will come through chang-es in the productivity and location of the fish stocks that are the current and future tar-gets of those fisheries (See Section 3 for more detail on these effects). Climate change can directly affect the growth rates, reproductive capacity, and mortality of fish populations (Brander, 2010; Section 3). Climate change can also affect the marine ecosystems that sup-port those populations, by altering primary productivity (Boyce et al., 2010; Sarmiento et al., 2004; Sumaila et al., 2011; Section 3) as well as the overall productivity, structure, and composition of the marine ecosystems on which fish depend (Brander, 2010; Sec-tion 3). In addition, ocean acidification, increased ocean temperatures, and changes in environmental conditions all strongly affect the spatial distributions of marine fish spe-cies (Allison et al., 2011; Cooley and Doney 2009; Doney et al., 2009; Gaines et al., 2003; Pauly, 2010; Pörtner and Knust, 2007; Sumaila et al., 2011). How these ecosystem effects will ultimately manifest in the fish stocks is uncertain in part because the complexity of trophic relationships makes predictions difficult (Brander, 2010).

At the level of an individual fishery, judging which species are likely to suffer significant adverse effects from climate change and which are likely to benefit is difficult. Fisheries that target stocks adversely affected by climate change may be able to target alternate stocks that benefit from climate change; thus, the ultimate impacts depend strongly on the capacity of particular fisheries to adapt (Brander, 2010). Similarly, fisheries that target fish stocks with evolving spatial locations will experience changes in the required amount of fuel and other fishing inputs, time at sea, and exposure to ice (Badjeck et al., 2010; Mahon, 2002), but whether these changes will be detrimental or beneficial depends on exactly how those locations change and the exact social and economic prices associated with available adaptation strategies on the part of fishermen, their families, their communities, and fisheries-dependent industries.

Economic effects on commercial fisheries and fishing-dependent communities

Sumaila et al. (2011) describe potential economic impacts of climate change on fisheries (refer to Appendix A for current levels of fishing pressure in each region of the U.S.). In particular, climate change can affect the quantity and quality of yields through biophysical impacts and the magnitude of these impacts will depend on responses to these changes by harvesting and processing sectors. These responses will be reflected in fish and seafood markets through changes in prices and yield values and also through changes in the costs of fishing, such as risingfuel prices. Taken together, the net value of fish (i.e., sales revenues minus costs) will determine the incomes of fishermen and the economic value of fish stocks to fishing-dependent communities. To assess the possible economic effects of climate change on fisheries, Sumaila et al. (2011) used the effects of El Niño-Southern Oscillation (ENSO)-induced climate variability on fisheries as a partial proxy for what could happen to individual fisheries in response to climate change (see Section 2 for details on ENSO).

In general, ENSO events are associated with a warming of sea surface temperature (SST) in the tropical Eastern Pacific. In a model of an ENSO event, a simulated 60 percent decrease in active sablefish (*Anoplopoma fimbria*) vessels in Monterey Bay was accompanied by a 25 percent decrease in ex-vessel price (Dalton, 2001). In addition to sablefish, a groundfish species, Dalton's (2001) analysis estimated effects on fisheries in Monterey Bay for Albacore tuna (*Thunnus alalunga*), Chinook salmon (*Oncorhynchus tshawytscha*), and market squid (*Loligo opalescens*). Together, these four fisheries accounted for approximately 50 percent of the ex-vessel revenues associated with landings at Monterey Bay ports during the period of analysis (1981 to 1999).

Results for the Albacore fishery showed temporary increases in active vessels and ex-vessel prices of approximately 20 percent in response to a major ENSO event. Results for the Chinook fishery showed no change in numbers of vessels but did exhibit a substantial decrease in ex-vessel prices. Results of the same ENSO model for the market squid fishery showed an increase in ex-vessel price together with a drastic decrease in the number of active vessels. In fact, no landings of market squid were reported at Monterey Bay ports during the major 1998 ENSO event, which had an SST anomaly of 1.9° Celsius. An impulse response function analysis was also conducted for the California sardine (*Sardinops sagax*) and Northern anchovy (*Engraulis mordax*) fisheries in Southern California and Monterey Bay to simulate economic effects of an SST anomaly of 1.2° Celsius on numbers of active vessels and ex-vessel prices (Dalton, 2001).

Except for effect on ex-vessel price of Northern anchovy at Southern California ports, the response to 1.2° Celsius increase in SST had positive effects on the numbers of vessels making landings and ex-vessel prices earned from those landings. These results demonstrate that climate change can be expected to have positive effects on some fisheries.

As reported below in Case Study 4-C for California sardine, these fish are more productive in warm water climate regimes. The effects of a single SST change are temporary but the persistence through time of the simulated SST changes described above shows the dynamic properties of the estimated models. In particular, Monterey Bay fisheries for Northern anchovy and Pacific sardine appear to be slightly more stable than those in Southern California.

Unlike ocean warming, which has a robust literature on fisheries impacts (see Sumaila et al., 2011 for a review of fisheries impacts), relatively few studies have assessed the wider impacts of ocean acidification on fisheries. Moore (2011) applies an integrated assessment model and an econometric estimation to U.S. mollusk production. Costs of ocean acidification to U.S. mollusk production are estimated to be approximately $10 million per year ($0.07 per U.S. household) in 2020 and to increase to almost $300 million per year ($1.78 per U.S. household) in 2100, with a cumulative cost of $734 million in net present value terms with a discount rate of 5 percent. This assessment of impacts on U.S. mollusk production was based on regression results from a study by Ries et al. (2009) for eighteen selected species of marine calcifiers. This study noted that ocean acidification will not always have negative impacts like that estimated for mollusks. Ries et al. (2009) found that blue crab, *Callinectes sapidus*, did not exhibit significant ocean acidification effects. However, some evidence already suggests that ocean acidification can impact other crab species (Walther et al., 2009, 2010), and that commercially important crab stocks in other areas, the North Pacific for example, are vulnerable. Furthermore, the small set of previous studies on ocean acidification impacts does not differentiate ocean acidification effects on different life-history stages, including early life stages that may be especially sensitive (Gazeau, 2010), nor does it apply to animals where demographic factors are a critical feature of population dynamics, which is clearly the case with many, and perhaps most, commercially important species.

In addition to the effects of climate change on fish stocks, both fishery operations and fishing-dependent communities are likely to be directly affected. Extreme weather events can disrupt fishing operations and damage the community-based infrastructure, such as landing sites, boats, and gear, that supports the fisheries (Badjeck et al., 2010; Jallow et al., 1999; Westlund et al., 2007). Changes in the variability of climate that manifest themselves in increased fluctuations in fishery production and income can affect communities through their choice of livelihoods and other social outcomes (Badjeck et al., 2010; Coulthard, 2008; Iwasaki et al., 2009; Sarch and Allison, 2000).

Fishing-dependent communities in the U.S. and elsewhere are economically and socially diverse. The effects of climate change on fisheries may be felt more acutely by those communities that are more dependent on fishing, including those with fewer alternative economic activities and/or higher reliance on fisheries, especially for subsistence, and those that are less diversified in their target fisheries and therefore more dependent on one or a few fish stocks (Phillips and Morrow, 2007).

Recent scientific concern about ocean acidification is turning to discussions of socio-economic security (NRC, 2010b) and particularly to the potential negative impacts of ocean acidification on certain commercial fisheries (Cooley and Doney, 2009; Cooley et al., 2009). According to a 2010 National Research Council report, "Ocean acidification may result in substantial losses and redistributions of economic benefits in commercial and recreational fisheries," adding that "Although fisheries make a relatively small contribution to the total economic activity at a national and international level, the impacts at the local and regional level and on particular user groups could be quite important" (NRC, 2010b, page 89). Furthermore, ocean acidification's impact on commercial fisheries is projected to generate increased job insecurity for workers employed in commercial and recreational fishing gear manufacture and sales, vessel construction, loading, and repair, fish processing, wholesale and retail, commercial docks, ice suppliers to commercial fishing vessels, and other support industries, possibly resulting in income decline and job loss. Workers in recreational fishing gear and vessel sales, recreational outfitting, marinas, and other recreational support industries could also be affected.

Regional effects of climate change on fisheries

Currently, relatively few socio-cultural or economic studies document climate change and existing economic studies (Markowski et al., 1999) do not measure all relevant climate-related effects on non-market benefits such as ecosystem health, species loss, and human amenity impacts. At this point, nothing definitive can be said about the net effect of climate change on the quality of life in the U.S.; there will be both positives and negatives. Grafton (2010), however, notes that although the specific effects of climate change on particular marine ecosystems and fish populations are difficult to predict, on a global and regional basis, sufficient research indicates that many, but not all, of these impacts will be negative. Some fish stocks are experiencing shifting distributions; for others, their overall abundance or population characteristics are fluctuating due to climate-induced shifts in marine ecosystems. The following presents a review of this research regarding the regional effects that are known or expected. Refer to Appendix A for a more detailed review of the extent of fishing in each region and further details on climate change impacts on specific fisheries.

SUBSISTENCE FISHING AND HUNTING IN THE NORTH PACIFIC. Alaskan communities and local economies depend on and are engaged in subsistence harvesting of marine resources more than any other region in the U.S. Regional climatic and environmental changes are already having a notable, although unpredictable and often non-linear, effect on subsistence activities in the ocean environment through changes in hydrology, seasonality, and phenology as well as fish and wildlife abundance and distributions (Loring and Gerlach, 2009; Loring et al., 2011; McNeeley 2009; Rattenbury et al., 2009; White et al., 2007). Residents of rural Alaska are already reporting unprecedented changes in the geographic distribution and abundance of fish and marine mammals, increases in the frequency and intensity of storm surges in the Bering Sea, changes in the distribution and thickness of sea ice, and increases in coastal erosion. When combined with ongoing social and economic change, climate, weather, and changes in the biophysical system interact in a complex web of feedbacks and interactions to make life in rural Alaska extremely challenging.

Climate-change-related changes in sea ice and weather patterns (see Section 2) are also creating numerous new environmental challenges for those who harvest marine species. The most striking change in the Arctic marine environment in recent years has been the rapid loss of summer sea ice (Perovich et al., 2011). In the Bering, Chukchi, and Beaufort Seas off Alaska's coast, this physical change has led to many ecological impacts (Moore and Gill, 2011; Mueter et al., 2011) and has altered physical access to the region (AMAP, 2008; AMSA, 2009), affecting human use of marine resources.

Despite the broadly-scaled, directional trends observed and projected for warming and drying in the region (Chapin et al., 2006), the smaller scale impacts of climate change are being experienced not directionally, but in terms of greater inter-annual and inter-seasonal variability (Bryant, 2009; Rattenbury et al., 2009; Wendler and Shulski, 2009). Uncertainty is high regarding how seasonal conditions will play out in the future (Lawler et al., 2010); for instance, the timing of the seasons, including fall freeze-up and spring break-up, is shifting in unpredictable ways from year to year (Mills et al., 2008; Mundy and Evenson, 2011). Further, winter ice is thinner and more unpredictable and variability in precipitation and snow pack will affect water levels in both the fall and spring (Euskirchen et al., 2007; Hunt et al., 2008; Wendler and Shulski, 2009).

As an example, residents of Alaska Native communities rely on sea ice to ease their travel to the hunting grounds for whales, ice seals, walrus, and polar bears (see Appendix A for a description of the extent of marine mammal subsistence hunting done in Alaska). Krupnik et al. (2010) identify numerous effects of climate change that challenge and threaten local adaptive strategies, including times and modes of travel for hunting, fishing, and foraging. In the Chukchi Sea, the loss of summer sea ice has reduced haul-out habitat for walrus, resulting in tens of thousands of walrus hauling out on land for the first time on record (Moore and Gill, 2011; see Section 3 for additional information) or in the memory of local hunters (Quakenbush and Huntington, 2010). As sea ice thins and retreats farther north, walrus, which rely on sea-ice as a resting place between foraging bouts over continental shelf waters, and polar bears, which need sea ice to hunt seals, will either be displaced from essential feeding areas or forced to expend additional energy swimming to land-based haul-outs or remaining in the feeding areas (Callaway et al., 1999; Laidre et al., 2008; Stirling et al., 1999). In addition to the stress on marine mammal and polar bear populations, hunters will have to travel farther and longer to reach haul-outs and will have to travel over open water for greater distances, both of which will increase the risk of hunting for subsistence-dependent populations (Gearhead et al., 2006). Fuel and vessel maintenance costs associated with subsistence hunting will also increase as hunters have to travel greater distances (Callaway et al., 1999).

The impacts of climate change on Alaskans are also seen in shifts in the abundance and distribution of culturally important species (described in Section 3). Salmon, which has been described as the cultural keystone food of Alaska, has likewise become a less dependable subsistence resource than in the past, which has direct implications for food security (Loring and Gerlach, 2010). A closure of the king salmon fishery on the Yukon River in 2009, for example, resulted in empty storage facilities, empty smokehouses, and barren fish racks from Stevens Village up through Fort Yukon and above. The 2009 closure produced a "perfect storm" for a food security crisis, especially in combination with low harvest rates of moose and other terrestrial resources in some areas, the high price of fuel, and climate-driven changes in hydrology and water resources (Loring and Gerlach, 2010).

However, climate change is not the first or even most important challenge facing people and communities of the North Pacific (Fazzino and Loring, 2009; Gerlach et al., 2011; Lynch and Brunner, 2007; Martin et al., 2008; Meadow et al., 2009). Rural Alaskans grapple with many difficult issues every day, including high and rising prices of food and fuel; rapid changes to the landscape and weather; fisheries closures and other management actions that keep freezers and smokehouses empty; social and political debates and conflicts regarding the development of land; and troubling health trends, such as increases in diabetes and heart disease, depression, and alcoholism. Each of these challenges may indeed be linked to climate change in various ways, which local people understand perfectly well. Nevertheless, these near-term challenges will likely take precedent for action over long-term climate change impacts.

CLIMATE CHANGE EFFECTS ON COMMERCIAL FISHING IN THE NORTH PACIFIC. Polar regions are expected to be affected by climate-related changes earlier and more extremely than other regions. The Eastern Bering Sea groundfish fishery, from which 14 percent of the total value of the fisheries of the U.S. is generated, is conducted north of the Alaskan Peninsula and Aleutian Islands (Hiatt et al., 2010). Climate change-related shifts in atmospheric conditions, ocean properties, and ecosystem interactions have the potential to greatly affect this multi-billion dollar industry; however, little concrete information is available regarding how these fisheries will be affected. Table 4-3 shows the expected impacts on North Pacific fisheries that have been published by a number of researchers.

Specific studies have been done on how Pacific Salmon, Walleye Pollock, and Pacific cod populations are expected to react (Case studies 4-A and 4-B); however, overall, climate change impacts on fisheries are not well studied in the North Pacific. Even still, the general consensus is that the impacts will be significant. Further information on North Pacific fisheries and predictions on how they may be affected by climate change is presented in Appendix A.

CLIMATE CHANGE EFFECTS ON SUBSISTENCE FISHING ON THE WEST COAST. For tribes in the Northwest, questions have been raised recently about how climate change will affect the maintenance and reproduction of indigenous rights for salmon and other marine species whose distributions may change with a changing ocean environment (Colombi, 2009). The right to harvest marine resources on traditional fishing grounds is guaranteed to these tribes through government-to-government treaties (NOAA Fisheries, 2009a); however, allocation of catch is based upon allowable catch quantities and treaty rights to harvest have referred to tribal "usual and accustomed" fishing areas. Because some predictions of climate change impacts involve target-species range shifts (Mantua et al., 2010), the implications for geographically-bounded tribal fishing rights are uncertain and of great to concern to Northwest "treaty tribes" as water temperatures and prey ranges shift (refer to Section 3 for more detail). Further information on West Coast fisheries and predictions on how they may be affected by climate change is presented in Appendix A.

Similarly, non-tribal fishermen relying on personal use of marine resources are potentially impacted by changes in the abundance and range habitats of targeted nearshore species as well as by climate-based shoreline shifts that may disrupt shoreside infra-structure. For example, 13 percent of pier fishermen surveyed on Los Angeles County piers identified food scarcity and security as concerns (Pitchon and Norman, in review). In Los Angeles County in general, more than three-quarters of a million low-income adults live with hunger or make daily decisions about whether to eat or pay for other essential needs such as shelter or clothing (Harrison et al., 2007). To the extent that urban extraction of locally-caught seafood represents a coping strategy for such food insecuri-ty, potential nearshore climate change impacts present livelihood and nutritional issues for a number of pier-based fishermen. Further information on West Coast fisheries and predictions of how they may be affected by climate change are presented in Appendix A.

Table 4-3: Known or expected direction of social and economic impacts on some major northeast commercial and recreational species

Species	Direction of impact
Walleye pollock and Pacific cod	Ambiguous
Pacific salmon	Ambiguous
Pacific herring	Ambiguous
Flatfish and rockfish	Ambiguous but perhaps positive
Pacific halibut	Ambiguous
King and tanner Crabs	Ambiguous

Based on Hare et al., 2010; Fogarty et al., 2008; Frumhoff et al., 2007.

CLIMATE CHANGE EFFECTS ON COMMERCIAL FISHING ON THE WEST COAST. Little has been documented as to how climate change is affecting fisheries along the West Coast of the U.S., but the largest effects on fisheries will likely be due to changes in the distributions and abundance of stocks, as is documented in other regions. Below we provide a case study of known effects of climate variability on the California sardine fishery as well as how fisheries managers have developed actions to adapt to these ef-fects (Case Study 4-C). Further description of expected climate change impacts to com-mercial fisheries along the West Coast is also provided in Appendix A.

CLIMATE CHANGE EFFECTS ON SUBSISTENCE AND COMMERCIAL FISHING IN THE PACIFIC ISLANDS. Climate change could have both direct and indirect ef-fects on all types of fishing opportunities in the Western Pacific Ocean. Much fishing effort in the region depends on species and habitat associated with coral reefs (Bell et al., 2011a); climate change impacts have been identified as one of the greatest global threats to coral reef ecosystems (http://coralreef.noaa.gov/threats/climate; see Section 3).

Coral reef habitats, and therefore the fish species that depend on them, are threatened by changes to water temperature, acidification of the ocean, and sea-level rise, as well as possibly more severe cyclones and storms (Bell et al., 2011a). Projected changes in SSTs, precipitation, and sea level are not uniform over the Pacific Ocean, but, in general, both surface temperatures and sea level are expected to be greater than they are now by the end of the 21st century. The loss of live corals results in local extinctions and a reduced number of reef fish species (Karl et al., 2009). Declining coral reefs will impact coastal communities, tourism, fisheries, and overall marine biodiversity; abundance of commercially-important shellfish species may decline, and negative impacts on finfish may occur (Fletcher, 2010). Reduced catches of reef-associated fish will widen the expected gap

Case Study 4-A
The effects of climate change on Pacific salmon

Five species of Pacific salmon exist in the Gulf of Alaska, Bering Sea and Aleutian Islands. Each species spends most of its ocean life in the off-shore, pelagic environment, briefly utilizing coastal areas as rearing juveniles and returning adults. A scenario with oceanic and atmospheric warming shows increases in snow melt and water flows causing fall/winter floods that could affect salmon eggs laid in gravel beds (Low 2008). In the summer, higher average temperatures could diminish the oxygen content of the water in streams where juvenile fish live, thus increasing mortality. Warmer temperatures could also affect smolt (emigrant juveniles) timing of marine entry, leading to lower survivals due to mismatches with their zooplankton-prey base (Taylor 2008). Salmon survival is likely determined early in marine life during coastal rearing where high mortality typically sets year class strength (Beamish et al. 2004). Evidence of the actual impacts at these various life cycle stages is accumulating slowly, because establishing impacts requires long time series information on abundance binned by latitude across the exceptionally wide latitudinal and geographical ranges of the five species. Although Alaskan salmon production in commercial fisheries increased from the mid-1970s to the late 1990s and has since been at or near historical highs (Low 2008), evidence from southern Alaska indicates that potentially detrimental changes in the migratory timings of juveniles and adults is occurring in concert with warming of the terrestrial and nearshore environments (Kovach et al. 2013).

Chinook Salmon (Source: NOAA).

Case Study 4-B

The effects of climate change on pollock and Pacific cod

The walleye pollock and Pacific cod fisheries are both directly affected by several biological characteristics of the target species. These include the total fishable biomass, the distribution of biomass throughout the fishing grounds, and the relative abundance of different sizes or ages of fish.

Studies have found that although a northward shift in the distribution of pollock and cod fishing has been seen in recent years (2006-2009), the northward shifts are associated with colder than average years in the Bering Sea (Haynie and Pfeiffer 2012; Pfeiffer and Haynie, 2012; pers. comm.). A large ice and cold pool extent concentrates fish populations in the northern region of the fishing grounds, giving fisherman in the north an advantage over those in the south. The redistribution has occurred in both the winter and summer seasons of the Pacific cod fishery and in the summer pollock fishery. However, the winter pollock fishery, which is driven by the pursuit of valuable roe-bearing fish that spawn in the southern part of the Eastern Bering Sea, has had little redistribution of effort. This large difference in value per fish in the roe fishery means that harvesters are unlikely to shift to the north for marginal increases in catchability.

Uncertainty remains as to the impacts of warming temperatures on the distribution of pollock fishing, because the relationship between climate and biomass complicates both retrospective analyses and predictions. Decreases in pollock biomass have been associated with warmer temperature regimes (Mueter et al., 2011), but the effects on fishable abundance are lagged by several years. For the time periods used in Haynie and Pfeiffer 2012 and Pfeiffer and Haynie's 2012 research, warm periods did not correspond to years with low fishable biomass.

This complicates the separation of the direct ice and cold pool effects from the effects of climate on abundance, which is necessary for prediction. Their work highlights the importance of considering the economic, institutional, and ecological characteristics of a fishery for improving our understanding of the effects of climate change on fisheries.

Local Cod Depletion Study (Source: NOAA)..

Case Study 4-C
The effects of climate change on Pacific sardine

Relationships between climate and productivity of the Pacific sardine stock have been previously examined (Norton and Mason 2003, 2004, 2005; Herrick et al., 2007). During a warm regime, biomass of Pacific sardine increases; conversely, a cold-water regime results in a decrease in abundance of the sardine stock and reduces its distribution to almost exclusively off Southern California, U.S. and Baja California. In response to fluctuations in productivity due to climate variability, the U.S. Pacific sardine fishery is managed using an environmentally-based harvest control rule to determine the annual harvest level. The harvest control rule is intended to prevent over-fishing, sustain consistent yield levels (Herrick et al., 2006), and reduce the exploitation rate if stock biomass decreases or if ocean conditions become cooler and less favorable for the stock.

Given the general lack of existing research and modeling capabilities on socio-economic impacts of climate change on fisheries, the impacts on Pacific sardine fisheries are uncertain. To this end, two scenarios are considered here out of the many potential scenarios that could occur in the Eastern Pacific Ocean. The first assumes that the increase in ocean temperature due to climate change is consistent with that experienced during a warm water regime in the California Current Ecosystem, which is favorable to sardine productivity. This results in an increase in the northern sardine stock

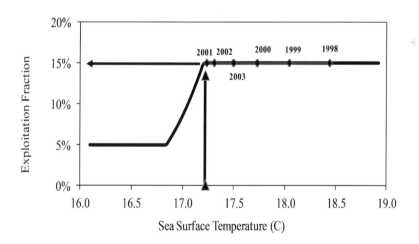

Sardine Harvest Control Rule:
Harvest = Fraction(SST)•(Biomass – Cutoff)

Pacific sardine harvest-control rule implements a decreasing exploitation fraction in cool years based on a 3-year moving average of SST at Scripps Pier, San Diego, California. 'Harvest' is the guideline harvest level in metric tons (mt), 'Biomass' is current biomass estimate, 'Cutoff' is the lowest level of estimated biomass at which harvest is allowed (150,000 mt), and 'Fraction (SST)' is the environmentally-based percentage of biomass above the cutoff that can be harvested (Source: M. Dalton, pers. comm.).

Case Study 4-C (Continued)

biomass in the U.S. EEZ off the West Coast, particularly off the Pacific Northwest, increased biomass in Canada's EEZ, and, depending on the extent of ocean warming, potentially harvestable biomass in the U.S. EEZ off Alaska. This is essentially the status quo scenario, with perhaps a slight increase in U.S. Pacific sardine fishing opportunities and a corresponding increase in economic activity and the economic value that would ensue.

The second scenario assumes that an increase in ocean temperature results in a northerly expansion of the entire subtropical marine biota. This would include an increase in abundance and a northerly shift of all Pacific sardine stocks, allowing sardine fisheries and conceivably fisheries that focus on their predators to expand along the entire Northeast Pacific coast. This scenario is suggestive of what has been observed during ENSO events in the eastern Pacific in which tropical tunas, which U.S. sardine vessels are capable of harvesting, typically become more available off the Southern California coast. Under these circumstances, a significant increase in fishing opportunities for U.S. vessels that target Pacific sardine and a resultant increase in fishing-related economic activity and economic value are expected.

between the availability of fish and the protein needed for food security.

A recent report published by the Secretariat of the Pacific Community (SPC) assessed the vulnerability of tropical pacific fisheries and aquaculture to climate change (Bell et al., 2011a).[1] Across the region, fish provide 51-94 percent of the animal protein in the diet in rural areas and 27-83 percent in urban areas. The great majority of fish for food security in the region is derived from coastal subsistence fishing; in 14 of the countries and territories, 52-91 percent of the fish eaten in rural areas is caught from coral reefs and other coastal habitats by the household and high levels of subsistence fishing are common in urban areas on many of the smaller island areas (Pratchett et al., 2011). Subsistence fishing in coastal and freshwater habitats produced three times as much fish as commercial fishing in coastal waters in 2007 (Bell et al., 2011b).

Nearly 70 percent of the world's annual tuna harvest, approximately 3.2 million tons, comes from the Pacific Ocean. Climate change is projected to cause a decline in tuna stocks and an eastward shift in their location, affecting the catch of certain countries (Karl et al., 2009). Thus, there are likely to be winners and losers; tuna catches are expected to be higher around islands in the eastern tropical Pacific Ocean and lower in the west (Bell et al., 2011b). However, the outlook appears less severe for the tuna on which the Pacific region depends, especially as a source of economic revenue as well as food (Bell et al., 2011b). Further description of expected climate change impacts to

---

1 The assessment did not include Hawaii but covered 22 Pacific Island countries and territories (PICTs): American Samoa, Cook Islands, Federated States of Micronesia, Fiji, French Polynesia, Guam, Kiribati, Marshall Islands, Nauru, New Caledonia, Niue, Commonwealth of the Northern Mariana Islands, Palau, Papua New Guinea, Pitcairn Islands, Samoa, Solomon Islands, Tokelau, Tonga, Tuvalu, Vanuatu, and Wallis and Futuna.

commercial fisheries in the Pacific Islands is provided in Appendix A.

On balance, the Pacific Island countries and territories, including the U.S., appear to be in a better position than nations in other regions to cope with the implications of climate change for fisheries and aquaculture. Expected effects for the region as a whole are among the better possible outcomes worldwide. In particular, the Pacific Island Commonwealths and Territories (PICTs) with the greatest dependence on tuna, such as Kiribati, Nauru, Tuvalu, and Tokelau, are likely to receive greater benefits as the fish move east, whereas the projected decreases in production occur in those PICTs where industrial fishing and processing make only modest contributions to GDP and government.

However, projections that storms, including cyclones, hurricanes and typhoons, could become progressively more intense would pose increased risk of damage to shore-based facilities and fleets for domestic tuna fishing and processing (Bell et al., 2011a) as well as increased risk to human safety at sea. The increased costs associated with repairing and relocating shore-based facilities and addressing increased risks to fishermen's safety could affect the profitability of domestic fishing operations.

Aquaculture impacts could also occur. Changing patterns of precipitation and more intense storms could damage aquaculture ponds and/or make small pond farming more difficult due to more frequent droughts (Bell et al., 2011a). Financial risks associated with coastal aquaculture could be higher as a result of greater damage to infrastructure from rising sea levels and more severe storms.

For island fisheries sustained by healthy coral reef and marine ecosystems, climate change impacts could exacerbate stresses such as overfishing, affecting both fisheries and tourism that depend on abundant and diverse reef fish (Karl et al., 2009). This context suggests that how society responds and adapts to the impacts of climate change may reduce some effects but could also exacerbate others, especially in the short-term. One approach to addressing climate change impacts to subsistence fisheries is to reduce other commercial and recreational fishing. Using this approach, subsistence fishing access and effort are already being affected by the recent establishment of annual catch limits for managed species in commercial fisheries.

Another likely response to reducing climate change effects on coral reefs and associated fish species will be the establishment of marine protected area (MPA) networks (Mumby et al., 2011a), which can enhance the resilience of marine resources to climate change and could protect certain areas from additional fishing pressure (see Section 6). Climate change will certainly add momentum to the existing public pressures to establish new MPAs. When climate change poses risks to protected species such as monk seals (Baker et al., 2006) or loggerhead turtles (Van Houtan and Halley, 2011), the resulting measures could potentially include reduced access to fisheries.

CLIMATE CHANGE EFFECTS ON FISHERIES IN THE SOUTHEAST. Given limitations on the current knowledge of the biophysical effects of climate change in the region, little is known about how the ocean services provided by the South Atlantic and Gulf of Mexico are or will be impacted. However, one of the most pronounced effects of climate change in the Gulf of Mexico is likely to be the increased intensity of hurricanes, which can lead to the loss of wetlands and barrier islands that protect nursery grounds for marine resources. The Gulf Coast represents the region with the highest potential for annual hurricane seasons that disrupt all types of fishing (NOAA Fisheries, 2009a). Loss of these habitats has been well documented (Ingles and McIlvaine-Newsad, 2007) and impacts are expected to be worse in the future given projected increases in barrier island and wetland losses. In addition, with the possible increases in hurricane intensity, storm surge, and high winds, communities that rely on coastal marine resources for subsistence are likely to be more and more limited in their ability to undertake harvesting activities.

There is also increasing concern for and modeling of the potential effect of harmful algal bloom (HAB) events on fish stocks in the Gulf of Mexico. For example, one of the first examinations of the issue reported: "Given the systematic decline in all indices between the years 2005-2006, there was a general concern the extensive red tide that occurred during 2005 on the West Florida Shelf may have substantially affected the red grouper population" (SEDAR Update Assessment, 2009, pages 24-25).

With regard to recreational fishing, Carter and Letson (2009) found that climate activity like ENSO had a moderate influence on the headboat fishery for red snapper in the Gulf of Mexico, with cooler, wetter weather in El Niño years corresponding to slightly increased headboat activity.

Further description of expected climate change impacts to fisheries in the Southeast is provided in Appendix A.

CLIMATE CHANGE EFFECTS ON SUBSISTENCE AND COMMERCIAL FISHING IN THE NORTHEAST. Most social and economic impacts of climate change on fisheries flow from biological and ecological changes related to issues such as water temperature and acidity (Cooley and Doney, 2009; Felthoven et al., 2009; Hannesson, 2007). For Northeast region fishermen and the families, households, firms and communities that depend on them, the most relevant changes are those occurring to the ocean of the Northeast U.S. (NEUS) shelf ecosystem and its denizens. Water temperatures are rising, surface seawater pH is decreasing, precipitation is increasing, salinities are decreasing, and stratification is increasing (EAP, 2012). All of these changes impact marine life (Table 4-4).

As described in the Case Study 4-D that follows, some Northeast species such as Atlantic cod (*Gadus morhua*) will likely lose juvenile habitat, thus lowering biomass, and will potentially move into Canadian waters and out of the range of Northeast fishermen due to warming water temperatures (Fogarty et al., 2008). Others, such as Atlantic croaker (*Micropogonias undulatus*), will likely see an increase in biomass as well as a range shift northward from the Mid-Atlantic into southern New England, thus providing New England fishermen with a larger stock to fish on but leaving Mid-Atlantic fishermen with less access (Hare and Able, 2007; Hare et al., 2010). Still other species, such as lobster, will likely also see their ranges move northward, leaving the waters of New York and Rhode Island and increasing their presence in Maine while warmer waters may also lead to increases in "lobster shell disease" (Frumhoff et al., 2007), making the impact on fishermen more difficult to judge. Increased acidity could affect shellfish, including scallops, lobsters, and blue crab, which are three of the Northeast's highest-value species, so economic and social impacts are potentially high (Cooley and Doney, 2009; McCay et al., 2011). (See Table 4-2 for summary.)

However, marine ecological changes are not the only climate change issues affecting

Table 4-4: Known or expected direction of social and economic impacts on some major northeast commercial and recreational species	
Species	**Direction of impact**
Atlantic cod (*Gadus morhua*)	Negative
Atlantic croaker (*Micropogonias undulatus*)	Positive
Atlantic lobster (*Homarus americanus*)	Ambiguous, but perhaps more negative
Atlantic sea scallop (*Placopecten magellanicus*)	Negative
Blue crab (*Callinectes sapidus*)	Negative

Based on Fogarty et al., 2008; Frumhoff et al., 2007; Hare et al., 2010.

fishermen and fishing-dependent industries. Many shellfish, including lobster, will suffer from sea level rise if the coastal wetlands necessary to their juvenile stages are flooded (Frumhoff et al., 2007, page 28). Sea level rise will also flood coastal infrastructure, especially docks and other fishing-related structures that are on the edge of the current coastline. In the Northeast, many smaller ports have already lost infrastructure to gentrification (Colburn and Jepson, 2012; Gale, 1991), among other causes. With vital infrastructure including boat repair facilities concentrated in fewer ports, loss in any of the remaining hubs could have important negative impacts on the entire region's fishing fleet (NOAA, 1997a, b; Robinson and The Gloucester Community Panel, 2003, 2005).

Many challenges remain for assessing the social and economic impacts of climate change on fisheries. For many Northeast species, studies have not yet been conducted to assess the impacts of climate change. In addition, the link between ocean conditions and fisheries is complex and poorly understood. Major uncertainties and gaps in understanding make it particularly difficult to quantify the effect of global climate change on the stocks of commercially important fish species (Markowski et al., 1999). Because many fishermen in the Northeast fish multiple species at a time or in sequence over the year, fully understanding likely social and economic impacts will be difficult without a clearer picture of biological impacts. Further description of expected climate change impacts to fisheries in the Northeast is provided in Appendix A.

Fisheries and communities adapting to climate change

Fisheries in the U.S. are managed through a process that uses benchmarks based on the productivity of fish stocks. Predicting how management might change in response to climate change can be explored through scenarios that link climate change to changes in these benchmarks (Hare et al., 2010). Such an approach has been used to show that the maximum sustainable yield of Atlantic croaker is likely to increase with predicted climate change along the mid-Atlantic coast (Hare et al., 2010), while a similar exercise shows that Atlantic cod is likely to suffer a decrease in productivity and its associated benchmarks (Fogarty et al., 2008). A major challenge will come from uncertainty regarding

Case Study 4-D
The effects of climate change on Atlantic cod and croaker

Atlantic cod, primarily found in New England from Maine to Connecticut, is likely to experience negative impacts from climate change, while Atlantic croaker, primarily found in the Mid-Atlantic from New York to Virginia, is likely to experience positive impacts.

Cod are sensitive to increases in ocean temperature (Fogarty et al., 2008) although the level of impacts varies. Some concern has also been voiced that certain prey species may not move in synch with cod, creating further difficulties in the Northeast Region (Murawski 1993). If cod move north, likely off of Georges Bank and potentially even completely out of the Gulf of Maine (Fogarty et al., 2008; Nye et al., 2009), U.S. commercial fishermen will need to substitute other species because they cannot follow the cod north into Canadian waters.

Even where climate change has a positive effect on a species, social and economic impacts can occur. Atlantic croaker are found from the Gulf of Maine to Argentina (ASMFC 2011), but in the Northeast, their primary range is from Hudson Canyon off the coasts of New York and New Jersey south to Cape Hatteras, North Carolina (Hare et al., 2010). Croaker is not one of the top commercial species, representing only 0.7 percent of total Mid-Atlantic landings revenue and 2 percent of Mid-Atlantic landed pounds. but it is one of the key recreational species, with 15 million fish caught.

Hare et al. (2010, page 452) estimate that, as ocean temperatures warm, "[a]t current levels of fishing, the average (2010–2100) spawning biomass of the population is forecast to increase by 60–100 percent." Similarly, the center of the population is forecast to shift 50–100 km northward.

Commercial fishermen in the Mid-Atlantic will likely be able to increase their landings of croaker, while fishermen further north will gain a new species (i.e., croaker) with a high biomass, potentially raising the importance of croaker as a commercial species. Further, Mendelsohn and Markowski (1999) and Loomis and Crespi (1999) note that recreational fishing and boating activities may significantly increase with warming. Tournament fishermen in New England (see NOAA Fisheries, 2009a, for examples of New England tournaments) will likely be able to add croaker to their repertoire. Subsistence fishermen in southern New England may add croaker to their landings, providing an additional species at a time when they may be able to collect fewer local shellfish (Steinback et al., 2009) because shellfish are expected to be negatively impacted by increasing ocean acidification.

the speed, magnitude, and location of effects brought on by climate change, which will make them difficult to distinguish from "normal" climatic variation (Coulthard, 2009). Not all stocks will be affected in the same manner; some stocks will benefit while others will be adversely affected, making general rules relating climate change to desirable management changes problematic. Part of the challenge will also be putting into place management structures that can adapt to climate change while recognizing, for example, that fish distribution will change over time. In such cases, systems such as quotas and protected areas that tie management choices to particular geographic areas may be ineffective if the distribution upon which protective measures were based change over time (OECD, 2010).

As an example, in American Samoa, researchers are conducting a two-phase project to assess villagers' perceptions of climate change, past experiences with climate and weather events, vulnerability and resiliency, and plans for adaptation. The first phase, conducted in 2009 by Supin Wongbusarakum and partners at the Pacific Regional Integrated Sciences and Assessments Program, consisted of two focus groups and a survey of households in two villages: Amouli on the main island of Tutuila and the smaller village of Ofu located in the more-rural Manua islands. The majority of household leaders in both villages believed that climate change was occurring and would affect village lives in many ways. Village residents were asked about their past experiences with many types of events expected to increase or intensify with climate change, such as sea-level rise and flooding, coastal erosion, saltwater intrusion, severe storms, droughts, and coral bleaching.

In particular, 86 percent of respondents in Ofu, compared to 62 percent in Amouli, believed that their household economy was sensitive to climate conditions. In Ofu, both fishing and farming were mentioned as the most important means of livelihood, while both were less important to Amouli households. 62 percent of natural-resource-dependent households in Ofu and 31 percent in Amouli sold fish and marine products, including fish, giant clam, octopus, lobster, shellfish, coconut crabs, and sea cucumbers. Fish was the most frequently mentioned type of food eaten in the households of both villages, with more fish self-caught than bought, and high rates of sharing both fish and farm products.

Socio-economic factors, including age, gender, education, occupation, livelihoods, income, availability of alternative livelihoods, and access to lifelines were related to perceptions of their household's vulnerability to certain types of extreme climate events. For example, a higher proportion of people with a livelihood alternative believed their household had a medium level of vulnerability across nearly all event types except for sea level rise, while the proportion of high vulnerability was comparable to those in households without alternative livelihoods.

Respondents were split roughly evenly between those who felt that their households were able to cope well with such events and those who felt they were not. Socio-economic variables did not uniformly explain these perceptions across all types of events. More women perceived higher household resilience across all types of extreme events when compared to their male counterparts. Except for tropical storms, for which all age groups seemed to assess their coping ability on a similarly high level, those who were over 60 years old were most likely to report that they were able to adjust to and recover from extreme events even though they agreed that they were more vulnerable. These respondents have more confidence in their household's resilience because they have experienced and successfully coped with extreme events. This is consistent with other research that has found that past experience in coping with economic, social, or environmental stressors can increase resilience. People whose livelihood activities depend on both fishing and farming are identified as one of the groups most vulnerable to extreme events.

Phase 2 of the project, ongoing as of 2012, is a collaboration with the University of Hawai'i Coastal Geography Group to develop detailed maps and models of local impacts of sea-level-rise scenarios and extreme-weather events in Amouli and another village on Tutuila. The maps and models will then be presented to village residents in Participatory Learning Assessment (PLA) workshops. Through involvement with all levels of the community including women, youth, and village leaders, the project aims to develop locally relevant, socially feasible, and sustainable solutions that can result in more climate-resilient communities.

4.3 Implications of Climate Change for Aquaculture

According to the NMFS Office of Aquaculture (http://www.nmfs.noaa.gov/aquaculture.html), the U.S. is a major consumer of aquaculture products but only a minor producer. The U.S. imports 86 percent of its seafood, approximately half of which comes from aquaculture production. Two-thirds of marine aquaculture is molluscan shellfish, such as oysters, clams, and mussels. The remainder is shrimp and salmon, with lesser amounts of barramundi, seabass, seabream, and other species.

The impacts of climate change on global aquaculture are not yet fully known. The potential impacts of climate change on North America may include rising sea surface temperatures; sea-level rise that will influence deltaic regions, increase saline water intrusion, and bring about major biotic changes; increasing ocean acidification; higher incidence of extreme weather events that will result in physical destruction of aquaculture facilities, loss of stock and spread of disease; increasing risks of transboundary pests and diseases; and altered rainfall patterns and river flows (FAO, 2010). Unlike capture fisheries, organisms being reared in captivity are subject to more controlled environments that may allow for adaptation to changing climate conditions. In fact, using aquaculture may reduce some of the impacts of climate change on wild stocks and their ecosystems; for example, the culture of seaweeds and, to a lesser extent, filter-feeding shellfish represent a net removal of CO_2 from the oceans. In addition, the application of hatchery stocking and selection could be used to increase the adaptive rate of key organisms to predicted changes in temperature, pH, and salinity.

As in commercial, recreational, and subsistence capture fisheries, climate impacts are likely to be both positive and negative, arising from direct physical and physiological processes and indirect impacts on natural resources required for aquaculture (De Silva and Soto, 2009). Examples of direct impacts on aquaculture due to an increase in sea surface temperature include the following.

Direct impacts of climate change

A rise in sea-surface temperatures may trigger the growth of HABs that can extend the spatial or temporal scope of a bloom or release toxins into the water that kill cultured fish and shellfish, particularly for fish in cage-based aquaculture systems and shellfish beds. Higher water temperatures may also result in increased disease incidence and parasites, which may develop more rapidly in warmer waters and higher salinities (see Section 3) and threaten the aquaculture sector. Species cultured in temperate regions, predominantly salmon and cod species, have a relatively narrow range of optimal temperatures for growth. Temperatures over 17 ºC, at which feed intake drops and feed utilization efficacy is reduced, would be detrimental to the salmon-farming sector (De

Silva and Soto, 2009).

Certain aspects of aquaculture may benefit from climate-related changes. Higher water temperatures may also increase the availability of new culture sites, especially in areas previously too cold to support aquaculture. An increase in water temperature may also have a positive effect on metabolism and stimulate growth of cultured species, as long as the change happens gradually and stays within the thermal tolerance of the species.

Aquaculture does have some advantages for dealing with climate-related impacts. For aquaculture species that are produced from a hatchery, selective breeding will likely help farmersstay ahead of changes in optimal temperature and pH. Improvements of 10 percent per generation are not uncommon for selective breeding programs (Gjøen and Bentsen, 1997). Most aquaculture species still contain a great deal of genetic diversity, which means they should be adaptable to the direct impacts of climate change. Likewise, aquaculture organisms can be treated for parasites and diseases (Dipnet, 2007; Moffitt et al., 1998). Vaccination, selective breeding, and better nutrition have all improved the resistance of farmed fish to wild diseases, a trend that is likely to continue (Torrisson et al., 2011).

Indirect impacts of climate change

The dependence of aquaculture on fishmeal and fish oil becomes an important issue under most climate change scenarios (De Silva and Soto, 2009). Because capture fisheries are a major source of protein and lipids for aquaculture, changes in fisheries caused by global climate change will impact aquaculture systems. Tacon et al. (2011) estimated that in 2008, the aquaculture sector on a global basis consumed 3.72 million metric tons of fish meal, or 60.8 percent of the total global fish meal production, and 0.78 million metric tons of fish oil, or 73.8 percent of the total reported global fish oil production in 2008. Industrial fishmeal and fish oil production is typically based on a few, fast-growing, short-lived, productive stocks of small pelagic fish in the subtropical and temperate regions. The major stocks that contribute to this global industry are the Peruvian anchovy, capelin, sand eel, and sardines. The U.S. is a net fish meal and oil exporter, with the largest source being menhaden and the second-largest source coming from trimmings associated with fish caught for human consumption. As aquaculture develops, trimmings from this sector are also becoming a source of fish meal and oil.

Due to climate change, biological productivity in the North Atlantic is predicted to decrease by 50 percent and ocean productivity worldwide by 20 percent (Schmittner, 2005). This would greatly impact the availability of the small pelagics for fishmeal and oil. Predicted changes in ocean circulation patterns might also result in the occurrence of ENSO influences becoming more frequent (Section 2), with impacts on the reliability of stocks of the small pelagics utilized for fishmeal production. Changes in the productivity of fisheries that cater to the fishmeal and oil industry, and particularly the main fisheries on which fishmeal and fish oil production is based, could limit the raw material available.

The reduction in forage fish is likely to have a greater impact on the wild fisheries that depend on them for forage than on fish reared in aquaculture. Because aquaculture organisms have no nutritional requirement for fishmeal or oil (Rust, 2002), alternative dietary ingredients are being developed (Barrows et al., 2008; Gatlin et al., 2007). All fish, whether herbivores or carnivores, require approximately 40 nutrients that are contained in fishmeal and oil in correct ratios, but these nutrients can also be assembled from other protein and lipid sources (Rust, 2002). Already, commercial salmon diets contain a fraction of the fishmeal and oil that they used to (Rust et al., 2010; Torrisson et al., 2011) and further replacement is occurring as the price for fishmeal and oil increases relative to other protein and lipid sources. Numerous carnivorous fish species have been fed experimentally on diets containing no fishmeal, with production rates equal to control groups being fed diets with fishmeal (Rust et al., 2010). Further, fish are more efficient with feed resources than land animals (Hall et al., 2011) causing some to shift resources away from livestock production and toward fish production.

Ocean acidification and aquaculture

In North America, ocean acidification is currently considered a serious near-term threat because of its potential to alter ocean foodwebs in a relatively short period (De Silva and Soto, 2009). For aquaculture, ocean acidification particularly influences shell formation and affects filter-feeding shellfish (see Section 3). Protecting vulnerable marine organisms grown in aquaculture facilities from the effects of ocean acidification may be possible in theory, but it presents practical challenges. Aquaculture is often conducted in tanks or ponds on land that are filled with coastal seawater or within coastal ocean pens. Adjusting seawater chemistry before supplying culture tanks on land would require equipment and monitoring that might increase the overhead of aquaculture operations and aquacultured animals in nearshore operations cannot be shielded from ocean acidification (Cooley and Doney, 2009). Research that focuses on monitoring and mitigating ocean acidification events could greatly benefit from international cooperation and coordination.

Seaweed aquaculture could be one of the few human activities that has the potential to mitigate some global climate change impacts and provide a net reduction in ocean acidification. Seaweed production from aquaculture could be used to make ethanol, protein concentrates, and other chemicals. Seaweed production, and to a lesser extent, shellfish aquaculture, has the additional benefit of binding CO_2 and nutrients from the ocean and returning them to the land. Although the scope of this potential is unknown at this time, it merits further investigation.

Social impacts of climate change on aquaculture

Likely the greatest social impact of climate change that must be addressed by the aquaculture industry is on human health. Seafood consumption may have a number of health benefits, including improved cardio-vascular function, reduced inflammatory disease, reduced macular degeneration, reduced mental depression, and higher IQ, among others (Institute of Medicine, 2006; FAO/WHO, 2011). Only aquaculture has the potential to supply the increased volume of seafood needed to support U.S. per capita consumption at historic or increased levels. If global climate change reduces wild harvest, then the production from aquaculture will have to be that much greater.

Finally, some social impacts of climate change on capture fisheries, such as damage to physical capital and impacts on transportation and marketing systems and channels, are likely to have similar effects on aquaculture (Cochrane et al., 2009).

4.4 Offshore Energy Development

Oil and gas

Offshore oil and gas development has been increasing in recent years. The oil and gas industry and its consumers are facing compulsory adaptation to climate changes that they contributed to generating. Figure 4-2 illustrates the perspective of adaptive government regulations in Alaska, which have been changing with warmer temperatures (ACIA, 2004). The main climate-related effects on these industries are anticipated to be increased pressure on water sources, failure of infrastructure that was not designed to withstand new climatic conditions, and changes in financial assets and resources extracted as energy production moves from the traditional oil and gas industry to renewable sources of energy. How the industry reacts to these changes depends on whether they are international companies, small independents, or state-owned; their geographical location; local policies and regulations; the company's ethics; and their combined performance on the national and international markets. These issues have to be analyzed and understood in a broad context given that the oil and gas industry delivers oil and gas as well as providing jobs, economic development, and research and development. Industry

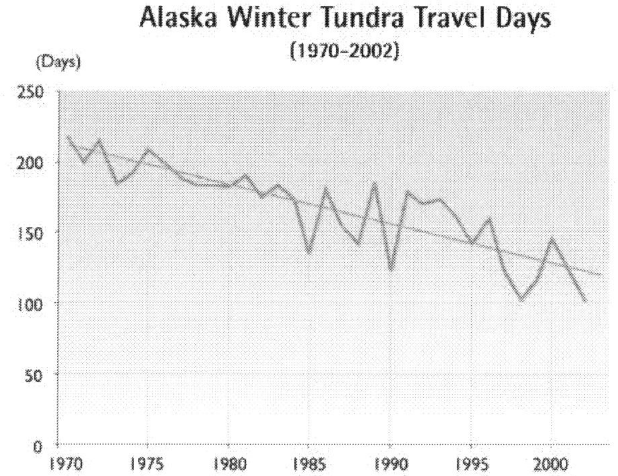

Alaska Winter Tundra Travel Days
(1970–2002)

Figure 4-2 The number of days in which oil exploration activities on the tundra are allowed under the Alaska Department of Natural Resources standards has halved over the past 30 years due to permafrost thaw, which is disrupting transportation; damaging buildings and assets particularly pipelines; and increasing the risk of pollution. Operational costs are increasing for oil and gas companies (ACIA, 2004)

research includes developing technologies that reduce greenhouse gas (GHG) emissions by investing in renewable energy sources and/or more efficient materials, machinery, and approaches to different problems.

CURRENT EXPLORATION EFFORTS AND PLANS. Offshore energy development planning has been proceeding rapidly in recent years. In late January 2011, the Obama Administration announced a proposed Central Gulf of Mexico oil and gas and lease sale. In the fall of 2011, several lease sales were announced by the Bureau of Ocean Energy Management (BOEM) in the Central and Western Gulf of Mexico. In addition, in

Table 4-5: Bureau of Ocean Energy Management lease sales schedule, 2012-2017		
Sale number	**Area**	**Year**
229	Western Gulf of Mexico	2012
227	Central Gulf of Mexico	2013
233	Western Gulf of Mexico	2013
244	Cook Inlet	2013
225	Eastern Gulf of Mexico	2014
231	Central Gulf of Mexico	2014
238	Western Gulf of Mexico	2014
235	Central Gulf of Mexico	2015
242	Beaufort Sea	2015
246	Western Gulf of Mexico	2015
226	Eastern Gulf of Mexico	2016
241	Central Gulf of Mexico	2016
237	Chukchi Sea	2016
248	Western Gulf of Mexico	2016
247	Central Gulf of Mexico	2017

the Arctic, Shell Oil Company is likely to move towards offshore energy development in the Chukchi Sea off the northern coast of Alaska. The continuation of this develoment is shown in BOEM's recently announced lease sale schedule for 2012-2017 in Table 4-5.

IMPACT FACTORS. Currently, five impacts of climate change have been associated with recent offshore oil and gas exploration (Acclimatise, 2009b):

1. Increased Pressure on Water Resources: Changing rainfall amounts, availability of potable water, and droughts will all increase the demand for water, which is key in sustaining the production of oil and gas.

2. Physical Asset Failure: Several types of existing equipment are old (e.g., shallow water oil platforms in the Gulf of Mexico) in addition to being designed to function under climate conditions typical of 20 to 40 years ago. Included in this category are energy supplies such as generators and batteries, off-site utilities, and waste- and water-treatment technologies. As an example of the impact of climate change on these types of equipment, changes in ambient air temperature and their impact on turbine and generator performance can have grave consequences once threshold values are reached.

3. Employee Health and Safety Risks: Not only are environmental conditions changing at most locations on the globe, but the oil and gas industry is also exploring potential oil and gas reserves in areas such as ultra-deep waters and the Arctic Ocean where ambient conditions are significantly more extreme and dangerous for industry workers than current areas under use. Consequently, insurance costs, salaries, and other operational costs will increase.

4. Drop in Value of Financial Assets: To meet the growing demand for energy, oil and gas companies need to continue to secure investment for new exploration, production, and manufacturing. Potential investors and stakeholders are placing greater importance on the business impacts of climate change because the risks impact cost and revenue drivers. Beyond the safety-driven increases noted above, insurance costs could potentially rise because of greater risk of physical plant damage due to weather events, an issue recognized by only 10 percent of respondents to the Acclimatise survey. The current reported value of proven petroleum reserves may also be affected by companies failing to take into account the full impact of climate change, which could have major financial implications.

5. Damage to Corporate Reputation: As knowledge and awareness of climate change grows, any failure to monitor and report the impacts of climate change on social and ecological resources is increasingly likely to harm oil and gas companies' reputations. Contractual relationships that do not adequately foresee and manage risks driven by climate change may damage a company's reputation with stakeholders, increasing the risk of parties turning to litigation.

CHANGES IN REGULATIONS. Governments will variably impact the oil and gas industry based on their particular climate change policies and regulations. Fish and marine mammal species foreign to the Arctic Ocean just five years ago are being sighted for the first time off the northern shore of Alaska (Acclimatise, 2009a). These new inhabitants of the Chukchi and Beaufort Seas will likely trigger new protective measures by state and federal regulatory agencies. Additional regulations will certainly increase operational costs, although in the Arctic, these setbacks will also be accompanied by opportunities. Reduced sea ice coverage, for instance, will lead to the opening of new shipping lanes to facilitate the transport of crude oil between the Atlantic and the Pacific Oceans. New or amended regulatory regimes may thus be required for these areas.

At lower latitudes, in the Gulf of Mexico area, it is likely that future tropical weather events, such as typhoons and hurricanes, will become more intense (Ulbrich et al., 2008). However, these modeling projections carry uncertainties which make preparing for possible impacts through regulatory changes even more difficult. Despite this, some companies are attempting to respond to potential changes in the regulatory environment; for example, in 2006, following the extreme 2005 hurricane season, the American Petroleum Institute launched a process for reviewing design and safety standards for offshore oil platforms.

IMPACTS ON INDEPENDENT AND NATIONAL OIL AND GAS COMPANIES. Independent oil companies may be more vulnerable than nationally-owned oil companies because they own less than 10 percent of the world's oil and gas reserves, but they are buffered in their vulnerability by the fact that 20 percent of global production is conducted through contractual arrangement with the nationally-owned oil companies (Acclimatise, 2009b). Industry is currently witnessing a trend toward national oil companies limiting access to reserves, resulting in investments by independent oil companies in regions with harsh operational environments and climatic conditions as well as higher geo-political risks.

OVERALL FINANCIAL IMPACT OF CLIMATE CHANGE ON THE OIL AND GAS INDUSTRY. Different companies are impacted differently by climate change. Companies may feel direct impacts, such as on-site changes of environmental conditions, or indirect impacts, such as pressures exerted by the public and governments. The financial impact of climate policies and environmental protections that restrict access to reserves,

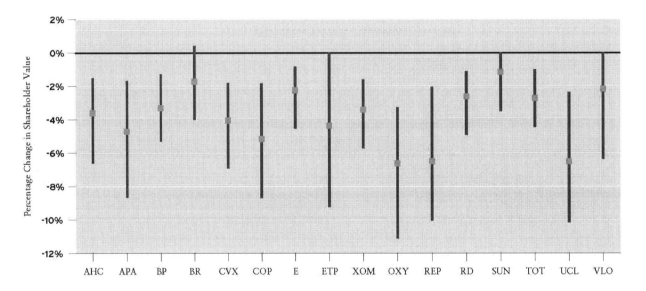

Figure 4-3 Combined financial impact of climate policies and restricted access to oil and gas reserves (range of possible outcomes and most likely impacts) (Source: Austin and Sauer, 2002).

in addition to the development of resources in less accessible locations every year, is estimated to reduce shareholder value by between 1 and 7 percent, depending on the company. These impacts are summarized in Figure 4-3 below (Austin and Sauer, 2002).

DRIVERS FOR CHANGE AND PROJECTIONS. Three main drivers are associated with change in oil and gas industry: a) cost and revenue, b) pressure by stakeholders, and c) new regulations by state and federal governments. A better understanding of readiness for compliance with new climate change regulations will ease the pressure mentioned in b) above. As an example, companies following regulations that will require them to invest in alternatives to fossil fuels and develop cleaner and more sustainable energy sources will see those changes reflected in their profits in the mid- and long-term. Oil companies would thus need to adapt by first conducting a high-level assessment of how climate change can impact their business models and then by identifing and analyzing individual areas within their operation that could generate the greatest impact on performance (e.g., Asset Lifecycle Management; Acclimatise, 2009b). Also, companies that adapt reporting and performance management to incorporate risks resulting from climate change will have a greater chance of remaining successful. Researchers agree that the adoption of carbon sequestration procedures, combined with the inclusion of renewable energy production, will transform the current oil and gas industry into a new version (Lovell, 2010) that will have the capacity to deliver energy products obtained from both renewable and non-renewable sources while simultaneously reducing emissions to comply with national and international regulations.

Historically, oil and gas companies have developed new innovations to keep up with fluctuating markets but it is unclear how the industry will be impacted in the future by the opposing trends of a) increased energy demands due to a projected increase of the U.S. and global populations, and b) increased use of alternatives sources of energy. In turn, the global economy is expected to recover and lead to increasing levels of prosperity, which would imply higher consumption of energy. In this context, operational costs are expected to increase significantly because of compliance with new regulations, the need for equipment able to perform under new environmental conditions, and higher insurance premiums, among other factors.

THE FUTURE OF THE INDUSTRY. The great dilemma that society and oil and gas companies face today is based on the relationship between limiting greenhouse gas emissions and desires for increased energy consumption and company profits (Van den Hove et al., 2002). Different companies have taken different approaches to this dilemma (Van den Hove et al., 2002). For instance, ExxonMobil was the first to argue that risking climate change to have a profitable oil and gas industry that can boost the economy and technological development is better than trying to reduce the risk and diminishing profits for the industry, because the reduced profits would presumably impact the entire economy (Button, 1992; Van den Hove et al., 2002). Different companies have given different weights and justifications to both sides of the problem: profits and the well-being of the economy versus climate change impacts on society and ultimately the economy as well. Some or all of these companies may change their strategies to focus on lowering emissions, especially if they are convinced that this approach will be the one that maximizes their profits.

Oil and gas companies could more fully assess and manage the risks and opportunities arising from climate change in addition to taking essential action to reduce emissions (Acclimatise, 2009b). Although our knowledge of the extent and rate of future climate change is uncertain, sufficient information exists to assess its impact on business models and to enable robust decisions to limit detrimental effects (Acclimatise, 2009b).

Oil and gas companies are beginning to act on clear signals that climate change is underway. From a purely financial perspective, the average impact across all companies is a loss of approximately 4 percent in shareholder value (Acclimatise, 2009b). Given this, climate and resource-access issues will play important roles in the future profitability of the industry. Lastly, given the influence of the oil and gas industry in our society, and our society's dependence on that industry, these issues deserve attention. Our economy and society will be detrimentally affected by either: a) any adverse impact on the oil and gas industry or b) any adverse impact of the oil and gas industry on climate change. Some companies seem to be expanding their reach from oil and gas to renewable energy while also changing their operational procedures (e.g., carbon sequestration approaches). These tendencies seem to be shaping the "new" oil and gas companies.

Renewable energy (wind, ocean waves, and currents)

Although coastal and marine environments do not currently host commercial facilities that generate electricity, several projects have been proposed and pilot projects are being tested. The possible types of renewable energy that may be developed in coastal and marine environments include wind, wave, ocean current, tidal, hydrogen generation, and solar. The Bureau of Ocean Energy Management (then the Minerals Management Service) prepared a Programmatic Environmental Impact Statement (PEIS; BOEM, 2007) that examined the potential environmental consequences of alternative energy development. The PEIS analysis determined that wind, wave, and ocean current technologies were the most advanced and likely to be developed on the outer continental shelf. Consequently, many projects are under way to test new technologies related to the harvest of renewable energy. The Roosevelt Island Tidal Energy Project demonstrates important new tidal technology in Long Island Sound, where tidal forces are strong where land constrictions exist (Verdant Power, 2012). Tidal energy is also under consideration in Cook Inlet, Alaska (Nelson, 2011). Along the Atlantic Coast, one wind facility has been approved for development and several others have been proposed. Wave energy is most intense along the Pacific Coast, where technology testing is underway. The most favorable area for ocean current development is the Gulf Stream along the southeast coast of Florida, where one pilot project is already in development.

With respect to climate impacts, coastal and marine renewable energy projects are evaluated as mitigation measures because they do not directly result in emissions of greenhouse gases. The effects of climate change on the renewable energy industry have not been assessed along the U.S. coast or elsewhere; however, as with the oil and gas industries, climate change is expected to affect the industry. Likely impacts include:

1. Damage to infrastructure from increased storm intensity through larger waves, stronger currents, or sediment erosion; and

2. Potential change in the resource extracted, including changes in wind speed, wave height, or ocean current intensity or direction. These changes could have either a positive or negative effect.

Model results of the future wave climate for the southwestern Baltic Sea suggest that the overall wave energy and direction could increase slightly (Dreier et al., 2011). Increases in wind speed have also been observed over the past two decades (Young et al., 2011).

The offshore renewable energy sector is nascent and, unlike the financially self-sustaining oil and gas industry, requires investment from the public sector, at least in the U.S., because construction of offshore facilities is more costly; specifically, offshore wind is more expensive than onshore. Uncertainty based on climate change projections could alter the evaluation of risk and potentially deter speculative investments in this emerging industry.

4.5 Tourism and Recreation

Tourism is an important part of the U.S. economy, contributing $1.8 trillion in economic output and supporting 14.1 million jobs in 2011 (U.S. Travel Association, 2012). Tourism is also one of the few sectors that has been growing during the current tentative economic recovery, with 101 million international tourist arrivals to North America last year, up 2.9 percent from 2010 (UNWTO, 2012). Nationally, 2.8 percent of gross domestic product, 7.52 million jobs and $1.11 trillion in travel and tourism total sales are supported by tourism (OTTI, 2011a, b).

Coastal tourism and recreation is used to describe all tourism, leisure, and recreationally-oriented activities that take place on the coast and in coastal waters. Main activities involved in coastal tourism and recreation include visiting beaches, diving and snorkeling, cruise tourism, boating and sailing, and bird and marine mammal watching. Additionally, infrastructure, including hotels, restaurants, vacation homes, marinas, dive shops, harbors, and beaches, is needed in coastal areas to support these tourism and recreational activities. Tourism statistics are difficult to disaggregate solely for coastal areas, but the most recent data from the Office of Travel and Tourism Industries show that, in 2009-2010, nine of the top ten states and U.S. territories visited by overseas travelers are coastal, including the Great Lakes, and seven of the top ten cities are located on the coast (OTTI, 2011a, b). The U.S. Travel Association publishes data based on overnight trips in paid accommodations or travel to destinations 50 or more miles from home, which amounted to 1.5 billion person-trips for leisure in 2010 (U.S. Travel Association, 2012).

In addition to international and domestic travelers, many U.S. residents live in coastal areas and participate in many forms of coastal and marine recreation (Jacob and Witman, 2006; Jepson, 2004; Shivlani, 2009). Participation in many types of coastal and marine recreational activities is rising throughout the U.S. and the U.S. population is increasing at a faster rate in coastal areas than on average (Year of the Oceans, 1998).

In the face of climate change, impacts to marine resource distribution, variable weather conditions, and extreme events such as typhoons and hurricanes are expected to pose the most significant impacts on the industry: positively in some regions of the country, negatively in others, and with mixed effects in others. The predicted impacts of climate change are expected to affect tourism and recreation industries and their associated infrastructure in a variety of ways (Moreno and Becken, 2009; Scott et al., 2004), some of which are described below; however, the science of assessing predicted impacts upon these industries is still in its early stages.

Many coastal and marine tourism and recreational activities depend upon favorable weather and climate; therefore, aspects of climate change associated with increasingly variable conditions are significant stressors upon tourism and recreation. Activities such as diving and snorkeling, and the associated boat travel to snorkeling and dive sites, rely on comfortable water and air temperatures and calm waters. Weather-related impacts such as changes in wind patterns and wave height and direction will affect activities such as sailing and surfing. As weather patterns change and air and sea surface temperatures rise, preferred locations for these and other types of recreation and tourism may change as well. For temperatures, this movement is expected to generally be poleward, with some destinations, such as Seattle, Washington, predicted to have a more desirable climate for more time throughout the year and others, such as Miami, Florida, to experience a decrease in desirability for tourism because of increased temperatures, except in the winter months (Yu et al., 2007). However, this means that even though current preferred locations may decrease in popularity, the popularity of new, more northern locations for coastal recreation and tourism are likely to increase. These implications are important for tourism and recreation activities as well as the infrastructure that supports these activities (Moreno and Becken, 2009; Scott et al., 2004).

Projections of changes to wind and wave patterns are not as clear (see Section 2); thus, changes in preferred locations affected by these changes are less predictable. Some activities are related to biophysical events, such as marine mammal watching, observing seabird migrations, or Arctic cruise tourism. With changing sea surface temperatures, marine mammal and seabird migrations are expected to change, affecting recreation involving watching or interacting with these animals (Lambert et al., 2010). Arctic cruise tourism is expected to increase with increasing sea surface temperatures and decreased sea-ice during Northern Hemisphere summers (Stewart et al., 2007), but researchers warn that climatic warming in the Arctic may change the distribution of sea ice, resulting in negative implications for tourist transits in the High Arctic and Northwest Passage regions (Stewart et al., 2007). Other biological changes that affect marine recreation and tourism may occur with climate change as well; an example is increased or prolonged algal blooms at beaches, which could deter people from visiting those locations (Hoagland et al., 2002).

Another component of marine tourism that is highly likely to be affected by climate change is whale watching. Whale- and other cetacean-watching activities rely on predictable movements of the species within a relatively small spatial area. Cetacean-watching operations are located in most coastal regions of the U.S. from Alaska to Florida. As sea surface temperatures increase, impacts upon cetaceans are predicted to include changes in species' distribution ranges, individual occurrence and abundance, migration timing and length, reproductive success, mortality levels, and community composition and structure. All of these impacts will affect where, when, how many, and how often cetaceans are likely to be observed, and thus the response of the industry itself (Lambert et al., 2010). In some cases, operators may choose to move their business to a more appropriate location based on where cetaceans are predicted to move, but impacts on specific species in specific locations are hard to foresee and moving operations may not be feasible. Although impacts to cetaceans in specific regions are difficult to predict, cetaceans in polar regions are likely to experience greater impacts because temperature changes there are predicted to be greater than in other regions. Cetacean species that cannot easily move as temperatures increase are expected to experience more severe impacts. Additionally, tours that focus on resident marine mammal populations that change their location and those that focus on migratory species are likely to be significantly affected because residency and migrations are likely to change with increasing sea surface temperatures.

Sea level rise may also impact coastal tourism and recreation in a variety of ways. Increased sea levels are likely to reduce the size of sandy beaches in some areas and possibly increase erosion rates (see Case Study 4-E; Yu et al., 2009). Additionally, higher sea levels could cause the landward migration or flooding of coastal lagoons and other coastal habitats for species that are attractive for wildlife viewing, such as seabirds (Bird, 1994). Furthermore, sea level rise directly and indirectly threatens coastal infrastructure such as marinas, boardwalks, hotels, and houses with increased inundation and erosion associated with increased sea levels (Scott et al., 2004).

Finally, HAB events have been shown to significantly reduce reported business revenues for the lodging and restaurant sectors in affected coastal communities with implications for local and state tax revenues. As a result, Morgan et al. (2010) estimated how and why participation in marine-based activities such as beach-going, fishing, and coastal restaurant patronage was affected during a red tide event, which are most common in Gulf of Mexico. The authors found that recreational activities of 63 and 70 percent of Southwest Florida residents who go saltwater fishing or go to beaches, respectively, were adversely affected by either cancellation, delays, or shortened or relocated trips by red tide events during the previous year. Given that the geographic and temporal scale of red tides have been shown to be affected by water temperature and potentially water quality, climate change could likely cause these impacts directly.

Although red-tide-monitoring information is available in offshore Gulf of Mexico areas that are more frequently impacted by red tide events, a recent study found that coastal tourism, measured by attendance at a large state park in coastal Southwest Florida with substantial beach areas and boating facilities, was more closely correlated with local newspaper articles about red tides than objective measures of HAB events. The implication of this finding is that media coverage of changing environmental conditions can cause behavioral responses that can negatively affect tourism before changes in the physical environment occur, indicating that investments in educational and media messages could be prudent expenditures.

4.6 Human Health

In addition to current and future climate change impacts on the biophysical and socio-economic dimensions of marine resources, broader knowledge of the human health dimensions of global climate change has grown. Climate change is expected to impact human health in a number of primary areas (Baer and Singer, 2009). In fact, according to the World Health Organization (WHO):

> *Our increasing understanding of climate change is transforming how we view the boundaries and determinants of human health. While our personal health may seem to relate mostly to prudent behavior, heredity, occupation, local environmental exposures, and health-care access, sustained population health requires the life-supporting "services" of the biosphere. Populations of all animal species depend on supplies of food and water, freedom from excess infectious disease, and the physical safety and comfort conferred by climatic stability. The world's climate system is fundamental to this life-support.* (McMichael et al., 2003, page 2)

The WHO is not alone in recognizing the known and potential human health consequences of global climate change. Public health scientists with the U.S. National Center for Environmental Health at the Centers for Disease Control and Prevention (CDC) have identified a number of primary areas in which climate change impacts human health and will likely exacerbate human vulnerability and sensitivity in the future (McGeehin,

Case Study 4-E
Economic impacts of the potential erosion of Waikiki Beach

Hawai'i's Waikiki Beach is recognized as a major tourism destination and a popular recreational spot for both visitors and residents. Waikiki Beach extends approximately two miles along the south shore of O'ahu, and is home to a number of major hotels. In 2007, approximately 4 million tourists visited Waikiki Beach to whom hotels sold 3.9 million room nights that generated approximately $1.2 billion in revenue (DBEDT, 2008). Waikiki represents approximately 88 percent of the O'ahu visitor market for hotel room demand.

Given the popularity and economic importance of Waikiki, the issue of beach erosion has been an ongoing concern. Recent recognition that sea level rise will exacerbate erosion has drawn increased attention to the economic contribution of the beach itself. Therefore, the Waikiki Improvement Association commissioned a study to estimate market and economic impacts using the scenario of complete erosion of Waikiki Beach.

Using reported visitor statistics (DBEDT, 2008), STR Global's (2012) hotel lodging and financial survey was able to calculate annual room demand for hotel rooms in Waikiki, total hotel revenue, visitor expenditure related to food and beverage, entertainment and recreation, retail and transportation but excluding lodging, and the associated General Excise Tax (GET) and Transient Accommodations Tax (TAT) for Waikiki (See Tables 1 and 2). To supplement the secondary data, the researchers subcontracted a visitor intercept survey to analyze the importance of beach availability and examine the economic importance of Waikiki Beach.

The visitor intercept survey, which focused on visitors from the U.S. (76 % of the sample) and Japan (24% of the sample) because they make up 84 percent of the tourism market for Oahu, was administered to 428 visitors at various locations in Waikiki. Over 90 percent of visitors surveyed reported that beach availability was very or somewhat important. 58 percent of U.S. mainland and 14 percent of Japanese respondents to the visitor intercept survey revealed that, if Waikiki Beach were completely eroded, they would not consider staying in Waikiki.

Based on these survey results, an estimated 38,000 daily visitors would not return to Waikiki if the beach were completely eroded. With this decrease in daily visitors, lost hotel room revenue is estimated to be $503.8 million. In addition to the potential loss in room revenue, other hotel revenues, such as those generated by food and beverage sales, spa visits, and parking, might also be

2007 Waikiki visitor expenditures and estimated losses with completely eroded beach

Visitor expenditure type	Revenue (2007)	Estimated loss (due to beach erosion)
Hotel room revenue	$1.2 billion	($504 million)
Total expenditure (excluding lodging)	$4 billion	($1.5 billion)
Total Waikiki visitor expenditures	$5.2 billion	($2 billion)

Case Study 4-E (Continued)

affected by declining room demand. Beach and water activities are important revenue generators in Waikiki. Such expenditures include surfboard, umbrella, chair, canoe, and snorkel gear rentals. An additional loss of $1.47 billion was estimated for expenditures other than lodging expenses. Applying the average daily expenditure per person per day for retail, food and beverages, transportation, entertainment, and recreation to the average number of visitors in Waikiki, complete erosion is estimated to cost nearly $2 billion per year in overall visitor expenditures. Additionally,

an estimated $66 million in tax revenue from GET and TAT taxes would be lost. Indirect effects could include hotel industry job loss of more than six thousand jobs per year.

These estimated impacts highlight the importance of beaches to the tourism industry in coastal states. Reducing these estimated losses in all regions should be a priority that could be addressed through a variety of approaches such as adjusting sales and marketing strategies, improving Waikiki's tourism products, or repositioning beach accommodations.

2007 Waikiki TAT and GET tax revenues and estimated losses with completely eroded beach

Type of tax	Tax revenue (2007)	Estimated loss
Transient Accommodations Tax (TAT) @ 7.25%	$86.6 million	($36.5 million)
General excise tax (GET) @ 4.50%	$69.8 million	($29.8 million)
Total	$156.5 million	($66.3 million)

2007). These include but are not limited to extreme weather-related injuries, morbidity, and mortality; decline in access to drinkable water; increased food insecurity and malnutrition; rising pollutant-related respiratory problems; and increased spread of infectious disease. This list of consequences illustrates the complex and varied ways in which the social impact of climate change in the U.S. extends beyond actuarial statistics and measures of economic loss to include broad, critical aspects of human health, well-being, and vulnerability (Brown 1999).

Health and vulnerability

Complex social and ecosystem conditions inform the reach and range of climate change effects on health, which is "not some absolute state of being but an elastic concept that must be evaluated in a larger socio-cultural context" (Baer et al., 2003:5). Although the environmental health effects of climate change on marine-resource users in particular are not widely known, recognition of human vulnerability and sensitivity in the wake of global climate change is growing. Vulnerability is a fundamental concept for assessing the role of climate change in determining health, especially because it merges theory and empirical findings from disaster studies in general and public health science, social, and economic analysis and risk assessment in particular (Baer and Singer, 2009). Disease vulnerability and environmental health risk may both increasingly become central issues in research exploring connections among climate change, marine resource contamination or decline, and poverty, especially since subjugated populations in coastal regions tend to be marine resource users.

Waterborne and foodborne diseases

The impact of warming oceans on waterborne pathogens that cause both seafood-related and direct-contact wound infections has generated growing concern in a time of climate change. Much attention has been directed at pathogens in the *Vibrio* family, especially *V. cholera* in global health research and intervention (Lipp et al., 2002). More recently, with regard to U.S. cases, there has been increased concern about other *Vibrio* species, including *V. parahaemolyticus* and *V. vulnificus*, both known sources of seafood-related acute gastroenteritis. In 2011, the CDC, which maintains a voluntary surveillance system of culture-confirmed Vibrio infections in the Gulf of Mexico region, estimated 45,000 annual cases of *V. parahaemolyticus* and 207 cases of *V. vulnificus* in the U.S. (Hlavsa et al., 2011). Figure 4-4 shows the CDC's Vibrio case reporting by state and region in 2009. The highest concentrations of *Vibrio* infections are in the Mid-Atlantic states that surround Chesapeake Bay, where 305 cases were reported in 2009. Given this high concentration in Chesapeake Bay, there is a high likelihood that these numbers could increase significantly if water temperatures in the Bay rise in the future.

Reported expansion of *V. parahaemolyticus* in the Pacific Northwest and Alaska also

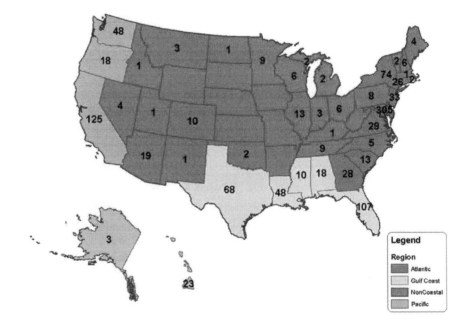

Figure 4-4 Number of cases of Vibrio infections by state and region, 2009 (Source: CDC, 2011).

closely corresponds with climate anomalies related to El Niño (CDC, 1998; Martinez et al., 2010; McLaughlin et al., 2005). In addition, in the days after Hurricane Katrina, 22 Vibrio wound infections were recorded, of which three were caused by *V. parahaemolyticus*, two of which led to the deaths of the infected individuals (CDC, 2005).

V. vulnificus is perhaps the most important pathogenic *Vibrio* in the U.S. because of its highly invasive nature and high fatality rate following infection (Case Study 4-F; Horseman and Surani, 2011). In recent years, *V. vulnificus* has come to be recognized as the most virulent foodborne pathogen in the U.S. with a fatality rate as high as 60 percent (Oliver, 2005) and has been responsible for the overwhelming majority of reported seafood-related deaths in the U.S. (Oliver and Kaper, 2007). *V. vulnificus* can be transmitted to humans by way of consumption and dermal exposure. The bacterium is frequently isolated from oysters and other shellfish in warm coastal waters during the summer months. Because it is naturally found in warm marine waters, people with open wounds can be exposed to *V. vulnificus* through direct contact with seawater. A review of this bacterium found that the 34.8 percent death rate among domestically acquired foodborne illness associated with *V. vulnificus* was significantly higher than the 0 to 17.3 percent rate associated with any of the other 31 foodborne pathogens assessed (Scallan et al., 2011). Additionally, *V. vulnificus* is becoming a significant and growing source of potentially fatal wound infections associated with recreational swimming, fishing-related cuts, and seafood handling (Weis et al., 2011). One study reported that almost 70 percent of infected individuals developed secondary lesions requiring tissue debridement or limb amputation (Oliver, 1989). *V. vulnificus* is most frequently found in water with a temperature above 20°C, which is especially important given the significant changes in temperature that are anticipated in coastal waters over the coming decades (see Section 2 and Case Study 4-F).

Furthermore, recent findings suggest an increase in global production of airborne dust and subsequent oceanic dust deposition that may stimulate growth of Vibrio species, including toxigenic *V. cholerae*, *V. vulnificus*, and *V. alginolyticus* (Lipp, 2011). Although no definitive link exists between the increasing incidence of Vibrio and climate change, sufficient data are available to warrant closer research attention (Greer et al., 2008).

In addition to members of the *Vibrio* family, a number of other marine pathogens also merit monitoring in a warming environment. *Aeromonas hydrophila* is a widely distributed inhabitant of both fresh and salt waters as well as a common fish pathogen. In humans, it causes gastroenteritis as well as a variety of extraintestinal infections, including endocarditis, pneumonia, conjunctivitis, and urinary tract infections, and is capable of causing localized wound infections in individuals with intact immune systems (Collier, 2002). *Myobacterium marinum* infections, of which approximately 200 are reported each year in the U.S., have been described as an emerging necrotizing mycobacteria-caused disease involving both marine and freshwater exposure during a water-related injury (Dobos et al., 1999). *Erysipelothrix rhusiopathiae* is found in diverse animal species including fish and shellfish. Successfully transferred to humans through cuts, this pathogen causes cutaneous eruptions on the hands or fingers. It is popularly referred to as "shrimp picker's disease" and "crab poisoning" in marine locations (Brooke and Riley, 1999). Increased rates of infection have been documented for these emerging diseases.

Case Study 4-F
Spread of Vibrio cases throughout the U.S.

The U.S. is experiencing increased cases of *V. vulnificus* infection (Jones and Oliver, 2009), the leading cause of death from seafood consumption in the U.S. and a pathogen scenario that showcases the growing relationships among climate change, marine environments, seafood safety, and human health. Based on the Foodborne Disease Active Surveillance Network (FoodNet), which involves population-based surveillance in ten states, the CDC reported an increase in *V. vulnificus* foodborne infections of 78 percent between 1996 and 2006, with the majority of cases reported in the Gulf of Mexico as a result of high consumption rates of raw oysters harvested in warm months (CDC, 2007; Shapiro et al., 1998). In subsequent surveillance-based reports in 2009 and 2010, the CDC reported that the incidence of Vibrio infection has continued to increase at a significant rate (85 percent) (CDC, 2009). Although the exact cause of the increase seen between 2001 and 2010 is still not established, heightened frequency of cases is associated with rising ocean water temperatures because *V. vulnificus* is most frequently found in water with a temperature above 20°C (Paz et al., 2007). Harvested oysters can be contaminated with *V. vulnificus* because the bacterium is naturally present in marine environments and its growth and distribution are related to warm seawater temperatures. Regional U.S. FDA specialists with expert knowledge about shellfish are assisting state officials in sampling harvest waters to discover possible sources of *V. vulnificus* and to make decisions on oyster bed closures when the pathogen is identified (McLaughlin et al., 2005).

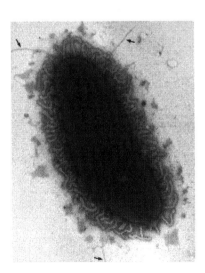

V. vulnificus **(Source NOAA).**

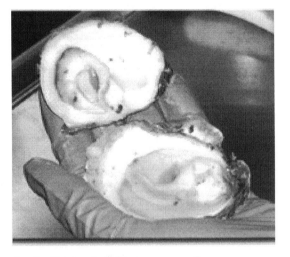

Analyzing oysters for presence of experimentally introduced *V. vulnificus* **(Source: NOAA).**

Like *V. vulnificus*, each of these pathogens has the potential for increased rates of infection as a result of warming ocean waters.

The cases noted here indicate that a range of other *Vibrio* and non-*Vibrio* pathogens should be monitored in association with warming ocean waters and other expressions of climate change such as extreme weather, flooding, and global desertification (CDC, 2005). In addition, focused development of hydrological and ecological modeling using existing pathogen knowledge, evidence on climate effects on pathogen concentrations, and varying climate change scenarios is needed (Hofstra, 2011).

Harmful algal blooms and climate change

HABs, which occur worldwide and in all U.S. states, have recently increased in duration and geographic range, and have involved new species and impacts (Anderson, 2012; Hallegraeff, 2010; Moore et al., 2008; also see Section 3). Many HABs produce potent toxins that can kill or sicken humans, fish, birds, turtles, marine mammals, and domestic animals and pets (Section 3). Human health is threatened through exposure to toxin-contaminated shellfish and fish, drinking water, or aerosols. Monitoring and adaptations, such as shellfish harvesting closures, beach closures, and drinking water treatment minimize threats to human health. HAB toxins threaten endangered and protected species, mariculture and aquaculture, fisheries, and ecosystem health. Non-toxic HABs also kill animals, discolor water, overgrow seagrass beds and coral reefs, and accumulate in stinking, unsightly piles on beaches (Figure 4-5). Control measures, although critical to protect public health, can reduce the availability of important sources of nutrition and/or income to communities that depend on the impacted resources. The economic consequence of HABs from a subset of events has been conservatively estimated at $82 million (Hoagland and Scatasta, 2006) a year, but only some of the impacts listed above are included in this estimate.

HAB occurrence may be altered by climate change impacts including increases in water temperature, stratification, and CO_2; alteration of currents or hydrology; and changes

Figure 4-5
Harmful algal bloom (Source: NOAA, n.d.).

in nutrient availability, due to upwelling or runoff (see Section 2 for details). However, HABs are a diverse group of organisms from many taxonomic groups with highly variable life habits, life cycles, physiology, and toxins, so their growth and toxicity will respond differently to changing environmental conditions (see Section 3 for details).

Increases in cyanobacterial blooms (CyanoHABs), many of which produce cyanotoxins linked to liver, digestive, skin, neurological illness, and even death, have already been well-documented and attributed to a combination of increased nutrients and climate change (Paerl and Paul, 2012). Massive toxic blooms threaten drinking water supplies and recreational use of water bodies, especially in areas experiencing droughts. CyanoHABs also threaten ecosystem health by altering the natural flora and fauna through increased turbidity, poor nutritional quality, and reduced bottom water oxygen. Although most common in fresh water, they can also overgrow coral reefs (Paerl and Paul, 2012) and have recently caused sea otter deaths in Monterey Bay, California (Miller et al., 2010b).

For other HABs, change in climate may increase the period of time when environmental conditions are suitable for blooms to occur. For example, in Puget Sound, as in many areas in the Northwest and Northeast U.S., shellfish harvesting is often closed for a period in the late spring or summer. The closure period corresponds to the window of opportunity when environmental conditions are optimal for blooms of *Alexandrium* (Moore et al., 2008, 2009, see Section 6). This toxic dinoflagellate produces potent neurotoxins that can cause illness and death in humans who eat contaminated shellfish. Modeling climate change scenarios, even when modest, indicate that the window of opportunity will increase substantially (Moore et al., 2011).

Climate change may also alter conditions so that they become unfavorable for HAB growth. Ciguatera fish poisoning (CFP) is caused by ciguatoxin produced by a benthic dinoflagellate, *Gambierdiscus*, living on macroalgae on tropical hard substrates, especially coral reefs. The toxins accumulate in higher trophic level fish, and, when humans consume those fish, cause ciguatera fish poisoning (CFP), a debilitating illness. CFP is the most common HAB-caused illness in the world, which deters fish consumption in many areas where protein is in short supply. A recent study showed that CFP incidence in the Caribbean is highest where water temperatures are highest and postulates that climate change may be one factor in recent outbreaks of CFP from fish caught near oil platforms in the more temperate Gulf of Mexico (Tester et al., 2010). However, data from the South Pacific suggest that waters may become too hot for the causative organism to grow (Llewellyn, 2010); therefore, the geographic distribution of CFP may change, but not the incidence.

Finally, some initial studies with three HABs from different taxonomic groups suggest that toxicity of HAB organisms may be differentially affected by climate change. In one case, toxicity decreased as temperature increased (Ono et al., 2000). In two other cases, it increased as CO_2 increased (Fu et al., 2010; Sun et al., 2011).

In conclusion, climate change will likely change the growth and distribution of many HAB species and thus affect their impacts on human, animal, and ecosystem health and local economies. Predicting those effects requires understanding the life cycle, physiological response, growth and toxicity of each HAB organism under a range of environmental conditions and the human responses to HABs such as the loss or change in tourism, as well as the application of this information in a specific regional ecosystem context.

Health risks related to climate impacts on marine zoonotic diseases

A global analysis of trends in infectious diseases found that emerging infectious disease events were increasing over time and were dominated by zoonotic diseases transmitted between animals and humans, with the majority of those diseases (72 percent) originating in wildlife (Jones et al., 2008). Climate change may impact infectious zoonotic diseases by prolonging the diseases' transmission periods and by changing geographic ranges of disease and animal reservoirs (Greer et al., 2008; see Section 3 for further discussion and examples of the potential impacts of climate change on disease in marine animals).

Zoonotic diseases that occur in marine animals are among those of concern for human health. Establishing a definitive link between increases in marine zoonotic disease and climate change has been difficult because of multiple contributing stressors (Burek et al., 2008; Wilcox and Gubler. 2005) and a lack of sufficient baseline data for some organisms (Burek et al., 2008), but examples of changes in latitudinal distributions of infectious organisms have been seen. For example, *Lacazia loboi* is a cutaneous fungus that has been reported to infect humans and dolphins in tropical and transitional tropical climates. The disease more recently has been diagnosed in dolphins off the coast of North Carolina, which represents a change in the latitudinal distribution of this fungus (Rotstein et al., 2009). To detect such changes, continued monitoring and assessments of disease in marine animals to establish baselines and identify trends is critical. Furthermore, an integrated monitoring and surveillance system will be important to provide early warnings and better public information for any emerging diseases that threaten human health.

Some coastal and tribal communities depend on marine animals as traditional sources of nutrition. In places like Alaska, access to animals has become more difficult following changing migration and reproduction patterns and ice flow shifts have impacted hunter and fishermen access (Ford, 2009; Gearhead et al., 2006). Communities such as these would be particularly vulnerable to outbreaks that impact their already at-risk food supply.

Health risks of extreme weather events

People living coastal environments might also be at greater health risk because of increases in extreme weather events resulting from global climate change (Greenough et al., 2001). The IPCC notes that warming would vary by region, but would be accompanied overall by changes in precipitation, the variability of climate, and the frequency and intensity of some extreme weather phenomena (IPCC, 2007a; see Section 2 for detail on expected changes in weather). The health risks of extreme weather are many, including but not limited to heat exhaustion and other heat-related illnesses (Bernard and McGeehin, 2004; Golden et al., 2008; Luber and McGeehin, 2008; McGeehin and Mirabelli, 2001; Semenza et al., 1999), mental health illnesses (Averhoff et al., 2006; Norris et al., 2006; Thienkura et al., 2006; van Griensven et al., 2006), and vector-borne and zoonotic diseases (Collinge et al., 2005; Eisen et al., 2007; Enscore et al., 2002; Gage et al., 2008; Glass et al., 2000; Parmenter et al.,1999).

Globalized seafood and emerging health risks

Due to a rise in both the globalization of seafood and demand in the U.S. market, the Food and Drug Administration (FDA) reports that the U.S. now imports more than 80 percent of its seafood supply, including wild caught fish and aquaculture fish (USDA, 2008). This seafood originates from over 13,000 suppliers in over 160 countries, with China being the largest exporter of seafood to the U.S. by volume (GAO, 2004). Despite the scale of the seafood import industry, the FDA, which is charged with promoting public health through regulation and supervision of food safety, excluding oversight of most meats, poultry, and processed egg products, directly inspects only a small percentage of the nation's imported seafood. In 2007, for example, the FDA reported almost 900,000 entries of imported seafood, of which only 14,000, or about 2 percent, were subject to FDA inspection and laboratory testing (USDA, 2008). The Government Accountability Office (GAO) has warned against the low level of testing conducted by the FDA and has cautioned that "imported foods have introduced new risks or increased the incidence of familiar illnesses" (Friedman 1996, page 2) in the U.S, but the FDA "cannot ensure that the growing volume of imported foods is safe for consumers" (GAO, 1998, pages 2-3). In 2004, the GAO again raised concerns about the adequacy of the FDA's seafood inspection program to protect U.S. consumers.

Imported seafood can be a source of health risk, involving multiple agents, especially bacteria such as *Salmonella, Campylobacter*, verotoxin producing *E. coli*, and *listeria*, parasites such as *Toxoplasma gondii, Cyclospora cayetanensis*, and *trichinella*, and viruses such as norovirus and hepatitis A virus, as well as rarer infectious agents and mycotoxins (Buisson et al., 2008). Between 1983 and 1992, seafood ranked third on the list of food products that caused foodborne disease (Lipp and Rose, 1997), a marked increase over the previous decade (Hui et al., 2001). In late 2006, two bacteria-caused outbreaks of fish poisoning were reported in the U.S., one in Louisiana and the other in Tennessee, involving tuna steaks imported from Indonesia and Vietnam, respectively (CDC, 2007). A likely source of the outbreaks was inadequate temperature control between catch and consumption. Direct testing also has yielded evidence of significant rates of the infection of imported seafood. For example, tested salmon, shrimp, and tilapia samples from three retail outlets in Baton Rouge, Louisiana that were imported from 12 countries found that 17.5 percent of the samples tested positive for *Salmonella*, 32.2 percent for *Shigella*, 4.1 percent for *Listeria* moncytogenes, and 9.4 percent for *Escherichia coli* (Wang et al., 2011). Similarly, tests of over 12,000 imported and domestic seafood samples over a nine-year period found that the incidence of *Salmonella* was 7.2 percent in imported seafood compared to 1.3 percent for domestic seafood (Heinitz et al., 2000). These study results suggest that a high potential for infection in imported seafood (Love et al., 2011).

In light of these findings, the critical question is whether climate change and the further globalization of seafood will contribute to additional jumps in the prevalence of infected seafood available to consumers in the American market. Climate change has the potential to adversely impact imported seafood in two ways. First, climate change is a risk to the extent that warming oceans and other changes to the marine environment increase rates of infection of various seafood stocks worldwide, including locations that export seafood to the U.S. Second, rising temperatures and changing weather patterns may result in inadequate cooling of seafood at various points in the import process, allowing the growth of infectious agents. These scenarios, combined with the low level of FDA testing of imported seafood, suggest the need for increased attention to this potential threat to U.S. public health in a time of climate change.

In many Alaskan communities, for example, seafood contamination has been a major source of environmental health concern (Loring and Gerlach, 2009). Many of these communities are already exposed to dangerous contaminants in their food and water, such as heavy metals like methyl mercury and persistent organic pollutants, as the Stockholm Convention addressed, from military dump sites and a variety of other sources (Herman et al., 2000). With climate change, some contaminant levels are projected to increase, though the details of these risks are still not well-understood (Godduhn and Duffy, 2003; Schiedek et al., 2007; Stockholm Convention, 2005).

Acidification and other unknown human health risks

Coupled with the current and emerging discussion of global temperature change is growing awareness of the problem of ocean acidification, which poses a fundamental challenge to marine life and marine resource users (NRC, 2010a). Although scientists are certain that increased acidification will impact particular marine species such as shellfish, the impacts of acidification on all seafoods that humans consume are unclear. Therefore, acidification poses a possible environmental health risk. Knowledge regarding human adaptations to ocean acidification has definite gaps, but the ecosystem and socio-economic impacts of increased ocean acidification are likely to negatively transform food systems and therefore introduce stressors that could lead to unknown public health risks.

According to a recent National Research Council report, "Communities in areas with [ocean acidification] affected marine resources may be highly dependent on them both for income and sustenance. There is thus a need to assess vulnerability and adaptation capabilities of these communities over different time frames" (NRC, 2010a, page 119). This emphasis on vulnerability will likely lead to greater linkages between studies of socio-economic and health-related vulnerabilities because ocean acidification will likely result in negative impacts on important food sources for communities in coastal regions that rely on marine resources for sustenance.

4.7 Maritime Security and Transportation

Security, transportation, and governance issues are also at play in how climate change may impact ocean services in the U.S. Climate shifts in the Arctic, namely sea ice coverage, are provoking discussion on the future of ocean governance, including marine resource and ecosystem-based management. Perhaps the most noteworthy issue in this arena is the increase in shipping accessibility in the Arctic. National security concerns and threats to national sovereignty have also been a recent focus of attention (Borgerson, 2008; Campbell et al., 2007; Lackenbauer, 2011). In general, the warming of the ocean has led to a redrawing of the biophysical map of Earth. This process will generate an expanded geopolitical discussion involving the relationship among politics, territory, and state sovereignty on local, national and international scales (Nuttall and Callaghan, 2000). Refer to Section 5.4.1 for further discussion on the international implications of climate change on these sectors.

4.8 Governance Challenges

Climate change is likely to lead to challenges for national governments and for international bodies and institutions. Among other things, new challenges will arise in the area of risk management. Many natural resource governance institutions have been built under assumptions of stable environmental conditions that are similar to observed historical experience (Peloso, 2010). In many instances of greater climate variability or climate change, these assumptions will be challenged or will no longer be valid. Changes in the distribution and accessibility of living natural resources and ecosystems will require changes in jurisdictional boundaries established by national or international management institutions in some cases. As a consequence, climate change will often result in the need to revise our current management approaches and, in some cases, to restructure governance systems for most ocean uses. While governments can employ technologies to achieve better climate change preparedness and response, it is important to keep in mind that those technologies are only as effective as the governments and institutions that implement them to deal assessing, planning and responding to negative impacts, like climate change (Dowty and Allen, 2011). In this regard, the National Academy of Sciences has repeatedly called for early, active, continuous, and transparent community involvement in risk management decisions rather than blind reliance on technology-based decisions (NRC, 1996, 2000a, b) orchestrated by government scientists and experts (Fischer, 2000). This broader, more inclusive governance approach is ever more important as marine resource users and coastal communities: 1) increasingly adapt and respond to changing security conditions, the restructuring of transportation networks, and a warming climate, and 2) demand further involvement in climate change discussions, especially those resulting in changes in marine resource management decisions and policies. The governance needs of ecosystem-based management may dovetail with this trend. In addition, novel territorial and jurisdictional issues will need to be addressed as as a result of climate change. These challenges are most apparent in the Arctic, where the reduction in sea ice will open up new ocean spaces to navigation and other maritime uses.

Fisheries management in the U.S.

The effects of climate change on fisheries present a number of governance challenges to fishery management institutions in the U.S. Federal fisheries management occurs mainly within the framework of the Magnuson-Stevens Fishery Conservation and Management Act (MSA). Within this framework, eight fishery management councils (Councils) develop fishing regulations for specific regions and fisheries in cooperation with the federal government, represented by the National Marine Fisheries Service (NMFS).

Management plans from the Councils are designed to meet ten National Standards (NSs) set by the MSA. Important among these standards is the requirement to prevent overfishing while achieving optimum yield. The optimum yield is specified by determining a stock's MSY, which is the harvest amount that will generate the largest long-term average yield (NS Guidelines, page 5, section 600.310(e)), as reduced by social, economic, or ecological factors (MSA, page 10, section 3(33)). This optimum yield is the basis for caps on total harvest or Annual Catch Limits (ACLs). ACLs are established by the Scientific and Statistical Committees (SSCs) of the Fishery Management Councils. The SSCs often establish uncertainty buffers to prevent overfishing. The SSCs are able to adjust annual harvest recommendations to fluctuations in stock size to prevent overfishing. Furthermore, regulations should "[M]anage individual stocks as a unit throughout their range, to the extent practicable; interrelated stocks shall be managed as a unit or in close coordination." This will require an investment in monitoring and assessment to accurately describe the ranges of and interrelationships among stocks as they change with climate conditions, but will require further increased coordination between Councils.

An additional factor that Councils must consider is changes in bycatch. When shifts in the distribution and abundance of stocks change bycatch patterns, management must adjust regulations and ACLs to account for the new patterns. For example, warming trends in the Bering Sea have caused increased overlap of pollock and salmon stocks, which has led to increased salmon bycatch in the pollock fishery (Stram and Evans, 2009). As a result, the North Pacific Fishery Management Council has taken steps to minimize salmon bycatch rates by limiting pollock fishing at certain times and in certain areas (see additional discussion in Section 6).

Increased uncertainty due to climate change will impact decisions made by fisheries management institutions. Climate change increases uncertainty about fisheries stocks in two ways. First, stock levels and spatial distribution may differ from observed, historical patterns in uncertain ways. Second, the year-to-year variability in stock levels and spatial distribution may increase as climatic patterns become more variable (See Section 4.2 above). Fishery regulations and ACLs are based on modeling- and data-intensive stock assessment methods. Climate change is expected to alter the basic population dynamics of some, and possibly many, fish stocks and is thus likely to impact the reliability of stock assessments and total allowable catch (TAC) limits and ACLs that are subsequently set. For example, the survey area upon which a stock assessment is based may no longer appropriately cover the range of the species, perhaps leading to an assessment that suggests a stock decline as opposed to a shift in location.

As noted elsewhere in this Report, the direction and magnitude of such changes are uncertain, and the changes could have negative or positive impacts on different fisheries. Thus, one could error in setting a TAC lower than it should be for a stock that is otherwise healthy. However, greater uncertainty should lead to greater caution in setting catch levels. Given this, the uncertainty about whether the stock has shifted or not is likely to be built into the setting of more conservative TACs and ACLs. In either case, the uncertainty generated by climate change exacts a cost through the TAC- and ACL-setting process even though the fish stock may not have changed. Case Study 4-G discusses some of this uncertainty in context with specific fisheries around the U.S.

An alternative to setting TACs and ACLs has occurred in the Arctic Management Area where the North Pacific Fishery Management Council closed large, previously unexploited areas to future fishing due to the high level of uncertainty regarding newly ice-free fishing grounds (see additional discussion in Section 6). This uncertainty stems from a lack of historical data on the balance of total catch revenues to the time and effort needed to explore these new Arctic areas and from imperfect knowledge about future climate (Kutil, 2011; Stram and Evans, 2009). In this way, current fisheries governance structures can incorporate increased uncertainty due to climate change, which therefore requires

Case Study 4-G
Fisheries management responses to climate change

U.S. fishing-dependent communities range from large urban centers to small rural outposts. They vary in mixes of species caught, size of vessels, types of gear used, and level of involvement in commercial, recreational, and/or subsistence fishing. Fishing-dependent communities may have fish buyers, processors, ice facilities, boat building facilities, seafood wholesalers, or retailers (NOAA Fisheries 2009b). Fishermen fish for multiple reasons, including income, adventure, family tradition, food, and religious rituals (Falk et al., 1989; Holland and Ditton 1992; Smith and Clay 2010; Fall and Koster 2011). In 2009, the U.S. seafood industry supported approximately 1 million full- and part-time jobs and generated $116 billion in sales, $32 billion in income, and $48 billion in value added. Recreational angler expenditures contributed $50 billion in sales to the U.S. economy and supported over 327,000 jobs (NOAA Fisheries 2010).

Many socio-economic impacts of climate change on fisheries flow from biological and ecological changes (Hannesson 2007; see Section 3). Potential effects of climate change on fisheries result from changes in the productivity and location of the fish stocks (Brander 2010) as well as the broader ecosystem (Sumaila et al., 2011). Due to strong temperature sensitivity (Fogarty et al., 2008), Atlantic cod (*Gadus morhua*) are expected to decrease in biomass and largely move north out of U.S. waters. Atlantic croaker (*Micropogonias undulatus*) are expected to increase in biomass and shift northward within U.S. waters (Hare et al., 2010). Rising temperatures may cause decreases in Alaska pollock (*Theragra chalcogramma*) biomass (Mueter et al., 2011), but a northward shift is unlikely (Haynie and Pfeiffer, 2012; Pfeiffer, 2012). Pacific sardine (*Sardinops sagax*) productivity is strongly influenced by climate variability; increases in biomass and northward extension are

expected, though whether those increases will be slight or more significant is unclear (Emmett et al., 2005).

With these changes, fishermen may need to change target species or move to follow the species they know. These have a variety of potential costs, including economic costs associated with buying new gear and rising fuel prices; social costs such as losing local support networks either onshore or off; cultural costs including losing a species critical to a spiritual ritual; and nutritional costs. Where stocks increase in biomass or extend in range, fishermen may increase household income, economic stability, food security, and opportunities for tournament-winning fish, if regulations allow. Increases in sea level may harm coastal fisheries infrastructure. These changes reverberate throughout the economy and society though rarely in a linear fashion. (Markowski et al., 1999; Loomis and Crespi 1999). Different groups such as fisheries and port communities are likely to have different levels of resilience and vulnerability.

Changes in marine target populations also affect fisheries management and thus fishermen (Pollnac et al., 2006). Managers must anticipate problems (Lubchenco and Petes 2010) and build flexibility into management plans; many current regulations tie fishermen to particular species in specific areas (OECD 2010). Further, significant discussion needs to take place on the societal value of fish stocks and those communities that depend on fishing to sustain their livelihoods. Transboundary and other jurisdictional issues will also emerge (Herrick et al., 2007).

greater investment in monitoring and assessment to reduce the level of uncertainty and requires scientists and regulators to recognize and to account for the uncertainty.

Finally, the issue of transboundary stocks will likely increase in importance given that international agreements on fishing shared stocks are based on stable, historical abundance and spatial distribution patterns (see Section 5). Miller and Munro (2003) use a theoretical model to illustrate potential problems with transboundary Pacific salmon stocks in the U.S. and Canada. Their results highlight the need to update existing and negotiate new, more flexible international agreements.

Offshore energy development

Climate change will present new opportunities and new challenges to offshore energy development. New areas, and especially the Arctic, will become accessible to development and to marine energy transport, but this will also bring industry activities into contact, and potentially conflict, with new environments and other uses such as subsistence and tourism. Operations in new and traditional areas, though brought on by changing climate, may face new climactic challenges with increasing storms, more severe operating conditions, and more sensitive species and ecosystems. In addition, routine operations, such as oil transport, are likely to carry increased risk in Arctic waters where lack of charts, extreme weather conditions, and poor oil spill response capabilities hamper safe operations. Additional regulations will certainly increase operational costs, although, in the Arctic, these risks will also be accompanied by opportunities because the reduced sea ice coverage will lead to the opening of new shipping lanes facilitating the transport of crude oil between the Atlantic and the Pacific Oceans. Both current fisheries and offshore energy policy and management regimes may need to be restructured, an effort that will include creating new protected areas for various marine resources and habitats. In addition, energy development companies will likely need to revisit their own internal policies and operations to cope with changing climate. An example of this is the American Petroleum Institute, which launched a review process of design and safety standards for offshore oil platforms in 2006 following the extreme 2005 hurricane season in the Gulf of Mexico.

Tourism and recreation

Tourism is seldom recognized as a single sector for policy and regulation. Policies in many different sectors including shipping, fisheries management, habitat protection, and the business sector affect tourism and the tourism industry. As such, changes made by the governance structures in each of these sectors in response to climate change are likely to impact tourism and recreation in the U.S. An important consideration is the safety of life at sea as tourists and cruise vessels venture into higher latitudes and uncharted waters far from established ports and rescue capabilities with decreasing summer sea-ice. The International Maritime Organization and the Arctic Council are already cooperating to increase governance measures in this area.

Human health

The consequences of climate change related to human health documented above showcase the complex and varied ways in which the social impact of climate change in the U.S. extends beyond actuarial statistics and measures of economic loss to broader and more critical aspects of human health, well-being, and vulnerability (Brown, 1999). Many agencies at the international, national, state, and local level in each country develop policy and regulations for issues including seafood safety, water quality, disaster response, and disease outbreak avoidance and containment and will need to develop adaptive means to respond to these changes.

Strategic planning

Institutions are re-evaluating their priorities and undertaking strategic planning to anticipate changes and develop strategies for responses to the present and future challenges of climate change. In the Pacific Islands, for example, where whole countries or territories are expected to disappear under a rising sea level, individual villages, communities, states, territories, and commonwealths are developing strategies for dealing with climate change and its effects on communities and their subsistence activities. Existing regional institutions, such as the Secretariat of the Pacific Community and the Western Pacific Regional Fishery Management Council, have adopted a climate change focus. In Alaska, 31 communities, many of which have already begun plans to relocate, have been identified as under imminent threat due to climate change effects (GAO, 2009). Similar changes are likely being seen in other regions.

4.9 Research and Monitoring Gaps

As this section has shown, research is critical for gaining insight into the effects of climate change on ocean services that are important to the economic, social, cultural, and personal well-being of the U.S. population and to the health of the ocean itself. These observed and predicted impacts of climate change on U.S. ocean services have raised awareness about potential socio-economic effects and highlighted areas where significant changes have not yet been observed but are expected to be seen in both the short and long term. Information needs are growing, both in regard to the socio-economic and biophysical effects of climate change and the ways in which U.S. industry, government, and civil society will react to such effects (Adger, 2003). Many socio-economic effects of climate change have yet to be studied.

The current body of case study literature documenting the socio-economic effects of climate change is far from robust, and as the scientific community moves forward, such research must be performed and incorporated into public and private sector responses locally, nationally, and internationally. The use of time series data for both social and economic indicators of human community vulnerability and resilience (Charles et al., 2009; Colburn and Jepson, 2012; Jepson and Jacob, 2007; Jacob and Jepson, 2009; Tuler et al., 2008), fishery performance (Clay et al., 2010; Kitts et al., 2011; Pollnac et al., 2006; Smith and Clay, 2010) and indicators of ecosystem function will help elucidate some of these impacts over time. However, interdisciplinary research that brings together scientists studying the biophysical effects of climate change with social scientists is still necessary to provide a much fuller understanding of both the biophysical and human dimensions of climate change.

This section has presented many important scientific findings and emerging areas of science that can or already are expanding our knowledge and understanding of how climate change is resulting in marine environment and human community-level changes and associated alterations to human uses of the ocean and ocean services. Researchers engaged in climate change research know that scientific assessments such as this help to identify knowledge gaps arising out of the difficulty in documenting and understanding complex natural and social systems. Greater attention in both the natural and social sciences is now being paid to the coupled nature or enmeshed reality of social, economic, and ecological systems. This section has showcased how different forms of science-based knowledge and expertise reveal the complex relations between climate change and marine environments and users as well as how knowledge gaps expose new pathways of scientific focus, practice, and public involvement (Button and Peterson, 2009), including a broader inclusion of traditional or local ecological knowledge (Acheson, 2003; Beaudreau et al., 2011; García-Quijano, 2007; Gasalla and Diegues, 2011; Nazarea, 2003).

Tools for filling some of these knowledge gaps are emerging to allow better understanding of social and economic impacts of climate change. Grafton (2010), for instance, has developed a risk and vulnerability assessment and management decision-making framework for adaptation. Existing hazards research (e.g., Cutter et al., 2008) can also provide insight. However, additional targeted research and case studies are also needed.

Socio-economic impacts for commercial and recreational fisheries

Currently, only a few examples exist of work that quantifies or predicts the socio-economic impacts of climate change on U.S. fisheries and fishing-dependent communities. In order to fill in this information gap, the following research topics should be addressed.

- The interconnections between climate change and marine-based socio-economic change. This type of research would benefit from integrations with emerging social science research on both the social and economic indicators of community resilience and vulnerability and of fishery performance.

- Case studies that examine the historical experiences of fisheries and fishing-dependent communities in response to changes in fishery conditions. Changes that are similar in type and magnitude to predicted climate change effects offer insight into the changes fisheries may undergo in the future (Crate and Nuttall, 2009).

- An interdisciplinary research effort and community-based risk assessment of the effects of ocean acidification. This would draw out the critical linkages between the biophysical and socio-economic consequences informed by ocean acidification scenarios in marine environments.

- Development of models of fleet dynamics, vessel operator income, and community social and economic impacts linked to biophysical models of climate-induced changes. This type of work can be performed within the framework of Integrated Ecosystem Assessment programs, even though these assessments are themselves simplified representations of complex social, economic, and ecological relations that call for greater interdisciplinary perspectives and policy based on human-environment relations (Nuttall, 2001; Nuttall and Callaghan, 2000).

- Development of dynamic models that incorporate risk and uncertainty to guide decisions on investment in adaptation and mitigation measures to prioritize expenditures related to adaptation and mitigation.

Subsistence fisheries

One domain of fisheries science where knowledge gaps persist is subsistence fisheries research. A variety of reports and workshops have been held to discuss these gaps (Callaway et al., 1999; United Nations, 2009) and some research has been conducted in certain areas, such as the Arctic (Moerlein and Carothers, 2012; Moller et al., 2004; Nuttall et al., 2004). However, significant gaps in knowledge still exist related to how climate change is going to affect subsistence uses of marine resources. These research gaps include the following.

- Identifying specific challenges and opportunities that subsistence-dependent communities have for adapting to climate change;
- Improving understanding of the ability of subsistence-dependent communities to predict local climate, social, biological, and economic trends;
- Understanding of the flexibility of subsistence living approaches and the capacity of subsistence communities to overcome potentially detrimental effects of climate change;
- Ethnographic research that collects and integrates traditional ecological knowledge (TEK), including the establishment of community-based monitoring projects;
- Development of protocols that ensure appropriate inclusion of TEK into biological and physical research projects and resource management decisions;
- Understanding the most effective means for developing co-management strategies that bring subsistence-resource users into discussions of climate change adaptation strategies; and
- Determining mechanisms for allowing resource management institutions to be flexible and able to respond rapidly to changes when setting annual limitations on subsistence harvests.

Offshore energy development

In order to advance our understanding on the impacts of climate change on the oil and gas industry in general, and on offshore production in particular, several reports and authors (e.g., Acclimatise, 2009a, b; Burkett 2011) suggest that research should be undertaken to develop knowledge related to a number of aspects of the industry, including:

- Understanding of the impact of climate change on oil and gas companies' business models to inform industry participants as they adapt to climate change;
- Addressing both the impacts of extreme events and the impacts of incremental change on offshore energy that may go unnoticed until thresholds are crossed (Acclimatise, 2009a);
- Evaluate the environmental, operational, and human safety risks of exploration

and production in new areas made accessible by climate change impacts;

- Conducting research at the level of individual corporations to evaluate appropriate risk management alternatives; and

- Consultation with local governments to anticipate possible changes in regulations and thus more easily incorporate change into operations.

Tourism and recreation

Researchers are beginning to explore how the major stressors associated with climate change might impact marine recreation and tourism. One significant contribution has been the development of indices, considered under projected climate change scenarios, to predict movements in preferred locations for tourism and recreation (Scott et al., 2004). In addition, Moreno and Becken (2009) developed a methodology to assess vulnerability of locations to the impacts of climate change. One suggestion identified by this research is for the institutions and infrastructure that support coastal tourism and recreation to be flexible in responding to climatic changes.

In many areas further information is needed to determine the impacts of climate change on ocean-based tourism and recreation in the U.S., including the following:

- Development of local-to-regional scale projections of climate impacts on ocean tourism and recreation (Scott et al., 2004; Lambert et al., 2010; Pallab et al., 2010);

- Development of a framework to adapt to the anticipated impacts of climate change through policy and management actions (Pallab et al., 2010);

- Development of agreements, institutions, and capabilities that ensure the safety of life at sea in more extreme weather conditions and new, more hostile operating environments; and

- Understanding of the impacts of climate change on cetacean movements.

Public health

Knowledge gaps remain in understanding climate-related public health impacts, risks, and human responses for marine systems. The U.S. will be better able to adapt to these changing conditions given better understanding of the following research areas:

- Improved understanding of ocean acidification impacts on human health;

- Enhanced protocols for testing domestic and imported seafood for toxins and infectious agents that may increase as a result of climate change;

- Development of integrated health warning systems through deployment of marine sensors for monitoring, updating public health surveillance systems, and developing early warning systems and risk communication strategies;

- Development of coupled socio-economic and biophysical models that can assist in the prediction of outbreaks of waterborne pathogens and harmful algal blooms; and

- Case studies of both individual species targeted for seafood poisoning and of decision-making processes of fishermen who struggle to negotiate the trade-offs they must make in choosing among different available species.

4.10 Conclusion

This section has shown how climate-related biophysical changes to marine resources lead to socio-economic, cultural, and governance effects. This section has also elaborated on questions that continue to dominate climate change discussions. What are the future effects of climate change? What effects are anticipated in the near future? What role will science and scientific assessments, in particular social scientific assessments, play in climate-change planning and policy? To meet the challenges of climate change in general, and climate change impacts on ocean services in particular, broad, interdisciplinary cooperation will become more critical in the years to come (Østreng, 2010).

Chapter 5

International Implications of Climate Change

Executive Summary

Climate change and marine ecosystems neither begin nor end at the U.S. border. Many marine organisms, such as fish, marine mammals, and seabirds, are highly migratory and do not remain in one jurisdictional boundary. As discussed in Section 3 and 4, we are currently observing and documenting widespread shifts in the timing, distribution, and abundance of many marine resources. Many of these species occupy the U.S. at some stage of their life cycle and are of conservation concern. As climatic changes become more apparent, and the rate of change potentially increases, habitats and species ranges will continue to shift significantly as species expand into countries where they had not previously lived. Current protected area networks may not match critical sites needed in the future. The focus of much conservation work has historically been on critically endangered species. In light of climate change, attention must be given to ensuring that other species and populations remain robust and resilient to the changes that are projected to occur throughout the marine biome (Simmonds and Eliott, 2009).

Multilateral regional fisheries management organizations (RFMOs) of which the U.S. is a member are aware of the issue of climate change. Flexible management strategies will be needed that will allow these organizations to manage fisheries sustainably. Only half of the 12 RFMOs that include U.S. fish species have taken or are taking actions to address climate change (Table 5.3). In addition, none of the six existing bilateral fisheries agreements between the U.S. and neighboring countries have formally addressed climate change issues. Bilateral fishing regimes, which tend to be re-negotiated periodically, will have to evolve over time in response to abundance and spatial and temporal changes in fish stocks brought about by climate change. The fishery governance process is much slower to adapt for an RFMO, where the process is built on existing environmental conditions and a stable decision environment such as a convention or treaty.

Security and transportation issues are at play in terms of expected climate change impacts to ocean services in the U.S. Most notably, climate shifts in the Arctic, especially sea ice coverage, are provoking discussion on the future of ocean governance, including marine resource and ecosystem-based management. Perhaps the most noteworthy issue in this arena is the increase in shipping accessibility in the Arctic. National security concerns and threats to national sovereignty have also been a recent focus of attention (Borgerson, 2008; Campbell et al., 2007; Lackenbauer, 2011). Ocean change will lead to an expanded geopolitical discussion involving the relationships among politics, territory, and state sovereignty on local, national, and international scales (Nuttall and Callaghan, 2000).

International forums that deal with species conservation have a major role to play in providing coordination and direction; therefore, international collaboration is both fundamental and foundational to understanding and managing climate change in the U.S. Strengthening existing international partnerships and developing new partnerships for knowledge sharing and strategy development will be necessary to understand and address climate change impacts on marine ecosystems and communities around the world. Working closely with partner countries to enhance the level of understanding related to climate change impacts is needed to build capacity to effectively plan and implement adaptation actions.

Key Findings

1. Many migratory species that span jurisdictional boundaries are exhibiting shifts in distribution and abundance.

   - Current policy and structured agreements assume species stationarity, which is no longer the case.

   - Many species will continue to shift significantly, expanding their ranges in countries where they previously have not lived. As a result, current protected area networks may need to be expanded to match critical sites needed in the future.

   - Several global conventions, as well as regional and bilateral migratory species agreements, are well-placed to support coordinated, multi-country climate change adaptation efforts on behalf of migratory species.

2. International partnerships will be necessary to ensure that monitoring protocols, management plans, and training programs are robust and coordinated for effective long-term implementation on shared marine resources.

   - Multinational large-scale and long-term work is needed to better understand the risks to oceans and marine resources as a result of climate change.

   - Strengthening synergies across existing treaties and conventions would provide better coordination and improved focus and facilitate the development of shared priorities.

3. Although aware of climate change, only half of the existing Regional Fisheries Management Organizations (RFMO) have addressed the issue.

   - Agreements, both multilateral and bilateral, will need flexibility to adapt to changing circumstances, particularly unanticipated, climate-driven changes in stock levels or distribution across EEZs or high-seas areas.

   - The potential for spatial displacement of aquatic resources and people as a result of climate change impacts will require existing regional structures and processes to be strengthened or enhanced.

4. Climate change will affect transportation and security issues in both the short and long term.

- Changes in available shipping lanes in the Arctic created by a loss of sea ice have generated an expanded geopolitical discussion involving the relationship among politics, territory, and state sovereignty on local, national, and international scales.

5. Accounting for the carbon sequestration value of coastal marine systems has the potential to be a transformational tool in the implementation of improved coastal policy and management.

- A number of countries including Indonesia, Costa Rica, and Ecuador have identified "Blue Carbon" as a priority issue and are currently developing strategies and approaches.

5.1 Implications of Climate Change in International Conventions and Treaties

A number of international treaties and conventions have been developed to aid in addressing ocean issues that affect multiple jurisdictions and countries. Many of these focus either primarily on marine resources or involve them in some fashion. Exploring and strengthening synergies between these treaties and conventions would be extremely beneficial for providing increased value, better coordination, and improved focus and facilitation of the development of key priorities (Robinson et al., 2005). The following discussion includes only a subset of the larger body of international conventions and treaties.

Convention on Migratory Species (CMS)

The Convention on Migratory Species (CMS) of Wild Animals is the only global, intergovernmental convention that is established exclusively for the conservation and management of migratory species (Robinson et al., 2005). The CMS recognizes that nations have a duty to protect migratory species that live within or pass through their jurisdictional boundaries, and acknowledges that, if effective management of these species is to occur, concerted actions will be required from all nations and range states in which a species spends any part of its life-cycle (Robinson et al., 2005). Species are listed under two CMS Appendices: Appendix I lists those species threatened with extinction and Appendix II lists species that would benefit from internationally-coordinated efforts. The U.S. is a range state for many of the marine species listed in these Appendices, including whales, seabirds, turtles, and sharks (see Table 5-1).

Cooperation between countries to tackle the impacts of climate change on specific migratory species is mandatory for effective management. The CMS provides an important opportunity to develop climate change strategies at the international level, a number of which have led to climate change actions that the CMS has undertaken already. A small working group called the Scientific Council Working Group on Climate Change reviewed the results of scientific work of other bodies, such as the Convention on Biological Diversity (CBD), the International Whaling Commission (IWC) and the Ramsar Convention, to ensure that the CMS had the most recent scientific information and to improve and strengthen links to these works. The sharing of information and expertise has been an important means of speeding up the responses of different countries to climate change issues and has given rise to a number of policy reports on species vulnerability as well as guidelines and incorporation of information into agreements and action plans (UNEP/CMS, 2011).

Table 5-1: Marine species with U.S. ranges listed in CMS Appendices

Appendix	Taxa	Species
I	Mammals	Humpback Whale (*Megaptera novaeangliae*), Bowhead Whale (*Balaena mysticetus*), Blue Whale (*Balaenoptera musculus*), Northern Atlantic Right Whale (*Eubalaena glacialis*), North Pacific Right Whale (*Eubalaena japonica*)
I	Birds	Short-tailed Albatross (*Phoebastria albatrus*), Bermuda Petrel (*Pterodroma cahow*), Hawaiian Petrel (*Pterodroma sandwichensis*), Pink-footed Shearwater (*Puffinus creatopus*)
I/II	Mammals	Sperm Whale (*Physeter macrocephalus*), Sei Whale (*Balaenoptera borealis*), Fin Whale (*Balaenoptera physalus*), West Indian Manatee (*Trichechus nanatus*)
I/II	Birds	Steller's Eider (*Polysticta stelleri*)
I/II	Reptiles	Green Turtle (*Chelonia mydas*), Loggerhead Turtle (*Caretta caretta*), Hawksbill Turtle (*Eretmochelys imbricate*), Kemp's Ridley Turtle (*Lepidochelys kempii*), Olive Ridley Turtle (*Lepidochelys olivacea*), Leatherback Turtle (*Dermochelys coriacea*)
I/II	Fish	Basking Shark (*Cetorhinus maximus*), Great White Shark (*Carcharodon carcharias*), Manta Ray (*Manta birostris*) Beluga Whale (*Delphinapterus leucas*)
II	Mammals	Narwal (*Monodon monoceros*), Pantropical Spotted Dolphin (*Stenella attenuate*), Spinner Dolphin (*Stenella longirostris*), Striped Dolphin (*Stenella coeruleoalba*), Killer Whale (*Orcinus orca*), Baird's Beaked Whale (*Berardius bairdii*), Northern Bottlenose Whale (*Hyperoodon ampullatus*), Bryde's Whale (*Balaenoptera edeni*), Dugong (*Dugong dugong*)
II	Birds	Black-footed Albatross (*Phoebastria nigripes*), Laysan Albatross (*Phoebastria immutabilis*), Black-browed Albatross (*Thalassarche melanophris*), Shy Albatross (*Thalassarche cauta*), Salvin's Albatross (*Thalassarche salvini*), White-chinned Petrel (*Procellaria aequinoctialis*), Spectacled Petrel (*Procellaria conspicillata*), Roseate Tern (*Sterna dougallii*), Arctic Tern (*Sterna paradisaea*), Little Tern (*Sterna albifrons*)
I/II	Fish	Whale Shark (*Rhincodon typus*), Shortfin Mako Shark (*Isurus oxyrinchus*), Longfin Mako Shark (*Isurus paucus*), Porbeagle (*Lamna nasus*), Spiny dogfish (*Squalus acanthias*), Green Sturgeon (*Acipenser medirostris*)

Figure 5-1 First short-tailed albatross chick to hatch outside Japan. (Source: Pete Leary).

The Zoological Society of London has also developed and tested a climate change vulnerability assessment method for the UNEP/CMS Secretariat on approximately half of the Appendix I species. Of these, approximately 50 percent are marine species. Results indicate that all of these species will be negatively impacted by climate change, including many species with ranges in the U.S. (Table 5-2). Key vulnerabilities and the main factors limiting adaptation have been identified for these species. Continuing this work in a more quantitative way is considered a key priority for the future.

A technical workshop held in June 2011 included experts from academia, non-governmental organizations (NGOs), inter-governmental organizations (IGOs), and government agencies that work on issues associated with migratory species and climate change. This workshop helped to charter a way forward by recommending key areas for future action and focus (UNEP/CMS/SCC17/Inf.12) under the CMS, including:

- Predicting how future range shifts should be considered;
- Establishing long-term datasets and baselines of species listed under CMS, as well as their critical prey items;
- Developing maps related to zoning, historical and shifting baselines, threats at spatio-temperal scales including transport routes, etc.,for use in the planning of renewable energy projects;
- Focusing on populations that are resilient and adaptive to climate change;
- Emphasizing ecological networks and design of protected areas;
- Highlighting that mitigation of climate change can be potentially more harmful to migratory species if sites are not carefully selected;
- Recognizing that tertiary effects, such as new shipping routes in the Arctic, are

increasing disturbance and exploitation of marine migrants, and acknowledging that coordinating responses with biodiversity-related Multilateral Environmental Agreements may be beneficial;

- Continuing to address research needs related to emerging issues including disease, invasive species, and ecosystem changes;
- Building capacity at the local level through climate change literacy training, participatory monitoring, and incentive creation for conservation among communities;
- Integrating climate change policies in more Multilateral Agreements and strengthening collaboration with the Convention on Biological Diversity (CBD), Ramsar, the Bern Convention, and UNFCCC.

Table 5-2: Marine species under CMS with U.S. Ranges vulnerable to climate change

Vulnerability	Taxa	Species
High	Reptiles	Green Turtle (*Chelonia mydas*), Hawksbill Turtle (*Eretmochelys imbricate*), Kemp's Ridley Turtle (*Lepidochelys kempii*), Loggerhead Turtle (*Caretta caretta*), Olive Ridley Turtle (*Lepidochelys olivacea*), Leatherback Turtle (*Dermochelys coriacea*)
High	Mammals	North Pacific Right Whale (*Eubalaena japonica*), Northern Atlantic Right Whale (*Eubalaena glacialis*), Bowhead Whale (*Balaena mysticetus*), Blue Whale (*Balaenoptera musculus*), Narwhal (*Monodon monoceros*)
High	Birds	Short-tailed Albatross (*Phoebastria albatrus*), Bermuda Petrel (*Pterodroma cahow*), Steller's Eider (*Polysticta stelleri*)
Medium	Mammals	Sperm Whale (*Physeter macrocephalus*), Sei Whale (*Balaenoptera borealis*), Humpback Whale (*Megaptera novaeangliae*)
Medium	Fish	Basking Shark (*Cetorhinus maximus*), Great White Shark (*Carcharodon carcharias*)

Multinational large-scale and long-term work is needed to better understand risks associated with ocean and marine resources as a result of climate change. Focus must be given to ensuring that species and populations other than just those critically endangered remain robust and resilient to the changes predicted to occur throughout the marine biome (Simmonds and Eliott, 2009).

Convention on Wetlands of International Importance (Ramsar)

The U.S. is a Contracting Party to Ramsar, which is an intergovernmental treaty that provides the framework for voluntary national action and international cooperation for the conservation and wise use of wetlands and their resources. Parties of Ramsar "work towards the wise use of all their wetlands through national land-use planning, appropriate policies and legislation, management actions, and public education; designate suitable wetlands for the List of Wetlands of International Importance ("Ramsar List") and ensure their effective management; and cooperate internationally concerning transboundary wetlands, shared wetland systems, shared species, and development projects that may affect wetlands" (http://www.ramsar.org). Nearly 2,000 designated Ramsar sites exist around the world, over 30 of which are in the U.S.

Marine/coastal wetlands are an identified wetland type recognized by Ramsar. Many of these are particularly important habitats for ocean and marine species, and include, among others: permanent shallow marine waters such as sea bays and straits; marine subtidal aquatic beds including kelp and sea-grass beds; tropical marine meadows; coral reefs; and rocky marine shores such as rocky offshore islands and sea cliffs). Similar to the CMS, the Ramsar Convention has explicitly recognized the threats caused by climate change and plays an important role in the future by understanding that additional work is required to understand the relationship between wetlands and climate changes.

Convention on International Trade in Endangered Species (CITES) of Wild Fauna and Flora

The U.S. is a Party to CITES, which regulates trade in species based on their conservation status. Appendix-I species are those threatened with extinction that are, or may be, affected by trade. Parties are not allowed to trade in Appendix I species for commercial purposes. Appendix-II species are not necessarily threatened with extinction but might be if trade is not regulated, and Parties may trade in these species as long as trade is not detrimental to the species' survival. Appendix III species are listed unilaterally by Parties seeking international cooperation in controlling trade.

Recently, CITES has begun to focus attention on the issue of climate change. At the UNFCCC Conference of the Parties (CoP) 15 in 2010, the Animals and Plants Committees were directed to identify the scientific aspects of the provisions of the CITES Convention and the Resolutions of the Conference of the Parties that are or are likely to be affected by climate change and to make recommendations for further action. A working group was created to draft recommendations to be finalized at the joint Animals and Plants Committees meeting in March 2012 and presented at the 62nd Standing Committee meeting in July 2012. The working group was tasked with addressing climate change in six CITES processes or mechanisms: species listing, non-detriment findings, Periodic Review of the Appendices, management of nationally stablished export quotas, Review of Significant Trade, and trade in alien invasive species.

The March 2012 joint meeting of the Animals and Plants Committee resulted in general consensus that the decision-making framework developed within CITES is flexible enough to accommodate the consideration of climate change in each of its six processes or mechanisms.

Inter-American Convention (IAC) for the Protection and Conservation of Sea Turtles

The IAC is focused on marine turtles, which are particularly vulnerable to climate change (see Section 3). It promotes the protection, conservation, and recovery of the populations of marine turtles and those habitats on which they depend on the basis of the best available data that take into consideration the environmental, socio-economic, and cultural characteristics of the Parties (Article II, Text of the Convention). These actions should cover both nesting beaches and the Parties' territorial waters. International collaboration is essential to marine turtle conservation because turtles that breed on U.S. beaches or migrate through U.S. waters need continued protection once they leave U.S. jurisdiction. The IAC is an intergovernmental treaty that provides the legal framework for countries in the Western Hemisphere to take actions for the benefit of these species. Continued active involvement of the U.S. in international expert groups and initiatives provides key support to emerging collective adaptation action for marine turtles.

In 2009, the Parties agreed to a number of actions to address the impacts of climate change on marine turtles specifically, including management actions as well as research and monitoring activities (http://www.iacseaturtle.org). According to annual reports submitted by the Parties, overall performance against these goals has been fair; however, climate adaptation concerns are often overtaken by more immediate priorities such as bycatch and non-climate impacts on nesting beaches.

Convention on Biological Diversity (CBD)

The objectives of the CBD are threefold: 1) the conservation of biological diversity, 2) the sustainable use of its components, and 3) the fair and equitable sharing of the benefits arising out of the utilization of genetic resources. The CBD affirms that conservation of biodiversity is a common concern of humankind and reaffirms that nations have sovereign rights over their own biological resources. The CBD covers both terrestrial and marine biota and Parties are explicitly required to implement the CBD consistent with the rights and obligations of States under the United National Convention on the Law of the Sea. The United Nations Convention on Environment and Development in Rio de Janeiro, June 1992 opened the CBD for signature and entered it into force on December 29, 1993. The U.S. has signed the CBD but has not yet ratified it.

Climate change and biodiversity is one of the CBD's major crosscutting issues. According to the Millenium Ecosystem Assessment[2], climate change is likely to become one of the most significant drivers of biodiversity loss by the end of the century. Climate change is already forcing biodiversity to adapt either through shifting habitat, changing life cycles, or the development of new physical traits. Conserving and restoring marine ecosystems, including their genetic and species diversity, is essential for the overall goals of both the CBD and the UNFCCC and plays a key role in the global carbon cycle and in adapting to climate change.

The CBD Conference of the Parties has made over 40 decisions regarding biodiversity

---

2 The Millenium Ecosystem Assessment was called for by UN Secretary General Kofi Annan in 2000. The objective of the Assessment was to assess the consequences of ecosystem change for human well-being and the scientific basis for action needed to enhance the conservation and sustainable use of those systems and their contribution to human well-being. The Assessment involved the work of more than 1,360 experts worldwide. Their findings provide a state-of-the-art scientific appraisal of the condition and trends in the world's ecosystems; the services they provide; and the options to restore, conserve, or enhance the sustainable use of ecosystems.

and climate change since it entered into force. Decision X/33 (*Biodiversity and Climate Change*) (http://www.cbd.int/climate/doc/cop-10-dec-33-en.pdf), adopted at the 10th meeting of the CBD in Nagoya, Japan in 2010, encourages the Parties, among other things, to "Enhance the conservation, sustainable use and restoration of marine and coastal habitats that are vulnerable to the effects of climate change or which contribute to climate-change mitigation, such as mangroves, peatlands, tidal salt marshes, kelp forests and seagrass beds, as a contribution to achieving the objectives of the United Nation Framework Convention on Climate Change, the United Nations Convention to Combat Desertification, the Ramsar Convention on Wetlands and the Convention on Biological Diversity."

5.2 Climate Change Considerations in Other International Organizations

Agreement for the Conservation of Albatross and Petrels (ACAP)

The Agreement for the Conservation of Albatross and Petrels (ACAP) is an international multilateral agreement that seeks to reduce known threats to albatrosses and petrels through coordination of international activity. The development of ACAP began in 1999 under the auspices of the Convention on the Conservation of Migratory Species of Wild Animals (CMS). Twenty-two species of albatrosses and seven species of petrels are currently listed under ACAP.

Since 2008, the ACAP Advisory Committee has had a standing agenda item titled *Impacts of Global Climate Change*. Recently, ACAP Parties have acknowledged growing scientific evidence that present climate change is already affecting marine ecosystems at all levels of the food webs, and projection of future change suggests that these effects will increase considerably. For this reason, the Parties recognized the importance of reviewing the potential impact of global climate variability and change on the conservation status of albatrosses and petrels. Despite this, published studies to date are limited to just a few species in the Indian Ocean; therefore, the Committee has recommended that Parties and Range States encourage further analyses on the combined impacts of environmental change and fisheries on albatross and petrel population trends.

International Whaling Commission (IWC)

The IWC is the body charged with the proper conservation of the world's whale stocks to make possible the orderly development of the whaling industry. The main duty of the IWC is to review and revise as necessary the measures laid down in the Schedule to the Convention that govern the conduct of whaling organizations throughout the world. Among other things, these measures: provide for the complete protection of certain species; designate specified areas as whale sanctuaries; set limits on the numbers and size of whales that may be taken; prescribe open and closed seasons and areas for whaling; and prohibit the capture of suckling calves and female whales accompanied by calves. The compilation of catch reports, as well as other statistical and biological records, is also required.

Figure 5-2
Sperm whale
rolling onto its
side (Source:
Stephen Tuttle).

Climate change and its impacts on cetacean species have been highlighted in discussions at the IWC Scientific Committee, which has considerable expertise in understanding and modeling climate impacts. The IWC passed Resolution IWC/61/16 entitled *Consensus Resolution on Climate and Other Environmental Changes and Cetaceans* at the 2009 IWC Annual Meeting in Madeira, Portugal. This resolution requested Contracting Governments to incorporate climate change considerations into existing conservation and management plans; directed the Scientific Committee to continue its work on studies of climate change and the impacts of other environmental changes on cetaceans as appropriate; and appealed to all Contracting Governments to take urgent action to reduce the rate and extent of climate change. The IWC has also hosted a number of workshops to enhance collaborations among various experts in cetacean biology, marine ecosystems, modeling, and climate change, as well as improving the conservation outcomes for cetaceans under climate change scenarios.

Future climate change-related challenges facing whale stocks require innovative, large-scale, long-term, and multinational response from scientists, conservation managers, and decision makers. Moreover, the reactions to emerging developments and changes will need to be swift (Simmonds and Eliott, 2009).

Commission for the Conservation of Antarctic Marine Living Resources (CCAMLR)

The CCAMLR was established mainly in response to concerns that an increase in krill catches in the Southern Ocean could have a serious effect on populations of krill and other marine life, particularly birds, marine mammals, and fish, that depend on krill for food. Climate change is on the agenda of CCAMLR's Scientific Committee, which reports on this item to the CCAMLR. Climate change is also a factor the U.S. is considering in the development of a proposal for a marine protected area in the Ross Sea. The U.S. would likely leave one area open to fishing and close an equivalent area so that the impacts of climate can be differentiated from the impacts of fishing. The CCAMLR welcomed the Scientific Committee's deliberation on climate change and, in particular, noted the recommendation of the EU/Netherlands-sponsored workshop on "Antarctic Krill and Climate Change" (SC-CAMLR-XXX/BG/3).

North Pacific Marine Science Organization (PICES)

The primary role of the Convention for a North Pacific Marine Science Organization (PICES) is to promote and coordinate marine research undertaken by the Parties (Canada, Japan, China, Korea, Russia, and the U.S.) in the temperate and sub-Arctic region of the North Pacific Ocean and its adjacent seas; to advance scientific knowledge about the ocean environment, global weather and climate change, living resources and their ecosystems, and the impacts of human activities; and to promote the collection and rapid exchange of scientific information on these issues. PICES provides an international forum to promote greater understanding of the biological and oceanographic processes of the North Pacific Ocean and its role in the global environment.

PICES is on the forefront of climate change issues in the North Pacific. PICES has published numerous scientific reports on the impacts of climate and climate change on fish species in the North Pacific and forecasting climate impacts on future production of commercially exploited fish and shellfish (Beamish, 2008; Hollowed et al., 2008). Since 2002, PICES has also hosted approximately 11 international symposia, 15 workshops, and 5 Special Sessions with climate change-related themes (see http://www.pices.int/publications/default.aspx).

Wider Caribbean Sea Turtle Conservation Network (WIDECAST)

WIDECAST is an active network of biologists, managers, community leaders, and educators in more than 40 nations and territories including the U.S. committed to an integrated, regional capacity that ensures the recovery and sustainable management of depleted marine turtle populations. WIDECAST has conducted workshops geared towards marine turtle conservationists and marine protected area practitioners, covering climate-related topics including monitoring, vulnerability assessment, selecting and prioritizing adaptation options, and communicating climate change.

5.3: Climate Change Considerations by Regional Fisheries Management Organizations and Living Marine Resource Conservation Organizations

On December 6, 2011, the United Nations General Assembly adopted draft resolution A/66/L.22, which stated, "Sustainable fisheries, including through the 1995 Agreement for the Implementation of the Provisions of the United Nations Convention on the Law

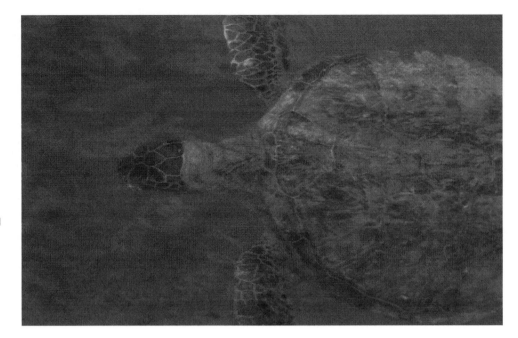

Figure 5-3 Green turtle (Source: David Patte).

of the Sea of 10 December 1982 relating to the Conservation and Management of Strad-dling Fish Stocks and Highly Migratory Fish Stocks, and related instruments" (Reso-lution 66/68). Resolution A/66/L.22 expressed concern over the current and projected adverse effects of climate change on food security and the sustainability of fisheries. The resolution urged nations, either directly or through appropriate subregional, regional, or global organizations or arrangements, to intensify efforts to assess and address, as appropriate, the impacts of global climate change on the sustainability of fish stocks and the habitats that support them, in particular the most affected ones.

As discussed in Sections 3 and 4, climate-related shifts in species geographic range will affect the distribution and composition of fisheries resources, which, in turn, may affect fishing operations, the allocation of catch shares, and the effectiveness of fisher-ies management measures (Sumaila et al., 2011). Multilateral regional fisheries manage-ment organizations (RFMOs) in which the U.S. is a member are cognizant of the issue of climate change, but, with few exceptions, most have done little to reduce the pos-sible effects or develop contingency plans despite the growing body of fisheries research on climate change. Only half of the 12 RFMOs that include U.S. fishery resources have taken or are taking actions to address climate change (Table 5.3). In addition, none of the six existing bilateral fisheries agreements between the U.S. and neighboring countries— five with Canada (albacore tuna, Pacific salmon, Pacific hake, Pacific halibut, and Great Lakes fisheries) and one with Russia (North Pacific and Bering Sea fisheries)—have for-mally addressed climate change issues. The common assumption is that bilateral fishing regimes, which tend to be renegotiated periodically, will have to evolve over time in response to abundance and spatial and temporal changes in fish stocks brought about by climate change; in other words, the effects of climate change are de facto taken into consideration in these negotiations. The fishery governance process is much slower to adapt for an RFMO, where the process is built on existing environmental conditions and a stable decision environment such as a convention or treaty. In addition, many RFMOs rely on consensus decision-making, which is an additional complicating factor, especially if the RFMO has many member countries.

Table 5-3: Primary RFMOs and arrangements that include U.S. living marine resources, by organization/membership, mission, relevant species, and climate change actions, 2012

Organization/ U.S. membership status	Mission	Relevant species	Climate change actions
Commission for the Conservation of Antarctic Marine Living Resources (CCAMLR)/ Member	Protect and conserve the marine living resources in the waters surrounding Antarctica	Fish, mollusks, crustaceans, and all other species of living organisms, including birds	CCAMLR includes climate change on the agenda of its Scientific Committee, which reports on this item to the Commission. Climate is also a factor considered in the development of a proposal for a marine protected area in the Ross Sea.
North Atlantic Salmon Conservation Organization (NASCO)/ Member	Promote scientific research and the conservation, restoration, enhancement, and rational management of salmon stocks in the North Atlantic Ocean	Atlantic salmon (*Salmo salar*)	NASCO is concerned about the potential impacts of climate change on wild Atlantic salmon and has requested the International Council for the Exploration of the Sea (ICES), which provides scientific advice to the organization, to advise it on the potential implications of climate change for salmon management at the 29th NASCO Annual Meeting to be held in Edinburgh, Scotland on June 5-8, 2012. NASCO has not published any studies directly addressing climate change and salmon to date.
Northwest Atlantic Fisheries Organization (NAFO)/ Member	Study, conserve and manage fishery resources in the NAFO Regulatory Area in the North Atlantic Ocean beyond 200-mile zones of member states	Cod, flounders, redfish, capelin, hake, skates, shrimp	The NAFO Scientific Council Standing Committee on Fisheries Environment has been discussing change patterns including climate change for nearly 50 years. Beginning in 1964, NAFO conducted four symposia on decadal reviews (1950-1959, 1960-1969, 1970-1979, 1980s-1990s) of environmental conditions in the Northwest Atlantic and their influence on fish stocks.

Table 5-3 Primary RFMOs and arrangements that include U.S. living marine resources, by organization/membership, mission, relevant species, and climate change actions, 2012 (Continued)

Organization/ U.S. membership status	Mission	Relevant species	Climate change actions
North Pacific Anadromous Fish Commission (NPAFC)/ Member	Promote the conservation of anadromous stocks and ecologically-related species in the high seas areas of the North Pacific Ocean	Pacific salmon (chum, coho, pink, sockeye, chinook, cherry, and steelhead)	The Bering-Aleutian Salmon International Survey-II (BASIS-II) is NPAFC's coordinated program of cooperative research on Pacific salmon in the Bering Sea designed to clarify the mechanisms of biological response by salmon to the conditions caused by climate change. Climate change and its impact on salmon have been discussed in a Symposium and two Special Publications: 1) a report on understanding impacts of future climate and ocean changes on the population dynamics of Pacific salmon (Beamish et al., 2009) and 2) a bibliography of literature associated with climate and ocean change impacts on Pacific salmon (Beamish et al., 2010)

The overarching theme of the NPAFC 2011-2015 Science Plan is "Forecast of Pacific Salmon Production in the Ocean Ecosystems under Changing Climate." |
| Western and Central Pacific Fisheries Commission (WCPFC)/ Member | Ensure, through effective management, the long-term conservation and sustainable use of highly migratory fish stocks in the western and central Pacific Ocean in accordance with the 1982 United Nations Convention on the Law of the Sea and the 1995 UN Fish Stocks Agreement. | All fish stocks of the species listed in Annex 1 of the 1982 Convention on the Law of the Sea occurring in the Convention Area, and other species of fish that the Commission may determine necessary to cover | The WCPFC Science Committee is becoming aware that the impact of oceanographic and climate variability is a key area of uncertainty that should be integrated in future stock assessments (Summary Report of the Seventh Regular Session of the Scientific Committee, 21 September 2011). |

Straddling fish stocks

In some cases, changes in the location of straddling fish stocks may lead to challenges in international fisheries management. The United Nations defines straddling stocks as stocks of fish that migrate between or occur in both the EEZ of one or more states and the high seas (Figure 5-4). One example of such a challenge occurred within the U.S.–Canada Pacific Salmon Commission (McIlgorm, 2010). Pacific salmon are anadromous fish that cross state and international boundaries in their oceanic migrations. Fish spawned in the rivers of one jurisdiction are vulnerable to harvest in other jurisdictions. The turbulent history of U.S. and Canadian cooperative management of their respective salmon harvests suggests that environmental variability may complicate the management of such shared resources. For six years beginning in 1993, the U.S. and Canada were unable to agree on a full set of salmon fishing regimes under the terms of the *Treaty between the Government of Canada and the Government of the United States of America concerning Pacific Salmon*. The breakdown in cooperation was fueled by strongly divergent trends in Alaskan and southern salmon abundance and a consequent change in the balance of each nation's interceptions of salmon spawned in the other nation's rivers. Although several natural and anthropogenic factors contributed to these trends, evidence shows that changing ocean conditions, notably the warm phase of the PDO, played a role. The period of high productivity in Alaska contributed to increased Alaskan interceptions of British Columbia salmon at a time when Pacific Northwest coho and chinook salmon could least withstand retaliatory actions by the Canadian salmon fleet. The mounting crisis led to a fundamental shift in the approach taken by the two nations to determine their respective salmon harvest shares. On June 30, 1999, Canada and the U.S. signed a 10-year agreement that laid the groundwork for a more sustainable, cooperative, abundance-based management regime (Miller et al., 2000). In 2009, that agreement was renegotiated and extended for another 10 years. The latest agreement improved modeling capabilities, and refined harvest rate calculations, and resulted in substantial investments in cooperative research programs.

Types of Fish Stocks

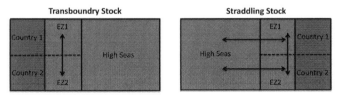

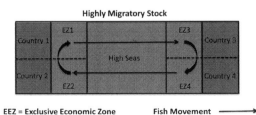

Figure 5-4 Types of fish stocks (Source: Munro et al., 2004).

EEZ = Exclusive Economic Zone   Fish Movement ⟶

Future climate change will continue to impact fisheries governance under the Pacific Salmon Treaty in the future as fisheries management science tries to assess the impact of climate change on Pacific salmon (McIlgorm, 2010).

Transboundary fish stocks

Climate change is also likely to impact the spatial distribution of transboundary fish stocks. These are stocks that range in the EEZs of at least two countries (Figure 5-4), such as hake stocks off the U.S. Atlantic and Pacific Coasts that straddle the U.S.-Canada border (Helser and Alade, 2012). Pacific hake spawn off the coast of California and forage off the coasts of Oregon, Washington, and Canada. As discussed in Sections 3 and 4, ocean conditions influence the latitudinal extent of Pacific hake and silver hake foraging migrations (Agostini et al., 2007; Nye et al., 2009). In addition, the age structure of the population also has a strong effect on the northerly distribution of Pacific hake; in other words, older fish are found farther north in warm years. In 2003, the U.S. and Canada signed a treaty to establish national shares of the coast-wide Pacific hake stock. Shares were initially determined by the percentage of the hake stock found in each country's waters during the summer fishing season. Shifts in spatial distribution of the foraging distribution will impact the stock assessment and may necessitate changes to current management agreements for these stocks.

Another consideration is the lack of formal conservation and management agreements for transboundary fish stocks. Pacific sardines, which migrate across international boundaries, are a case in point. Currently, no international management agreement exists for Pacific sardines, but scientists and members of industry from the U.S., Mexico, and Canada informally meet at the annual Trinational Sardine Forum, where research results and ideas are exchanged. In view of the combined impact of fishing and ocean climate variability on the sardine stock, an important emerging issue is the need for stable transboundary management, given potential changes in the stock's availability within the affected countries' EEZs. The question becomes whether cooperative management of the stock will result in economic and biological gains to all three countries. If cooperative conservation and management is a positive sum game, a related concern then becomes whether cooperative management will be stable in the face of climate change (Herrick et al., 2007; see Section 4 for more details).

Similar situations exist for a number of West Coast transboundary groundfish stocks such as sablefish, petrale sole, and numerous rockfish species. These are currently managed separately by the U.S. Pacific Fishery Management Council and Canada's Department of Fisheries and Oceans in British Columbia.

Highly migratory fish stocks[3]

---

3 The term "highly migratory species" comes from Article 64 of the United Nations Convention on the Law of the Sea (UNCLOS). Although the Convention does not provide an operational definition of the term, UN-CLOS Annex 1 lists the species considered highly migratory by Parties to the Convention. The list includes: tuna species (albacore, bluefin, bigeye, skipjack, yellowfin, blackfin, little tunny, southern bluefin, and bullet), and tuna-like species (pomfret, marlin, sailfish, swordfish, saury and ocean going sharks, dolphins, and other cetaceans).

Some fish populations migrate over long distances, passing through multiple territorial waters (Figure 5-4). The stability and success of RFMOs that govern the harvests of straddling and highly migratory fish stocks will depend, in part, on how effectively they can maintain member nations' incentives to cooperate despite the uncertainties and shifting opportunities that may result from climate-driven changes in productivity, migratory behavior, or catchability of the fish stocks governed by the RFMO (Miller, 2007).

The reliance of island nations on tuna fisheries, and the potential adverse effects of climate change on this resource, emphasizes the need for precautionary approaches to management. The Western and Central Pacific is a complex ocean fishery with island nations and foreign fishers including the U.S. taking purse seine and longline tuna catches in areas to the north and east of Papua New Guinea. Independent states in the Pacific have several regional organizations such as the South Pacific Forum Fisheries Agency (FFA) and the Western and Central Pacific Fishery Commission. The majority of fishery production is by distant water-fishing nations under access agreements to FFA member states. Gaining economic benefits from domestic tuna fishery processing by island states is a priority to supplement income from access license fees.

Tuna are relatively sensitive to sea temperature and move quickly to areas of preferred temperature. Consequently, the total stock of tuna does not necessarily change dramatically if the sea temperature changes, but the spatial distribution may shift substantially (Aaheim and Sygna, 2000). Climate change will impact all industrial fishers and processors due to changes in the location of fishing sites, with vessels spending several months of the year inside a national EEZ as fish move to the high seas or an adjacent EEZ. This movement has implications for licensing of foreign fishing vessels and tuna canneries using local suppliers. The increased risk of fish availability will have implications for future capital investment and labor requirements. Fishery governance systems need to be aware of potential climate change impacts on annual catches and the location of fish schools, both of which increase the variability in an already complex system (McIlgorm, 2010, see Section 4).

Arctic

Climate change is expected to have profound impacts in the Arctic; some of these changes are already being observed (see Sections 2, 3, 4). The loss of sea ice is expected to impact the timing of seasonal production and extend the growing season. Expansion or movement of some sub-Arctic species into the Arctic may occur over time and the rates of expansion and movement will differ by species in relation to their vulnerability to climate change (see Section 3). Fishing may expand in response to periods of reduced ice cover in the Chukchi and Beaufort seas. In recognition of this possibility, the North Pacific Fishery Management Council acted swiftly to close the U.S. Arctic Exclusive EZ to commercial fishing until sufficient information is obtained to manage the stocks sustainably (Wilson and Ormseth, 2009; see Section 6). Although this action protects stocks within the U.S. Arctic EEZ, the Arctic highseas waters are international waters in which international fishing agreements may be needed for fisheries sustainability.

On a regional fisheries level, the potential for spatial displacement of aquatic resources and people as a result of climate change impacts will require existing regional structures and processes to be strengthened or enhanced. Agreements, both multilateral and bilateral, will need flexibility to adapt to changing circumstances, particularly un-anticipated, climate-driven changes in stock levels or distribution across EEZs or high seas areas.

5.4 Climate Change and Other International Issues

Maritime transportation and security

Climate change will likely affect maritime security, transportation, and governance. For example, the warming of the Arctic is making regions of the ocean more passable by vessels than in the past, which has provoked discussion on the future of ocean governance, including marine resource and ecosystem-based management. According to some researchers, the Arctic region could slide into a new era featuring jurisdictional conflicts and increasingly severe clashes over the extraction of natural resources among the global powers (Berkman and Young, 2009). National security concerns and threats to national sovereignty have also been a recent focus of attention (Borgerson, 2008; Campbell et al., 2007; Lackenbauer, 2011). Others are confident that existing bilateral and international arrangements are robust enough to handle the new challenges. In any case, given that the effects of climate change will vary across regions of the U.S. and the world, diverse and novel governance and security challenges will likely emerge.

Climate change adaptation and mitigation actions often extend beyond regional scales and regional governance and security concerns. According to the Energy, Environment, and Development Programme (EEDP) of the United Nations, "Climate change is no longer an environmental protection' issue but one intimately connected with a wider world. Given the scale and urgency of the challenge, many of the decisions critical for global climate security and the effective transition to a low-carbon, high-efficiency economy will take place *outside* the field of climate change. It is the decisions made in the areas of foreign and trade policy, security and geopolitics, energy policy and investment that will have an influence on the global response to climate change" (Paskal, 2007, page 11; also see Section 5).

SECURITY. The topic of security is a growing theme of global environmental change discussions, especially those focusing on climate change (Barnett, 2006). In general, the U.S. is unlikely to escape the plausible adverse consequences of global climate change, and marine-dependent communities are in an especially vulnerable position. If a substantial rise in sea levels occurs, humans will be forced to move away from the coast, leading to a reorganization of populations in U.S. coastal regions, which could result in radical socio-economic changes to marine resource-based communities. The central role of the U.S. in emerging discussions of the relationship between climate change and security response strategies is largely due to the strong likelihood that: "In the immediate aftermath of any natural disaster, whether caused by climate change or other factors, the international community will look to the U.S., with its unique world role and response assets (including those in the U.S. military), to assume a leading role in organizing the relief operation" (Campbell et al., 2007:108).

U.S. foreign policy and national security agencies, and the government and population they serve and represent, face numerous challenges in this and coming decades.

In addition to energy security, global trade, terrorism, nuclear non-proliferation, and global poverty, global climate change may become a significant foreign policy and national security challenge as it complicates and exacerbates many more traditional security issues.

TRANSPORTATION. According to the Arctic Council's 2009 Arctic Marine Shipping Assessment (AMSA) Report, "The Arctic is now experiencing some of the most rapid and severe climate change on earth... Of direct relevance to future Arctic marine activity, and to the AMSA, is that potentially accelerating Arctic sea ice retreat improves marine access throughout the Arctic Ocean" (Arctic Council, 2009, page 26). With sea ice receding in the Arctic as a result of warming temperatures (see Section 2; Figure 5-5), global shipping patterns are already changing and will continue to change considerably in the decades to come (Berkman and Young, 2009; Cressey, 2007; Khon et al., 2010; Stewart et al., 2007; Figure 4-14).

As the Northern Sea Route and Northwest Passage routes become more passable by vessels because of melting Arctic sea ice, these regions are experiencing greater maritime travel (Khon et al., 2010) and sailors are witnessing "an age-old dream come true" (Kerr, 2002, page 1490). This regional transformation has global economic implications. The international shipping industry influences much of current world trade, suggesting that increasing the capacity of maritime transport in the Arctic may highly affect the import and export of goods throughout the global economy. Furthermore, world seaborne trade is increasing, which is tightening the linkages among economic growth, trade, and demand for maritime transport services (Kitagawa, 2008). Given this state of affairs, climate change impacts on marine resources and users will likely involve direct and indirect challenges and opportunities for the U.S. seaborne transportation sector.

Many of these challenges and opportunities will also be manifested in coastal areas, and particularly in and around ports and harbors, by climate change-related effects such as sea level rise. These issues are considered in depth in the Coastal Impacts, Adaptation, and Vulnerabilities Technical Input for the 2013 National Climate Assessment.

Blue carbon

Blue carbon is a term to describe the carbon dioxide sequestered by seagrasses, tidal marshes, and mangroves and stored in large quantities in both the plants and in the sediment of coastal and marine ecosystems (as discussed in section 2.9). Accounting for the carbon sequestration value of coastal marine systems has the potential to be a transformational tool in the implementation of improved coastal policy and management. Currently no policy, financing, management, or other system specifically values the role of marine and coastal ecosystems in sequestering greenhouse gases or the potential emissions that result from degradation or conversion of these systems. However, attention on carbon sequestration in blue carbon systems is increasing both within the U.S. and internationally. A number of policy and financing mechanisms currently exist that support nature-based climate change mitigation solutions such as blue carbon:

- The United Nations Framework Convention on Climate Change (UNFCCC): Reducing Emissions from Deforestation and Forest Degradation (REDD+), National Appropriate Mitigation Actions (NAMAs), and the Clean Development

Mechanisms (CDM) and Land-Use, Land-Use Change and Forestry (LULUCF) (Climate Focus, 2011). These mechanisms provide incentives and financial support for national-level accounting and project-level activities including conservation, restoration, and sustainable use of natural systems such as forests and peatlands. Coastal ecosystems can be integrated into these existing UNFCCC-supported mechanisms.

- The UNFCCC Subsidiary Body for Scientific and Technological Advice (SBSTA) has recognized the need for increasing policy-maker awareness and capacity on coastal carbon and has invited the submission of information on emissions from coastal and marine ecosystems to the body (UNFCCC SBSTA, 2011).

- Under the UNFCCC, the IPCC has established an expert working group to update the 2006 IPCC Guidelines for including wetlands in National Green-house Gas Inventories. This revision will include a chapter on Coastal Wetlands that was previously absent (IPCC, 2011).

- The Blue Carbon Initiative is a global agenda to maintain the blue carbon stored in coastal ecosystems and to avoid emissions from their destruction. The initiative, coordinated by Conservation International (CI), the International Union for Conservation of Nature (IUCN), and the Intergovernmental Oceanic

Figure 5-5 U.S. Coast Guard Vessel in Arctic Ocean (Source: http://www.msnbc.msn.com/id/39394645/ns/world_news-world_environment/t/ships-take-arctic-ocean-sea-ice-melts/).

Commission (IOC) of UNESCO, has established international expert working groups to develop: 1) the necessary scientific basis and tools and 2) the international- and national-level policy framework needed to support blue-carbon-based policy, management, conservation, and science globally. Field-based demonstration projects have been identified by the working groups as a current priority to test the viability of blue carbon projects and to facilitate the development of practical, science-based methodologies and building capacity in target countries.

- The Verified Carbon Standard, an international carbon registry, is updating its requirements to provide for the inclusion of eligible wetlands projects for carbon financing.

- A number of countries including Indonesia, Costa Rica, and Ecuador have identified blue carbon as a priority issue and are currently developing strategies and approaches. A number of U.S. federal agencies are currently investigating the integration of coastal blue carbon into their priority activities. These countries are in need of technical and resource support to complete this process and implement effective coastal-carbon-based management and policy.

- Snapshots, The Nature Conservancy's Coastal Resilience Tool for New York/Connecticut

Chapter 6

Ocean Management Challenges, Adaptation Approaches, and Opportunities in a Changing Climate

Executive Summary

Individuals, communities, resource managers, and governments across the U.S. are beginning to understand, plan for, and address the impacts of climate change on oceans. Although the practice of climate adaptation is relatively nascent, particularly for marine systems, strategies and actions are emerging. Adaptation planning requires access to the best available science, including long-term monitoring and assessment of environmental and societal change, to understand baselines, track changes through time, and evaluate the effectiveness of actions. Tools and services are currently being developed to meet the growing demand for user-friendly, science-based information that supports ocean adaptation efforts. Two-way communication between scientists and practitioners is critical for ensuring that information meets the needs of decision makers.

Climate change presents a challenge as well as an opportunity to revisit and improve existing plans and management strategies to make them more robust and forward-looking. Integration of climate change into management and stewardship efforts such as fishery management plans and the design of protected areas will enhance ocean resilience. A key strategy to enhance ecosystem function and resilience to climate change is to reduce non-climatic stressors such as pollution and habitat destruction. Existing legal and regulatory frameworks can also be leveraged to promote ocean adaptation efforts. Future success will depend on flexible, adaptive management that can accommodate uncertainty.

Key Findings

1. Climate change presents both challenges and opportunities for marine resource managers and decision makers.

   - In comparison to terrestrial, freshwater, and coastal systems, relatively few adaptation actions have been designed and implemented for marine systems.

   - Barriers to ocean adaptation currently exist and include a lack of usable scientific information, awareness, and institutional capacity.

   - Despite barriers, creative solutions are emerging for advancing adaptation planning and implementation for ocean systems.

2. Ocean-related climate information, tools, and services are being developed to address the needs of decision makers and managers.

   • Long-term observations and monitoring of ocean physical, ecological, social, and economic systems provide essential information on past and current trends as well as insight into future conditions. An impending challenge is ensuring that the provision and accessibility of high-quality information regarding resolutions is commensurate with the scales at which adaptation decisions are made.

   • User-friendly tools, guidance, and services are emerging to foster dialogue, grow communities of practice, and inform and support decisions to enhance ocean resilience in the face of climate change.

3. Opportunities for adaptation include incorporation of climate change into existing ocean policies, practices, and management efforts.

   • Climate considerations can be integrated into fisheries management and spatial planning, as well as the design of marine protected areas to enhance ocean resilience and adaptive capacity.

   • Application of existing legislative and regulatory frameworks can advance climate adaptation efforts in the marine environment.

4. Progress is being made across the U.S. to enhance ocean resilience to climate change through local, state, national, federal, and non-governmental adaptation frameworks and actions; however, much work still remains.

Key Science Gaps/Knowledge Needs:

A strategic, use-inspired, and integrated science agenda is necessary to inform and support efforts to anticipate and effectively adapt to a changing climate. Adaptation will require sustained dialogue, mutual information exchange, and feedback among scientists, decision makers, and information users throughout the research and implementation processes, as well as robust evaluation of the performance of adaptive actions. There is a growing need to:

   • Enhance and sustain long-term observations and monitoring of ocean physical, ecological, social, and economic systems to inform adaptive management;

   • Apply integrative indicators of ocean ecosystem health to foster holistic understanding, monitoring, assessment, and evaluation of change;

   • Advance assessments of climate risks, impacts, and vulnerabilities in marine systems, with an emphasis on local-to-regional scales, to inform adaptive actions;

   • Support research on relevant social, behavioral, and economic sciences to assess adaptation options and trade-offs, determine costs of action vs. inaction, design effective governance processes, identify thresholds and tipping points in social systems, investigate decision making under uncertainty, and improve understanding of human responses to change for marine systems; and

- Develop and implement methods for evaluating the effectiveness of adaptation actions in order to enable a flexible and responsive management approach.

6.1 Challenges and Opportunities for Adaptation in Marine Systems

Adaptation involves processes related to preparing for and building resilience to climate change, as well as responding to unavoidable impacts (IPCC, 2007b; NRC, 2010a). The climate impacts observed today will continue to increase at least into the next several decades, regardless of whether greenhouse gas emissions are limited (NRC, 2010a); therefore, climate change adaptation is a critical component of society's effort to foster a more sustainable future through enhancing the social, economic, and ecological resilience of ocean systems.

Although a diversity of adaptation actions exist, most processes use the following general approach, as articulated by Glick et al. (2011):

- Assess vulnerability to climate change;
- Identify, design, and implement management, planning, and/or regulatory actions and policies to reduce the vulnerabilities identified;
- Design and implement monitoring programs to assess change and evaluate effectiveness; and
- Create an iterative process by which adaptation actions can be re-evaluated and redesigned as necessary.

Tenets of adaptation for ecosystem managers include protecting adequate and appropriate space, reducing non-climatic stressors that interact with the impacts of climate change, and adopting adaptive management practices (Hansen et al., 2010; Glick et al., 2011).

Compared to terrestrial, aquatic, and coastal systems, relatively few adaptation actions have been designed and implemented for marine systems. To better understand perceived and real barriers to action, a survey was conducted of North American coastal and marine managers, who articulated the following as barriers (Gregg et al., 2011):

- Lack of economic resources and budgetary constraints;
- Lack of institutional support, governance, and mandates to take adaptation action;
- Lack of institutional capacity and guidance on how to take action;
- Lack of key information on locally and regionally specific climate projections, and tools to support assessments and monitoring;
- Uncertainty about risk and vulnerability; and
- Lack of awareness, stakeholder support, and engagement.

These constraints have been echoed by others (e.g., CCSP, 2008; Glick et al., 2009; IPCC, 2007a). Fortunately, solutions exist for overcoming many of these barriers, including enhanced provision of information, tools, and services that support ocean-related adaptation decisions and integration of climate change into existing policies, practices, and management. This section synthesizes and assesses progress, strategies, and solutions for ocean adaptation in the U.S.

6.2 Information, Tools, and Services to Support Ocean Adaptation

Indicators, tools, and services are currently being developed to meet the growing demand for user-friendly, science-based information that supports ocean adaptation efforts. Incorporation of science is essential for successful adaptation planning, implementation, and evaluation. Decision makers rely on science to assess vulnerability and risk to a plausible range of climate change futures, understand potential impacts, inform adaptive actions, and evaluate the effectiveness of response options. Sustained interaction and feedback between scientists and information users such as adaptation practitioners can help to ensure that the information provided is accessible, understandable, and relevant. In addition, these interactions can grow scientists' awareness of decision makers' key information needs.

Importance of long-term observations and monitoring for management

The establishment of current baselines and trends are a core element of adaptation approaches (CCSP, 2008). Long-term data are essential for enhancing understanding of how ecosystems and human communities respond and adapt to climate change (Heinz Center, 2008; Peterson and Baringer, 2009). A range of observations on physical, ecological, social, and economic systems are needed to provide information on past and current trends as well as to gain insight into future conditions. These observations detect changes and help communicate trends to managers and the public in addition to assessing risks, developing meaningful climate indices, and supporting adaptive management. Observations and monitoring data can provide critical insight into the relative contributions of anthropogenic change versus natural variability in ocean systems. In addition, long-term data can inform the development of more accurate and higher-resolution climate models that enhance the predictive capacity of managers and other decision makers.

Key variables that inform the development of ocean adaptation actions include but are not limited to:

- Physical parameters, such as surface and subsurface water temperature, water quality, sea level, salinity, pH, currents, and solar radiation (NCA, 2010a);
- Ecological parameters such as phenology (e.g., timing of the spring phytoplankton blooms and life-history events), species abundance, distribution, diversity, and primary productivity (NCA, 2010b); and
- Socio-economic parameters, such as demographics, food supplies, social and economic well-being, and public health (NCA, 2011).

This information can help support assessments of vulnerability as well as the development of strategies for minimizing climate risks to ecosystems and communities. One opportunity is to leverage existing observation and monitoring systems, including those in marine protected areas (Case Study 6-A), to establish sentinel sites for understanding and managing for climate change (National Ocean Council, 2012).

Case Study 6-A

Marine protected areas as sentinel sites for monitoring, understanding, and managing climate change

Under the 1999 Marine Life Protection Act, and through a public-private partnership, California is implementing a statewide, 1,100-mile network of marine protected areas (MPAs) to protect marine life, habitats, and ecosystems. The task of monitoring and evaluating these MPAs to inform adaptive management of the network is significant. Recognizing this, the California Ocean Protection Council has invested over $20M to conduct baseline characterizations of the ecosystem and to develop a novel approach for objective, scientifically rigorous, and cost-effective MPA monitoring. Monitoring now underway in California is establishing a benchmark against which future assessments of ecosystem health and MPA performance can be evaluated.

The emerging network also provides a unique opportunity to advance our understanding of climate change effects on temperate marine ecosystems. A statewide network of MPAs, in which other anthropogenic stressors are reduced, provides a large-scale natural laboratory to understand how climate changes manifest in ocean ecosystems. The innovative approaches to MPA monitoring being developed in the state also provide a framework that can be applied to inform the climate change management dialogue.

Through a partnership with EcoAdapt, California has recently developed a new framework for climate change monitoring (MPA Monitoring Enterprise, 2012), leveraging the opportunity presented by the MPA network. This approach efficiently and effectively augments MPA monitoring with metrics that can track climate change effects on habitats and species, understand the effects on MPA performance, and evaluate climate change adaptation measures.

Sea Ranch, California (Photo Credit: J.J. Meyer).

A challenge ahead is to ensure that coastal and ocean resource managers have access to high-quality information at resolutions commensurate with the scales at which decisions are made (Fauver, 2008; National Ocean Council, 2012). For example, oyster growers in the Pacific Northwest depend on local observations to alert them to potentially harmful changes in ocean pH levels so that they can take proactive measures to protect young animals (see Case Study 3-A). Oceanographic models developed at the Woods Hole Oceanographic Institution and supported by long-term observations are now able to predict blooms of the toxic alga, *Alexandrium fundyense*, providing local public officials and harvesters with an early warning tool to minimize health risks and economic losses from tainted shellfish (Li et al., 2009). These examples illustrate the utility of accessible information at relevant decision scales.

Barriers remain in providing long-term information to support ocean adaptation decisions. Distilling and synthesizing large quantities of data into useful products is increasingly important for providing practical information to inform management. The lack of a systematic approach to sustained, high-resolution climate observations is currently constraining attempts to develop informative and meaningful climate indices (NCA, 2011). Efforts such as the Ocean Health Index (Case Study 6-B) offer a promising approach for synthesizing and improving accessibility of information on ocean stressors as well as the valuable ecosystem services on which humans depend. User-friendly climate tools, services, and products based on long-term data will be essential for understanding changes and informing adaptation measures (Heinz Center, 2008).

Tools and services for supporting ocean management in a changing climate

Efforts are underway to enhance the development and deployment of science in support of adaptation; to improve understanding and awareness of climate-related risks; and to enhance analytic capacity to translate understanding into planning and management activities (e.g., Moser and Luers, 2008). In order for science to be useful to decision makers, the information provided must be timely, accessible, relevant, and credible. Although critical knowledge gaps exist, a wealth of climate- and ocean-related science pertinent to adaptation can be found. However, the majority of this information is currently "inaccessible" to adaptation practitioners; it is either unavailable, too technical to be understood and applied by non-scientists, or does not address the specific needs of decision makers. To address this challenge, a diversity of user-friendly tools and services are emerging.

Below are some examples of accessible tools, guidance, and services developed to foster dialogue, grow communities of practice, and inform and support decisions to enhance ocean resilience in the face of climate change.

- The Sea Grant Climate Network is an online resource that includes adaptation-relevant information for coasts and oceans, a discussion forum, links to upcoming events, and social networking opportunities for the broader Sea Grant community, including extension agents (http://sgccnetwork.ning.com/).

- Coral bleaching is a significant climate challenge for marine resource managers. Several tools and services have been developed to help reef managers anticipate and respond to bleaching events. For example:
    - The Reef Manager's Guide to Coral Bleaching, produced by NOAA, the Australian Great Barrier Reef Marine Park Authority, and the International Union for the Conservation of Nature, provides information on the causes and impacts of coral bleaching, as well as a set of management strategies to help reef managers respond and enhance resilience to bleaching events. (http://coris.noaa.gov/activities/reef_managers_guide/)
    - NOAA's Coral Reef Watch (http://coralreefwatch.noaa.gov/satellite/) has developed several tools, including the Satellite Bleaching Alert System (http://coralreefwatch-satops.noaa.gov/SBA.html), an automated email alert system that notifies subscribers when thermal conditions become conducive to bleaching at select reef.

Case Study 6-B
The Ocean Health Index

Protecting or restoring healthy oceans represents a core goal of almost any marine resource management action, but exactly what is meant by "ocean health" is often vague and poorly defined. Even with a definition in hand, what is not measured cannot be managed. The Ocean Health Index was developed to solve both of these problems with the aim of providing a tool for guiding management decisions.

The Index defines a healthy ocean as one that can sustainably deliver a range of benefits to people now and in the future and measures this health along 10 widely held public goals for the ocean and coasts, such as clean water, food provision, livelihoods, and cultural values (McLeod et al., in review). As such, the Index provides assessments that are relevant to management objectives and mandates and allows systems that are sustainably used, rather than simply protected, to score highly. The Index converts into a common measure the disparate ways in which climate impacts, fisheries production, pollution, species protection, coastal jobs, and other factors are assessed.

The Index can be used to assess the impacts of climate change on each of the 10 public goals as well as the likely benefit to any given goal under management scenarios that target climate impact reduction versus improved fisheries management, land-based pollution regulations, or other measures. For example, climate change impacts that reduce the extent of coastal habitats (e.g., through sea level rise, increased storm intensity) will affect many goals including carbon storage, coastline protection, and biodiversity, in turn reducing overall ocean health. Management actions that increase resilience and reduce non-climatic pressures on coastal habitats should help ameliorate climate impacts and maintain or even improve ocean health in the face of climate change.

Global results for the Index have recently been calculated (Halpern et al., 2012), with the U.S. scoring above average but with ample room for improvement. Regional assessments for the California Current along the U.S. West Coast and the Mid-Atlantic Bight are currently underway.

- Scanning the Conservation Horizon: A Guide to Climate Change Vulnerability Assessment (Glick et al., 2011): This guide, produced by the National Wildlife Federation in partnership with the Department of Defense, the National Park Service, the U.S. Fish and Wildlife Service, the U.S. Forest Service, NOAA, and the U.S. Geological Survey, provides conservation practitioners and resource managers with methods and guidance for understanding and addressing the impacts of climate change on species and ecosystems (http://www.nwf.org/News-and-Magazines/Media-Center/Reports/Archive/2011/Scanning-the-Horizon.aspx).

- The Climate Adaptation Knowledge Exchange (CAKE), a joint project of Island Press and EcoAdapt, is aimed at building a shared knowledge base for managing natural systems including oceans in the face of climate change. CAKE helps users by vetting and organizing available information, communicating adaptation case studies, building a community via an interactive online platform, creating a directory of practitioners to share knowledge and strategies, and identifying and explaining data tools and information available from other sites (http://www.cakex.org/).

- The Coastal Climate Adaptation website of the NOAA Coastal Services Center provides a diversity of resources in support of coastal and ocean adaptation, such as state-level adaptation plans, case studies, climate communications information, risk and vulnerability assessments, and guidance and outreach materials (http://collaborate.csc.noaa.gov/climateadaptation/default.aspx).

These tools and guidance documents help decision makers and managers navigate the complex landscape of information as they work to enhance preparedness and response efforts to safeguard ocean resources in a changing climate. The existing and emerging efforts to coordinate and provide timely, useful, and relevant climate information, tools, and services serve as a critical platform. However, a great deal of work remains in providing accessible information to meet the diverse set of adaptation planning, implementation, and evaluation challenges faced by marine resource managers and practitioners. Creative partnerships will be required in the near future to improve multi-directional communication and ensure that information provided by the scientific community meets the needs of decision makers.

6.3 Opportunities for Integrating Climate Change into U.S. Ocean Policy and Management

Although climate change presents challenges to marine resource managers and other ocean decision makers, solutions exist for incorporating climate change into ocean management. For example, because climatic and non-climatic stressors interact, reducing stressors such as land-based pollution and habitat destructure over which there is more direct control can enhance resilience to climate change and ocean acidification (Kelly et al., 2011; Lubchenco and Petes, 2010). Below, we describe opportunities for reducing climate-related vulnerabilities of oceans by incorporating climate change considerations into marine spatial planning and marine protected area design, fisheries management, and application of existing legislative and regulatory frameworks.

Incorporating climate change into marine spatial planning and marine protected area (MPA) design

Both coastal and marine spatial planning (CMSP) and MPAs spatially allocate human uses of the ocean as a means to better protect and sustainably use marine resources. CMSP focuses on all human uses of the oceans and seeks to allocate those uses across the ocean in a way that minimizes ecological, social, cultural, and economic impacts while supporting and improving resource use and conservation goals (Ehler and Douvere, 2009). MPAs instead focus primarily on limiting access to some or, in the case of "no-take" marine reserves, all human uses within particular locations, typically for conservation or fisheries management purposes (Klein et al., 2008). Improving the enforcement and management of existing protected areas and refugia, and increasing connectivity and the amount of protected space, provide mechanisms for enhancing climate resilience (Glick et al., 2009). In addition, in order to enhance long-term effectiveness, CMSP and MPA processes must incorporate climate change into their planning, implementation, and evaluation efforts.

Accounting for the impacts of global climate change in CMSP and MPA planning may appear challenging because impacts can be diffuse and are often not under local control or management. However, at least three direct mechanisms exist for incorporating climate change into the design of management plans: 1) build resilience to climate impacts, 2) account for spatial patterns of climate impacts, and 3) anticipate future patterns of change.

Building climate resilience into spatial management remains the most commonly pursued approach of the three options (Halpern et al., 2008a; McLeod et al., 2009), in part because this mechanism is relatively straightforward and can leverage existing regulations and mandates. Targeted actions to limit or remove non-climatic stressors can help reduce the cumulative impact of total stressors, thus improving the ability of the ecosystem to cope with increases in climate-related stressors (Halpern et al., 2008b, 2010). For example, land-based pollution is the overwhelmingly dominant stressor to coastal areas off the Gulf of Mexico due to nutrient run-off from the human-dominated Mississippi River watershed, which drains into adjacent coastal areas (Halpern et al., 2009a). Efforts to reduce land-based stress would enhance resilience to climate impacts such as sea level rise by improving the health of coastal salt marshes and wetlands. Furthermore, the size of protected areas or zones for limited use can be increased, and the spacing between conservation patches decreased, as a means of buffering climate-related impacts and other increasing or catastrophic stressors (Allison et al., 2003; McLeod et al., 2009). For example, the large areas encompassed by the MPAs recently established in the Northwestern Hawaiian Islands, the Papahānaumokuākea Marine National Monument, and U.S. Pacific holdings, Marianas Trench, Rose Atoll, and Pacific Remote Islands National Marine Monuments, are protected from many human activities and therefore may be more resilient to climate impacts than are areas subject to higher levels of non-climatic stress from human activities.

Equally important are efforts to account for smaller-scale variation in climate impacts when designing spatially explicit resource management. Patterns of existing and projected impacts of changing sea surface temperature, ocean acidification, and sea level rise can be highly variable (Burrows et al., 2011; Halpern et al., 2008b). For management with a conservation goal, as in the case of MPAs, one potential strategy is to place protected areas in locations that exhibit high resilience to climate change. As assessments designed to inform CMSP and MPA planning processes engage and inform more sectors and more comprehensive planning, these efforts will be better able to address costs, benefits, and trade-offs across multiple sectors (White et al., 2012).

Anticipating future patterns of climate impacts and using these predictions in CMSP and MPA design is the most challenging of the three options, primarily because of the difficulty in predicting small-scale patterns of future climate impacts. Projected shifts in species ranges and ocean circulation patterns can be used to anticipate where species will exist in the future as well as the potential for population connectivity through larval transport. MPA planning and CMSP processes can be designed to both anticipate and facilitate the transition to new geographies through strategies such as the creation of stepping stone reserves that offer refuge or habitat to species as they migrate in response to warming waters (McLeod et al., 2009). In some cases, implementing networks of MPAs may help to diffuse climate risks by protecting multiple replicates of the full range of habitats and ecological communities within an ecosystem (CCSP, 2008).

Integrating climate change into fisheries management

Climate-related processes are affecting, and will continue to affect, the production of fisheries resources in marine ecosystems under U.S. jurisdiction and beyond (Cochran et al., 2009; Doney et al., 2012; see Sections 3, 4, and 5). Fish resources may respond to climate change in a variety of ways including changes in mortality, migration, and distribution patterns, and these changes can have important ramifications for fishery population dynamics, the ability to assess the status of fish populations, and the validity of future stock forecasts and rebuilding plans (Kraak et al., 2009).

The future sustainability and adaptation of fish resources in a changing climate depends on: 1) understanding past, current, and projected future climate impacts and 2) incorporating this information into the scientific bases of fishery management decisions so that decision-makers can effectively respond to impacts on existing fisheries and take advantage of new opportunities as conditions change (Link et al., 2010; Sumaila et al., 2011). Although some progress is being made, much work remains to ensure that fisheries management can effectively prepare for and adapt to the impacts of climate change on fish resources and the communities and economies that depend on them (Hare et al., 2010; Link et al., 2010).

Most of the progress to date is in understanding climate impacts on fisheries. For example, in some U.S. regions, oceanographic and fisheries observing systems are increasingly being mobilized to monitor and track the impacts of climate variability and change on fish and other living marine resources. These observation systems have been instrumental in producing the growing number of studies documenting persistent changes in spatial distribution of fishes attributable to large-scale changes in oceanographic processes (Link et al., 2010; Nye et al., 2009, 2011; Overholtz et al., 2011). The development of marine ecosystem status reports in some regions is also providing a key mechanism for compiling and assessing marine ecosystem conditions as a part of efforts to move towards ecosystem-based management (EAP, 2009).

An increasing number of efforts are underway to understand and project the risks and impacts of climate variability and change on fish populations as well as to advance modeling tools and their application to fisheries management (Hollowed et al., 2009; Stock et al., 2011). For example, Hollowed et al. (2009) developed and tested a framework for modeling fish and shellfish responses to future climate change. Hare et al. (2010) used an ensemble-based modeling approach to project increases in average spawning biomass of Atlantic croaker and a northward shift in population distribution that could translate into a 30-100 percent increase in maximum sustainable yield. Mueter et al. (2011) applied a variety of modeling techniques to project negative impacts of climate change such as reduced recruitment on walleye pollock populations in the Bering Sea. Another promising step is the development of ecosystem models to help explore the complex dynamics of marine ecosystems in a changing climate (Link et al., 2010).

In addition, the body of literature and tools for assessing the vulnerability of natural resources in a changing climate is growing (e.g., Glick et al., 2011). Although most of this work has historically focused on terrestrial or freshwater environments, some recent effort has focused on developing and conducting vulnerability assessments for marine species (Johnson and Welch, 2010).

Relatively few examples exist of fishery management efforts that have explicitly incorporated climate-related information, but these efforts are expected to increase as more information and tools on climate impacts and vulnerabilities become available. One of the key questions is how fisheries managers respond to changing climate, ecosystem, and fisheries conditions. In other words, what are the adaptation options available to fisheries managers, and how and when should they be applied? At present, little information and guidance exists to support fisheries management decisions along this path. Link et al. (2011, page 461) provide guidelines for incorporating distribution shifts into fisheries management, concluding that their approach is "feasible with existing information, and as such, fisheries managers should be able to begin addressing the role of changes in stock distribution in these fish stocks." Consideration of climate impacts on fishery resources will likely become more common with the development and application of ecosystem-based approaches through mechanisms such as integration of changing environmental and ecological conditions into fishery management plans.

Efforts to integrate climate considerations into existing legislative and regulatory frameworks

Recent years have witnessed an increase in awareness that climate strategies must include both efforts to limit and adapt to climate change (Lazarus, 2009; Ruhl, 2010). Regulatory and management frameworks must be able to operate and remain effective in the face of increased and potentially significant uncertainty and change (Craig. 2010; Gregg et al., 2011). Although no single piece of existing federal legislation directly targets climate change adaptation in the marine environment, several potential mechanisms have been developed, some of which are already being implemented, to incorporate climate change considerations into existing statutory and regulatory processes (Gregg et al., 2011; Sussman et al., 2010). As noted by the Government Accountability Office, federal resource management agencies "are generally authorized . . . to address changes in resource conditions resulting from climate change in their management activities" (GAO, 2007, page 2). Whether agencies are required to address climate change depends on their delegated statutory and regulatory authority.

Broadly applicable policy initiatives may enable climate change adaptation in the ocean and marine environment. For example, the National Ocean Council has developed draft implementation plans for two relevant national priority objectives: 1) "Resiliency and adaptation to climate change and ocean acidification," and 2) "Changing conditions in the Arctic" (National Ocean Council, 2012). The Council on Environmental Quality has also drafted guidance for federal agencies regarding the incorporation of consideration of greenhouse gas emissions and adaptation measures into environmental reviews conducted pursuant to the National Environmental Policy Act (NEPA; 42 U.S.C. § 4371 *et seq.*; CEQ, 2010).

In addition, the U.S. is undertaking specific efforts to address adaptation to climate change impacts in the marine environment through existing legislative and regulatory frameworks. The following describe efforts related to incorporating climate change considerations into regulation and management.

- **Ocean Acidification:** The Clean Water Act (CWA) has been cited as a potential mechanism for managing climate change impacts in U.S. waters (Craig, 2009; Kelly et al., 2011). The purpose of the CWA is to restore and maintain the chemical, physical, and biological integrity of U.S. waters (33 U.S.C. § 1251 *et seq.*). One of the statutory tools that may help manage ocean acidification is the designation of impaired waters, which are water bodies that fail to meet specified water quality standards. Pursuant to a settlement between the Center for Biological Diversity and the Environmental Protection Agency (EPA), the EPA solicited input on state approaches to determining whether waters are threatened or impaired by ocean acidification (CBD v. EPA 2009; Federal Register. 2010a). The result was the agency's reassertion that states should seek scientific information on impacts and, when sufficient information is available, list as impaired those water bodies for which pH is below the recommended range (CWA § 304(a), EPA, 2010; Kelly et al., 2011). States can also list as impaired water bodies that fail to meet biological water quality standards because of ocean acidification; that is, water bodies that fail to meet criteria established for coral reef ecosystems, bivalves, or other organisms protected under CWA aquatic life designated uses can be listed as impaired (Bradley et al., 2010).

- **Threatened and Endangered Species:** Climate change is adversely impacting a number of marine species as well as the habitats on which they depend (see Section 3). Public and private efforts have worked to ensure consideration of climate impacts through protected species management laws and listing decisions. In accordance with the Endangered Species Act (16 U.S.C. § 1531 *et seq.*), NOAA and the U.S. Fish and Wildlife Service (USFWS) designate, protect, and recover threatened and endangered species. These agencies cited climate change impacts such as increased sea surface temperatures, sea level rise, loss of sea ice, and ocean acidification as habitat stressors in their decisions to list as threatened elkhorn and staghorn coral species (Federal Register, 2006), polar bears (Federal Register, 2008), and the southern distinct population segment of the spotted seal (Federal Register, 2010b), as well as the finding that listing of the Pacific walrus as threatened or endangered is warranted (Federal Register, 2011). NOAA also identified climate change as a key factor in finding that 82 coral species may warrant a threatened or endangered listing (Federal Register, 2010c). In addition to listings, agencies could factor climate adaptation considerations into critical habitat designations, recovery plans, and consultations on proposed Federal actions (Craig, 2009: Kostyack and Rohlf, 2008; Owen, 2012).

- **Fisheries Management:** Climate change is affecting the abundance, distribution, and diversity of fish species as well as the species and habitats they rely upon (see Sections 3 and 4). The Magnuson Stevens Fishery Conservation and Management Act (16 U.S.C. §1801 *et seq.*) requires regional fishery management councils to develop fishery management plans (FMPs) that must be approved by the Secretary of Commerce. Although no requirements specifying consideration of climate change exist, at least one regional council has begun to consider such impacts. In 2009, the North Pacific Fishery Management Council (NPFMC) issued an FMP that closed Arctic fisheries to commercial harvesting (NPFMC, 2009). Recognizing that loss of sea ice could remove barriers to previously inaccessible fisheries, the Arctic FMP established a prohibition against commercial fishing until adequate information is available to support sustainable management (NPFMC, 2009; see Sections 4 and 5). This decision reflects a precautionary approach to managing and developing fisheries in the face of climate change.

Enhancing the resilience of the Nation's oceans to climate change will require action at all levels. The federal statutory, regulatory, and policy efforts noted above represent some of the existing approaches that agencies are taking to support adaptation to and management of climate change impacts in the marine environment. In addition, many state and local actors are taking steps to address climate change effects on oceans. For example, many state coastal management programs have already developed adaptation policies (CSO, 2007). Kelly et al. (2011) identify several actions that local and state governments may take to help reduce the causes and impacts of ocean acidification, including management of runoff and coastal erosion. The actions described in this section provide an illustrative but not comprehensive picture of how existing legal and policy frameworks can be used to undertake climate change adaptation efforts for oceans.

6.4 Emerging Frameworks and Actions for Ocean Adaptation

Although the science and practice of marine adaptation are relatively nascent, many individuals, communities, ecosystem managers, and governments across the U.S. are developing strategies for enhancing ocean resilience in the face of a changing climate. Marine systems, and especially coral reef systems, are the sites of some of the earliest efforts to develop adaptation strategies and policies (Marshall and Schuttenberg, 2006). Adaptation frameworks, including ocean-related adaptation efforts, are emerging across the U.S. Federal Government (Center for Climate and Energy Solutions, 2012; see Table 6-1 for examples). In addition, a diversity of frameworks for ocean adaptation action have been developed at national, regional, state, local, and non-governmental levels (see Table 6-1 for examples). These efforts provide a platform to inspire and support the planning and implementation of on-the-ground actions.

In addition to the development of adaptation frameworks, incorporation of climate change into forward-looking adaptive management actions is beginning to occur for marine systems throughout the U.S. For example, the North Pacific Climate Regimes and Ecosystem Productivity (NPCREP) (http://www.pmel.noaa.gov/foci/NPCREP/) project is working closely with resource managers to understand and address impacts of climate change on North Pacific and Bering Sea ecosystems. From 2000-2005, anomalously high sea surface temperatures in the Bering Sea coincided with low pollock recruitment. The NPCREP integrated this information into fisheries stock assessment models and projected poor pollock recruitment conditions, which were reported to the North Pacific Fishery Management Council (NPFMC)'s Science and Statistical Committee. Based on recommendations from the Committee, the NPFMC temporarily reduced pollock quotas between 2006 and 2010 until conditions became more favorable. This effort illustrates adaptive management based on changing environmental conditions.

Coral reefs are being impacted by both climatic and non-climatic stressors (see

Table 6-1: Examples of ocean-related climate adaptation frameworks in the U.S.

Adaptation project	Description
NATIONAL/FEDERAL	
Interagency Climate Change Adaptation Task Force	The Interagency Climate Change Adaptation Task Force (ICCATF) was initiated in Spring 2009 to determine progress on federal agency actions in support of national adaptation and to develop recommendations for additional actions. The ICCATF is composed of over 20 federal agencies and Executive branch offices and has involved over 300 federal employees. One outcome of this effort is a mandate under Executive Order 13514 for all federal agencies, including those with ocean-related responsibilities, to develop adaptation plans. *http://www.whitehouse.gov/administration/eop/ceq/initiatives/adaptation*
National Fish, Wildlife and Plants Climate Adaptation Strategy	The National Fish, Wildlife, and Plants Climate Adaptation Strategy, initiated through Congressional directive in 2009, is currently under development. The Strategy provides a nation-wide blueprint for coordinated action among federal, state, tribal, and non-governmental entities to safeguard the nation's valuable natural resources, including marine resources, against a changing climate. The draft Strategy was released in January 2012 for public review and input. http://www.wildlifeadaptationstrategy.gov/index.php
National Ocean Policy	The National Ocean Policy was created in July 2010 under Executive Order 13547 to create a comprehensive, integrated framework for the stewardship of the ocean, coasts, and Great Lakes of the U.S. In January 2012, a draft Implementation Strategy was released including a set of interagency actions and milestones focused on enhancing the resiliency of oceans to climate change and ocean acidification. *www.whitehouse.gov/administration/eop/oceans/policy*
REGIONAL	
West Coast Governors Alliance on Ocean Health	The U.S. West Coast Governors launched an Agreement on Ocean Health in 2006 to create a framework for regional collaboration on protection and management of ocean and coastal resources. The Alliance includes a Climate Change Action Coordination Team that is initially focusing on a West Coast assessment of shoreline change and anticipated impacts to coastal areas and communities due to climate change over the next several decades. This effort will inform adaptation to climate change and coastal hazards. *http://www.westcoastoceans.org/*

Table 6-1: Examples of ocean-related climate adaptation frameworks in the U.S. (Continued)

Adaptation project	Description
STATE	
State of California Climate Adaptation Strategy	In 2009, recognizing the need to prepare for climate change, the State of California released their Climate Adaptation Strategy. The Coastal and Ocean Resources Working Group is currently working to implement components of the strategy through assessing the risks of sea-level rise, mapping susceptible transportation areas, and conducting vulnerability assessments, among other actions. *http://resources.ca.gov/climate_adaptation/docs/Statewide_Adaptation_ Strategy.pdf* *http://www.climatechange.ca.gov/adaptation/*
Massachusetts Climate Change Adaptation Report	In response to the state's Global Warming Solutions Act of 2008, the Secretary of Energy and Environmental Affairs and the Massachusetts Climate Change Adaptation Advisory Committee produced the Massachusetts Climate Change Adaptation Report to describe state-level climate impacts, vulnerabilities, and adaptation strategies for key sectors, including Coastal Zone and Oceans. *http://www.mass.gov/eea/air-water-climate-change/climate-change/climate-change-adaptation-report.html* *http://www.mass.gov/eea/docs/eea/energy/cca/eea-climate-adaptation-chapter8.pdf*
NON-GOVERNMENTAL	
Alaskan Marine Arctic Conservation Action Plan for the Chukchi and Beaufort Seas	The Nature Conservancy's Alaska chapter developed the Conservation Action Plan to guide the organization's management and conservation efforts in the region. Climate change is identified in the Plan as the primary threat to the region's natural resources. An expert panel helped guide the selection of primary conservation targets including bowhead whales, ice-dependent marine mammals, seabirds, boulder patch communities, and benthic fauna and fish. Recommendations include promoting adaptation strategies and ecosystem-based management, investing in baseline and long-term data collection, and identifying and protecting climate refugia, among others. *http://www.nature.org/images/summary-of-arctic-cap-2.pdf*
A Climate Change Action Plan for the Florida Reef System (2010-2015)	The Florida Reef Resilience Program, a public-private partnership, released a Climate Change Action Plan in 2010. Florida reefs are subject to non-climatic and climatic stressors that are degrading overall ecosystem resilience. Priority actions identified in the Plan include expanding disturbance response monitoring throughout the Florida Reef Tract, developing a marine zoning plan to address non-climate stressors, decreasing negative user impacts from fishing and diving, mapping areas of high and low resilience to prioritize protection, and restoring resistant reefs. The Plan can be adopted by reef managers into existing management plans. *http://frrp.org/SLR%20documents/FL%20Reef%20Action%20Plan-WEB.pdf*

Section 3). To understand and address these threats, the U.S. Geological Survey's Coral Reef Ecosystem Studies (CREST) (http://coastal.er.usgs.gov/crest/) project is investigating drivers and trends of coral reef ecosystem change. CREST is conducting monitoring and research efforts in National Parks such as the Dry Tortugas, Virgin Islands, and Biscayne and areas of the Florida Keys National Marine Sanctuary. Projects include habitat mapping, assessment of calcification change related to ocean acidification, identification of diseases, and improving understanding of reef responses to sea-level change, among others. This work will improve understanding about coral health, advance the ability to forecast future change, and guide management decisions.

Oyster producers in the Pacific Northwest have been facing persistently low seed survival, in part due to acidified waters. These failures are resulting in low harvest rates and economic impacts to shellfish hatcheries. In response, the Pacific Coast Shellfish Growers Association and partners launched the Emergency Oyster Seed Project (http://www.seafoodchoices.com/whatwedo/profile_pcsgaOA.php) in Washington State to establish monitoring programs in key estuaries, develop solutions for enhancing hatchery production, and identify resilient oyster genotypes. Some hatcheries are already implementing adaptive management practices by coordinating water intake to avoid periods of high acidity.

These efforts represent only a handful of the emerging ocean adaptation activities underway in the U.S. They demonstrate that, although much work remains to be done, actions are taking place across a spectrum of adaptation-related efforts, from risk, impact, and vulnerability assessments, to the development of guidance and tools, to on-the-ground implementation. In general, most ocean-related adaptation activities are still in the phases of improving understanding of risks, impacts, and vulnerabilities; some are in planning phases, while a few are in implementation. However, this also means that we can expect to see more actions in the future as marine managers and decision makers advance through the process of adaptation.

Chapter 7

Sustaining the Assessment of Climate Impacts on Oceans and Marine Resources

Key Findings

1. Sustained assessment of climate impacts on U.S. ocean ecosystems is critical to understanding current impacts, preparing for future impacts, and taking action for effective adaptation to a changing climate.

2. Assessing what is known about past, current, and future impacts is a challenging task for a variety of reasons. Many uncertainties and gaps still exist in understanding about the current and future impacts of climate change and ocean acidification on marine ecosystems.

3. A number of steps could be taken in the near term to address these challenges and advance assessment of impacts of climate change on oceans and marine resources.

   • Identify and collect information on a set of core indicators of the condition of marine ecosystems that can specifically be used to track and assess the impacts of climate change and ocean acidification as well as the effectiveness of mitigation and adaptation efforts over time at regional to national scales.

   • Assess capacity and effectiveness of existing ocean-observing systems to detect, track and deliver useful information on these indicators as well as other physical, chemical, biological, and social/economic impacts of climate change on oceans and marine resources.

   • Increase capacity and coordination of existing observing systems to collect, synthesize, and deliver integrated information on physical, chemical, biological, and social/economic impacts of climate change on U.S. marine ecosystems.

   • Conduct regional-scale assessment of current and projected impacts of climate change and ocean acidification on ocean physical, chemical, and biological components and human uses.

   • Increase data, information, and capacity needed to assess and project impacts of climate change on marine ecosystems.

   • Build and support mechanisms for sustained coordination and communication between decision makers and science providers to ensure the most critical information needs related to impacts, vulnerabilities, mitigation, and adaptation of ocean ecosystems in a changing climate are being met.

- Build and support mechanisms for getting and sharing information and resources on impacts, vulnerabilities, and adaptation of U.S. ocean ecosystems in a changing climate.
- Build and support mechanisms with neighboring countries and other international partners for assessing and addressing impacts of climate change and ocean acidification on marine ecosystems of key interest to the U.S.

7.1 Challenges to Assessing Climate Impacts on Oceans and Marine Resources

Based on current understanding and projections, it is very likely that marine ecosystems under U.S. and other jurisdictions will continue to be affected by climate change through a suite of changes in ocean physical, chemical, biological, social, and economic systems (Doney et al., 2012; Hollowed et al., 2011; see previous sections). These climate-driven changes will have significant impacts on U.S oceans (see Sections 2 and 3), the communities and economies that depend on them (see Section 4), and U.S. international relations (see Section 5).

Despite this foundation of information, many uncertainties and gaps remain in understanding the current and future impacts of climate change and ocean acidification on marine ecosystems (see research gaps articulated in Sections 2, 3, 4, 5, 6 and Doney et al., 2012). Assessing what is known about past, current, and future impacts is a challenging task, but addressing these challenges is critical to preparing for and responding to impacts of climate change on U.S. ocean ecosystems. The following is a list of some of the major barriers and needs encountered in completing the work for this assessment.

- Lack of a clear set of indicators and information on the condition of marine ecosystems that could be used to track and assess: 1) the impacts of climate change and ocean acidification on U.S. oceans and 2) the effectiveness of adaptation efforts over time.
- Limited capacity and coordination of existing observation systems to collect, assess, integrate, and deliver information on physical, chemical, biological, social, and economic impacts of climate change on U.S. marine ecosystems at decision-relevant scales.
- Lack of regional-scale assessments of impacts of climate change and ocean acidification on U.S. ocean ecosystems.
- Lack of assessments and information on the implications of climate change for U.S. international responsibilities, agreements, and interests related to ocean ecosystems.
- Limited capacity for integrating and synthesizing data and information acquired at different levels of space, time, taxonomic resolution, etc. into useful forms for assessing impacts and decision making.
- Large gaps in key data, information, and capacity to assess and project impacts of climate change on marine ecosystems (see previous sections for lists of key science needs).

- Compared to terrestrial and coastal systems, relatively few strategies, tools, and examples for incorporating climate change information into ocean management processes.
- Relatively few ocean adaptation actions implemented to date.

7.2 Steps for Sustained Assessment of Climate Impacts on Oceans and Marine Resources

Sustained assessment of climate impacts on U.S. ocean ecosystems is critical to understanding current impacts, preparing for future impacts, and taking action for effective adaptation to a changing climate. Many key steps are needed to sustain and advance the capacity to assess and respond to climate impacts on ocean and marine resources (Doney et al., 2012; Fautin et al., 2010; Murawski, 2011; Osgood, 2008). The following are some near-term steps that would greatly aid in addressing the challenges for sustaining and advancing future assessments of climate change impacts on oceans and marine resources. This is not intended to be a comprehensive list, and items are not listed in priority order.

- Identify a set of core indicators of the condition of marine ecosystems that can specifically be used to track and assess: 1) the impacts of climate change and ocean acidification and 2) the effectiveness of mitigation and adaptation efforts over time at regional to national scales.

- Assess capacity and effectiveness of existing ocean-observation and monitoring systems to detect, track, and deliver useful information on the physical, chemical, biological, social, and economic impacts of climate change on oceans and marine resources.

- Enhance capacity and coordination of existing observing systems to collect, synthesize, and deliver integrated information on physical, chemical, biological, social, and economic impacts of climate change on U.S. marine ecosystems.

- Conduct regional-scale assessments of current and projected impacts of climate change and ocean acidification on ocean biological, physical, and chemical systems as well as ocean-dependent human systems.

- Increase data, information, and capacity needed to assess and project impacts of climate change on marine ecosystems, including (also see previous sections for lists of key science needs):
    - Information on how the various physical, chemical, biological and ocean use components of marine ecosystems respond to changes in climate and ocean acidification, how these responses are interconnected, and where tipping points or thresholds can be found that could produce major changes in marine ecosystems;
    - Information on past variability in climate and ocean conditions for use in understanding how anthropogenic climate changes may influence future ocean conditions on a variety of temporal and spatial scales;

   – Information and capacity to model and project impacts of climate change on ocean physical and chemical conditions at finer spatial (e.g., regional) and temporal (e.g., decadal) scales most relevant for decision makers; and

      – Information and capacity to couple models and projections of climate change impacts on physical, chemical, biological, and ocean use components of marine ecosystems.

- Build and support mechanisms for sustained coordination and communication between decision makers and science providers to ensure that the most critical information needs are being met related to impacts, vulnerabilities, and adaptation of ocean ecosystems in a changing climate.

- Build and support mechanisms for obtaining and sharing information and resources on impacts, vulnerabilities, and adaptation of U.S. ocean ecosystems in a changing climate.

- Build and support mechanisms with neighboring countries and other international partners for assessing and addressing impacts of climate change and ocean acidification on marine ecosystems of high relevance to the U.S.

A diversity of potential mechanisms could enhance the sustained coordination and assessment of climate impacts on ocean systems. For example:

- Regular communication and interaction between personnel involved with federal ocean and/or climate efforts such as the National Climate Assessment, National Ocean Policy, National Fish, Wildlife, and Plants Climate Adaptation Strategy, and Interagency Climate Change Adaptation Task Force. This would help to ensure that emerging knowledge needs of decision makers are addressed through a coordinated federal science agenda.

- Enlist a core team of experts to continuously work on the ocean climate assessment process for the National Climate Assessment. These individuals could draw in additional experts as needed and membership of the core team could be on a staggered rotational basis.

- Informally convene federal agencies that are developing ocean-related components for their agency adaptation plans mandated under EO13514 to share ideas and lessons learned and to look for opportunities for further collaboration.

- Hold a workshop on ocean climate impacts and adaptation to convene relevant scientists and local, state, tribal, federal, NGO, and private sector practitioners to discuss ocean adaptation challenges, approaches, and opportunities. This effort could potentially lead to a handbook, tool, or online community of practice.

- Improve regional coordination to connect climate information to the large marine ecosystem scale through informal dialogues between relevant regional climate and ocean entities such as NOAA Fisheries Science Centers, NOAA Regional Climate Service Directors, DOI Climate Science Centers, and DOI Landscape Conservation Cooperatives and their marine partners within the regions.

Appendix A

Status of and Climate Change Impacts to Commercial, Recreational, and Subsistence Fisheries in the U.S.

A.1 Commercial and Recreational Fisheries [4]

Commercial fisheries

In 2009, commercial fishermen in the U.S. harvested 7.9 billion pounds of finfish and shellfish. The value of this catch, measured as landings or ex-vessel revenue, was $3.9 billion. The leading species in terms of revenue were shrimp ($378 million), sea scallop (*Placopecten magellanicus*; $376 million), Pacific salmon ($370 million), and walleye pollock (*Theragra chalcogramma*; $308 million). In terms of pounds landed, the leading species were walleye pollock (1.9 billion pounds), menhaden (*Brevoortia tyrannus*; 1.4 billion), and Pacific salmon (705 million), which altogether comprised over half of total pounds landed in 2009. The U.S. seafood industry in 2009 supported approximately 1 million full- and part-time jobs and generated $116 billion in sales impacts, $32 billion in income impacts, and $48 billion in value added impacts. [5]

Landings, revenue, and economic impacts vary widely across individual regions of the U.S. (Figure A-1). For example, commercial fishermen in Alaska caught the most salmon (*Oncorhynchus spp.*; 671 million pounds) and earned $345 million for their catch in 2009. Landings of tuna in Hawai'i totaled 15 million pounds and generated $48 million in revenue. On the East Coast, Maine fishermen landed the vast majority of American lobster (*Homarus americanus*) in 2009, earning $231 million for the 79 million pounds landed. In Massachusetts, sea scallop was a major contributor to total revenue, earning $197 million for 30 million pounds landed. Louisiana harvesters caught more blue crab (*Callinectes sapidus*; 51 million pounds) than any other state, earning over $36 million, and more than half of the U.S. menhaden harvest in 2009 (786 million pounds), earning $43 million in landings revenue. In the Gulf of Mexico, shrimp was a highly valued species, earning $131 million for the 2009 harvest of 90 million pounds by Texas fishermen. Although more pounds of shrimp were landed in Louisiana (114 million pounds), the total landings revenue was less ($121 million). The greatest employment impacts

---

4 Data reported in this subsection (Commercial Fisheries) are documented in NOAA Fisheries, 2010.

5 The seafood industry includes the commercial harvest sector, seafood processors and dealers, seafood wholesalers, and distributors, importers, and seafood retailers.

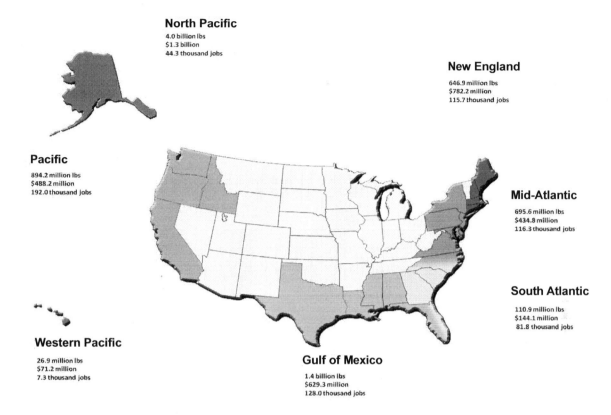

North Pacific
4.0 billion lbs
$1.3 billion
44.3 thousand jobs

New England
646.9 million lbs
$782.2 million
115.7 thousand jobs

Pacific
894.2 million lbs
$488.2 million
192.0 thousand jobs

Mid-Atlantic
695.6 million lbs
$434.8 million
116.3 thousand jobs

South Atlantic
110.9 million lbs
$144.1 million
81.8 thousand jobs

Western Pacific
26.9 million lbs
$71.2 million
7.3 thousand jobs

Gulf of Mexico
1.4 billion lbs
$629.3 million
128.0 thousand jobs

Figure A-1 U.S. Commercial Fisheries by region (2009 landings, revenue, and job impacts) (Source: National Marine Fisheries Service, 2010).

generated by the seafood industry were in California with 121,000 jobs, followed by Massachusetts with 78,000 jobs, Florida with 65,000 jobs, and Washington with 58,000 jobs. The lowest number of jobs was supported in Delaware, which had only 407 seafood-industry jobs. The import sector generated the greatest job impacts in California, Massachusetts, Florida, and Washington.

Recreational fisheries[6]

In 2009, approximately 11 million recreational anglers operated across the U.S. Approximately 9.4 million of these anglers were residents of a U.S. coastal county and 1.7 million anglers were residents of a non-coastal county. These anglers took 74 million saltwater fishing trips, spending $4.5 billion on those trips as well as $15 billion on durable fishing-related equipment. These expenditures contributed $50 billion in sales to the U.S. economy and supported over 327,000 jobs. Recreational anglers most often caught sea trout (*Cynoscion regalis*; 44 million fish) and Atlantic croaker and spot (*Micropogon undulatus* and *Leiostomus xanthurus*, respectively; 36 million fish).

---

6 Data reported in this subsection (Recreational Fisheries) are documented in NOAA Fisheries, 2010.

The Gulf of Mexico was the destination for 2.8 million anglers, who took 22 million trips there in 2009, with the Mid-Atlantic (2.6 million anglers, 17 million trips) and the South Atlantic (2.4 million anglers, 19 million trips) the next most popular regions. The Pacific (1.8 million anglers, 6.3 million trips), New England (1.4 million, 7.5 million trips), North Pacific (284,000 anglers, 914,000 fishing days), and the Western Pacific (246,000, 2.2 million trips) regions followed in terms of total anglers.

Anglers fishing in the Mid-Atlantic caught most of the Atlantic croaker (15 million fish) and summer flounder (*Paralichthys dentatus*; 24 million), while New England anglers caught most of the striped bass (*Morone saxatilis*; 9 million) in 2009. Most sea trout (35.5 million) were caught in the Gulf of Mexico. In the North Pacific region, salmon (Chinook (*Oncorhynchus tshawytscha*), chum (*Oncorhynchus keta*), coho (*Oncorhynchus kisutch*), pink (*Oncorhynchus gorbuscha*), and sockeye and Pacific halibut (*Oncorhynchus nerka* and *Hippoalossus stenolepis*, respectively) were the most commonly caught species or group in 2009 with 1.1 million fish and 761,000 fish caught, respectively. Rockfish and scorpion fish (2.7 million fish), mackerel (*Scomber scombrus*; 2 million fish), barracuda (*Sphyraena barracuda*), white sea bass (*Atractoscion nobilis*) and bonito (*Sarda chiliensis*) (1.6 million fish) were caught in high numbers in the Pacific region, while bigeye tuna (*Thunnus obesus*) and mackerel (1.1 million fish) were the most numerous fish caught by anglers in the Western Pacific.

A.2 Commercial and Recreational Fishing-Dependent Communities [7]

U.S. fishing-dependent communities and ports are located in coastal areas that span a wide range of climate zones. Many important fishing-dependent communities are located within the North Pacific arctic and polar zones as well as the temperate middle latitudes that characterize the New England and Mid-Atlantic regions, most of the Pacific region, and some of the South Atlantic region. The southern third of California, the coastal areas of the South Atlantic region's states of South Carolina, Georgia, and Florida are all subtropical as are the coastal areas of all the states in the Gulf of Mexico region. The tip of the Florida Keys, the Caribbean region, and the Western Pacific region are in the tropics. These differences affect local and regional fisheries.

The nation's top commercial fishing-dependent communities and ports, defined as the top ten commercial landings locations per state, range from subareas of major metropolitan centers such as Houston, Texas (pop. 1,953,631), San Diego, California (pop. 1,223,400), Honolulu, Hawai'i (pop. 876,156), and Jacksonville, Florida (pop. 735,617), to small villages such as Winter Harbor, Maine (pop. 988), Naknek, Alaska (pop. 678), La Push, Washington (pop. 371), Wachapreague, Virginia (pop. 236), and Valona, Georgia (pop. 123). These 222 top fishing-dependent communities in the U.S. have an average poverty rate of 10.1 percent, just above the national rate of 9.2 percent. Poverty rates in top commercial fishing-dependent communities range from 0 percent in Valona, Georgia to 33.7 percent in Crescent City, California, with the majority of communities falling

---

7 Data reported in this subsection (Commercial Fishing-dependent communities) are documented in NOAA Fisheries, 2009a.

between 2 percent and 10 percent. These fishing-dependent communities also exhibit a wide range in other important social characteristics. For example, the percentage of residents in fishing-dependent communities five years of age or older who speak a language other than English at home range from 0 percent of Crescent, Georgia's residents and 1 percent in Bowers Beach, Delaware to 87 percent in Brownsville, Texas and 93 percent in Ni'ihau, Hawai'i. Twenty-two percent, or 48 of 222, of the top fishing-dependent communities in the U.S. had a higher rate of such residents than the national rate. The national median household income was $42,000 according to the 2000 U.S. Census; top commercial fishing-dependent communities had median household incomes that ranged between $18,000 in Crisfield, Maryland and $146,755 in Darien, Connecticut. Thirty-eight percent, or 84 of 222, of the top commercial fishing-dependent communities in the U.S. had a higher median income than the national median, although this was not always primarily due to income from fishing.

A.3 Regional Involvement in Commercial and Recreational Fishing

Although described to some extent in the previous sections, a more detailed description of the socio-economic impacts and climate change implications of commercial and recreational fisheries in each region of the U.S. are provided below.

North Pacific

Three large marine ecosystems exist in the waters off Alaska: the Gulf of Alaska (GOA), the Bering Sea–Aleutian Islands (BSAI), and the Arctic ecosystems. The first two currently support major fisheries, and fishing is currently excluded in the third. The GOA is exposed to the ocean environment of the central North Pacific Ocean. The dominant circulation in the GOA is characterized by the cyclonic flow of the Alaska Gyre. The Bering Sea is a semi-enclosed high-latitude sea and its broad continental shelf is one of the most biologically productive areas in the world. An important feature is ice coverage over most of its eastern and northern continental shelf during winter and spring.

The main groundfish species harvested in the BSAI fisheries are walleye pollock (*Theragra chalcogramma*), Pacific cod (*Gadus macrocephalus*), yellowfin sole (*Limanda aspera*), northern rock sole (*Lepidopsetta polyxystra*), and Atka mackerel (*Pleurogrammus monopterygius*). Other species comprise a minor fraction of the total catch. Total groundfish catch in the BSAI region is limited to two million tons under the BSAI Groundfish Fishery Management Plan. In the GOA, the dominant species harvested are walleye pollock, Pacific cod, flatfishes, Pacific ocean perch (*Sebastes alutus*), various species of rockfish, and sablefish (*Anoplopoma fimbria*).

Groundfish fisheries in the Bering Sea and GOA are among the most productive in the world. From Hiatt et al. (2010), the commercial groundfish catch off of Alaska totaled 1.5 million tons of retained catch in 2010 with an ex-vessel value of $636 million, which accounted for 43 percent of the weight, and 14 percent of the ex-vessel value, of total U.S. domestic landings. The gross value after primary processing of the 2010 groundfish catch off of Alaska was approximately $1.9 billion, an increase of 11 percent from 2009. The groundfish fisheries accounted for the largest share (41 percent) of the ex-vessel value of all commercial fisheries off Alaska in 2010, while the Pacific salmon fishery was second with $506 million or 32 percent of the total Alaska ex-vessel value. The value of the shellfish fishery amounted to $206.3 million or 13 percent of the total for Alaska and exceeded the value of Pacific halibut by about $5.8 million.

Walleye pollock has been the dominant species in the commercial groundfish catch off Alaska. The 2010 pollock catch of 888,500 metric tons accounted for 56 percent of the total groundfish catch of 1.6 million tons. However the stock has gone through major changes recently. The total allowable catch (TAC) for Eastern Bering Sea pollock, the largest component, was close to 1.5 million metric tons from 2004-2007. It was reduced sharply in 2008 to 1.0 million metric tons and further reduced in 2009 and 2010 to a little over 800 thousand metric tons, but the final TAC for pollock in the Eastern Bering Sea for 2011 and 2012 was expanded to 1.2 million metric tons.

Haynie and Pfeiffer (2012, pers. comm.) found that although pollock fishing has shifted northward in recent years (2006-2009), the northward shifts are associated with colder than average years in the Bering Sea. A large ice and cold pool extent concentrates fish populations in the northern region of the fishing grounds, giving fisherman in the north an advantage over those in the south. The redistribution has occurred in the summer pollock fishery, but the winter Pollock fishery, which is driven by the pursuit of valuable roe-bearing fish that spawn in the southern part of the Eastern Bering Sea, has seen little redistribution of effort. This large difference in value per fish in the roe fishery means that harvesters are unlikely to shift to the north for marginal increases in catch.

Uncertainty remains as to the impacts of warming temperatures on the distribution of pollock fishing, because the relationship between climate and biomass complicates both retrospective analyses and predictions. Decreases in pollock biomass have been associated with warmer temperature regimes (Mueter et al., 2011), but the effects on fishable abundance are lagged by several years. For the time periods used in Haynie and Pfeiffer's (2012a, b) research, warm periods did not correspond to years with low fishable biomass. This complicates the separation of the direct ice and cold pool effects from the effects of climate on abundance, which is necessary for prediction. Their work highlights the importance of considering the economic, institutional, and ecological characteristics of a fishery to improve understanding of the effects of climate change on fisheries.

The 2010 catch of flatfish, which includes yellowfin sole (*Pleuronectes asper*), rock sole (*Pleuronectes bilineatus*), and arrowtooth flounder (*Atheresthes stomias*), was 291,800 tons or 18.3 percent of the total 2010 groundfish catch (Hiatt et al., 2011). The Pacific cod catch in 2010 accounted for 250,300 tons or 16 percent of the total 2010 groundfish catch, up about 9 percent from a year earlier. Pollock, Pacific cod, and flatfish comprised just under 90 percent of the total 2010 catch. Other important species are sablefish (*Anoplopoma fimbria*), rockfish (*Sebastes* and *Sebastolobus spp.*), and Atka mackerel (*Pleurogrammus monopterygius*).

Pacific halibut (*Hippoglossus stenolepis*) are managed by the International Pacific Halibut Commission separately from other groundfish species. The fishery has been managed under an Individual Transferable Quota system since 1995. The species is found throughout the coastal waters of the GOA and BSAI and concentrate in the central GOA between Kodiak Island and the Kenai Peninsula. In 2009, fishermen landed 57.7 million pounds of halibut valued at $134.6 million in Alaska, following only pollock, salmon, and crab landings in overall value of Alaskan commercial fisheries (NOAA Fisheries,

2010, page 20). In addition, a growing sport fishery with both private anglers and charter businesses targets halibut. In 2009, recreational anglers in the state kept or released 486,000 halibut.

Climate variability has been shown to affect the life history and distribution of Pacific halibut. During the 20th century, dramatic and persistent changes in the growth and recruitment of the species have occurred that cannot be readily explained by changes in stock size (Clark et al., 1999). Recent work has strongly suggested that halibut recruitment is driven primarily by the Pacific Decadal Oscillation (PDO) (see Section 2 for details) (Clark and Hare, 2002). The PDO has alternated between positive, or productive for halibut, and negative, or unproductive, phases every 25 to 35 years (Mantua et al., 1997).

Three species of king crab: red (*Paralithodes camtschatica*), blue (*P. platypus*) and golden or brown (*Lithodes aequispina*), and two species of Tanner crab, Tanner crab (*Chionoecetes bairdi*) and snow crab (*C. opilio*), have traditionally been harvested commercially off Alaska. Until 1967, Japanese and Russian fisheries dominated Bering Sea landings, but those fisheries were phased out by 1974 (Low, 2008). In the Bering Sea, domestic catches peaked in 1980 and then dropped precipitously in 1981. In the GOA, catches peaked in 1965 and varied at a relatively low level for the next decade before dropping lower still in 1983. Almost all GOA king crab fisheries have been closed since 1983. For Tanner crab, the 1965-1975 period was a developmental phase for the fisheries. Catch peaked in 1979 and then declined to 1984. After 1984, the catch increased, reaching an all-time high in 1991, and then decreased to 1997, when the Tanner crab fishery was closed. Abundance trends for Bering Sea stocks indicate that the Tanner crab stock declined from a relatively high level in the late 1970s to a low level in 1985. The stock recovered and then declined again subsequent to 1989 and is currently at a low level. Snow crab rebounded sharply from a low level in 1985, producing large catches in the 1990s, and then declined in 1999 to low levels (Low, 2008).

In 2007-2009, the most recent reporting year available, the Bristol Bay red king crab fishery produced an annual average of nearly 12 million pounds of finished products and was estimated to have generated real first-wholesale revenues of about $107 million dollars per year (all values in 2009 U.S. dollars; Garber-Yonts and Lee, 2011). In addition, the Eastern Bering Sea (EBS) snow crab fishery produced an annual average of about 26 million pounds of finished products and was estimated to have generated first-wholesale revenues of about $117 million per year. Catches are restricted by quotas, seasons, and size and sex limits, with landings limited to large male crabs. Bering Sea crab fisheries were rationalized during 2004-2005. Fishing seasons are set at periods of the year to avoid molting, mating, and soft-shell periods to both protect crab resources and maintain product quality. Because of poor biological conditions and other factors, the status of North Pacific crab stocks is currently vulnerable.

Recruitments of red king crab were estimated to have been relatively high during the 1970s but then declined sharply and have remained low since about 1985. In particular, recruitments were estimated to have been extremely low in the past five years (NPFMC, 2011, page 168). The EBS snow crab population was declared overfished by NOAA Fisheries in 1999, and a rebuilding plan with a maximum ten-year rebuild period was adopted. In 2009, NOAA Fisheries determined that EBS snow crab did not make adequate progress towards rebuilding within the specified rebuilding period and, therefore would not be rebuilt within the maximum ten-year period. EBS Tanner crab is another economically important fishery that has experienced dramatic swings in abundance and catch. The federal Tanner crab fishery was closed between 1997-2004 because of severely depressed stock conditions (NPFMC, 2011, page 301), and, based on the 2010 stock assessment, this stock was determined by NOAA Fisheries to be overfished (NPFMC, 2011, page 285).

The Aleutian Islands golden king crab fishery is also commercially important, with average annual finished products of more than two and a half million pounds and estimated first-whole revenues of almost $17 million per year. A population dynamics model is under development for this fishery to determine whether the stock is overfished and to establish catch limits under mandated harvest regimes. Catch in this fishery has not exceeded the overfishing limit in recent years, but an allowable biological catch was not established until this year. Overall, status of this stock will be extremely uncertain until a population dynamics model is available for an assessment.

Blue king crab stocks are in a depressed state and and cannot be commercial fished currently. Blue king crab population in the Pribilof Islands area have been considered overfished since the mature biomass fell below threshold levels in 2002 (Low, 2008). Abundance of mature biomass continued to decrease in 2004 and is now the lowest on record. Little or no recruitment is apparent in the population, which has been declining continuously since 1995. Continued warm conditions in the surrounding waters may be contributing to the decline. Blue king crab popoulations around St. Matthew Island are also considered overfished. That population has declined steeply since 1998 (Low, 2008).

The eastern Bering Sea Tanner crab population was high in the early 1980s and from 1988 to 1992 (Low, 2008). The population has been low since then and currently continues to decrease due to low recruitment. The mature biomass is below threshold level and the stock is therefore considered overfished. The fishery has been closed since 1996 (Low, 2008).

The mature biomass of EBS snow crab was moderate to high in the early 1980s and from 1987 to 1997 (Low, 2008). The biomass declined sharply from 1998 to 1999, and the stock is considered overfished. Snow crab recruitment was higher during 1979-1987 than in other years (Low, 2008). Low recruitment estimates since 1988 could be due to fishing, climate, and/or a northward shift in snow crab distribution (Low, 2008).

Most crab population fluctuations off Alaska are caused by recruitment variability, which is driven primarily by variability in the egg and larval stages when survival of the crabs is determined by oceanic circulation. Zheng and Kruse (2006) examined the effects of environmental factors on recruitment of six crab stocks in the Eastern Bering Sea. Among the crab stocks, Bristol Bay red king crab recruitment trends appeared to relate most strongly to decadal shifts in physical oceanography, particularly the Aleutian Low Pressure Index, hypothesized to affect food availability for these crab larvae (Zheng and Kruse, 2006). Loher (2001) hypothesized that changes in near-bottom temperatures associated with the 1976/77 regime shift were causes for spatial shifts of red king crab female distributions. These regime shifts slowed juvenile growth due to relatively low temperatures in the north (Stevens, 1990), thus affecting recruitment strength. Orensanz et al. (2004) proposed a similar hypothesis to explain spatial changes for snow crab. In addition, correlations exist between Tanner crab year classes and warm spring sea surface temperatures (Rosenkranz et al., 1998, 2001).

Five species of Pacific salmon exist in the Gulf of Alaska and Bering Sea and Aleutian Islands: pink (*Oncorhynchus gorbuscha*), sockeye (*O. nerka*), chum (*O. keta*), coho (*O. kisutch*), and Chinook (*O. tshawytscha*). Each species spends most of its ocean life in the offshore, pelagic environment, with brief periods of migration through coastal areas as juveniles and returning adults. A scenario with oceanic and atmospheric warming includes increases in snow melt and water flows including fall/winter floods, which would affect salmon eggs laid in gravel beds (Low, 2008). In the summer, higher average summer temperatures could diminish the oxygen content of the water in streams where smolts live. Warmer temperatures could also affect migration of smolts and lead to timing mismatches with their zooplankton prey base, timing that is critical as they enter salt water. Salmon survival is likely determined early in its marine life when mortality is high. Actual impacts at these various life phases are difficult to quantify because these fish occupy an exceptionally wide latitudinal and geographical range. Alaskan salmon production in commercial fisheries increased from the mid-1970s to the late 1990s and has been at or near historical highs recently. This may be attributed to changes in oceanic and atmospheric conditions that increased wild fish survival and hatchery success (Low, 2008).

The main areas where Pacific herring (*Clupea pallasii*) are harvested commercially include Prince William Sound (PWS), southeast Alaska, and the Togiak district in the Bering Sea, areas that have high recruitment variability. Pacific herring recruitment in PWS experiences high variability as a result of at least three main types of factors: large-scale environmental factors, smaller-scale environmental factors, and diseases such as viral hemorrhagic septicemia virus and parasite *Ichthyophonus hoferi* (Low, 2008; Moffitt, 2004). Additionally, the spawning biomass of Pacific herring in southeast Alaska has been variable. Since 1980, seven of the nine primary locations have exhibited long-term trends of slightly increasing biomass with major fluctuations of both increasing and decreasing biomass around these long-term trends (Low, 2008). Decadal-scale variability of recruitment is not apparent in the three widely-recognized climate regimes of 1978-1988, 1989-1998 and post-1998 in the North Pacific (Low, 2008). However, Togiak herring recruitment events are variable. A large-year class occurs once every 9 or 10 years, and these events drive the Togiak herring population. The larval stage of herring life history is the most important factor for determining year-class strength to which environmental conditions are critical, with temperature having the strongest influence on recruitment (Low, 2008; refer to Section 3 for more details).

West coast

The Pacific region includes California, Oregon, and Washington. Federal fisheries in this region include Pacific coast groundfish, Pacific coast salmon, coastal pelagic species, and West coast highly migratory species. The Pacific groundfish and salmon fisheries are subject to weak stock management in which access to the harvestable surplus of healthier stocks is often restricted to protect the weaker stocks with which they co-mingle in the ocean. These weaker stocks include eight rebuilding groundfish stocks and salmon listed under the Endangered Species Act as well as other non-listed stocks that also constrain the fishery. Salmon management is further complicated by the need to ensure equitable allocation of harvest among diverse user groups, including tribal governments that have treaty rights to harvest salmon, and to coordinate with other entities that have jurisdiction over other aspects of salmon management.

In 2009, commercial fishermen in the Pacific region landed roughly 894 million pounds of finfish and shellfish (NOAA Fisheries, 2010, page 29). Hake and squid had the highest annual landings in the Pacific region in 2009, with 253 million pounds and 204 million pounds, respectively. Although they accounted together for 51 percent of the total landings in the Pacific region, they accounted for only 14 percent of the $488 million in total landings revenue generated in 2009. Landings revenue was dominated by other shellfish ($129 million) and crab ($124 million). In terms of pounds landed, California contributed the most (372 million pounds), followed by Oregon (198 million pounds) and Washington (164 million pounds). Washington had the highest landings revenue in the region, with $228 million in 2009, followed by California ($150 million) and Oregon ($102 million). The Pacific region's seafood industry generated $20 billion in sales impacts in California, $1.1 billion in sales impacts in Oregon, and $7.3 billion in sales impacts in Washington. California also generated the largest value added income and employment impacts ($4.3 billion; $7.1 billion; 121,000 jobs) (NOAA Fisheries, 2010, page 29).

In 2009, almost 1.8 million recreational anglers took 6.3 million fishing trips in the Pacific region (NOAA Fisheries, 2010, page 30). Over 64 percent of these anglers were residents of a regional coastal county. In terms of the Pacific region's key species and species groups, rockfishes and scorpion fishes (2.7 million fish), mackerel (2 million fish), barracuda, bass and bonito (1.6 million fish) and surf perches (1.5 million fish) were the most often caught by anglers in 2009. Most of the rockfishes and scorpion fishes in the Pacific region were caught in California, while most of the salmon and other tunas were caught in Washington and Oregon. Employment impacts from recreational fishing expenditures in California were the highest in the region with almost 14,000 full- and part-time jobs generated in that state, with Washington (3,300 jobs) and Oregon (1,600 jobs) following. In addition to employment impacts, the contribution of recreational fishing activities to Pacific region's economy can be measured in terms of sales impacts and the contribution of these activities to gross domestic product (value-added impacts). In 2009, sales impacts were also the highest in California ($2 billion in sales impacts), followed by Washington ($347 million) and Oregon ($168 million) (NOAA Fisheries, 2010, page 30).

Pacific sardines (*Sardinops sagax*) off the West Coast of North America form three subpopulations or stocks: a northern subpopulation (northern Baja California, Mexico, to Alaska); a southern subpopulation (off Baja California); and a Gulf of California, Mexico subpopulation. The northern stock inhabits the California Current Ecosystem (CCE) and is a transboundary stock, occupying the EEZs of Mexico, the U.S., and Canada. The stock is fished and managed independently by each country in an arrangement that has worked reasonably well recently, while sardines have been relatively abundant in the CCE. The productivity of the sardine stock in the CCE is strongly influenced by climate variability on both inter-annual and decadal scales. Observed decadal-scale climate variability, such as the PDO (Mantua et al., 1997), affects both the sardine stock's abundance and its geographic range (Emmett et al., 2005; Herrick et al., 2006; Rodriguez-Sanchez et al., 2002); therefore, the PDO is probably the best indicator of how the Northeast Pacific commercial sardine fisheries might be impacted by global climate change (Figure A-2).

Relationships between climate and productivity of the Pacific sardine stock (e.g., SST anomalies vs. sardine abundance; Figure A-3; Hill et al., 2000) have been previously examined (Herrick et al., 2007; Norton and Mason, 2003, 2004, 2005). During a warm regime, biomass of Pacific sardine increases. Conversely, a cold-water regime results in a decrease in abundance of the sardine stock and reduces its distribution to an area almost exclusively off Southern California, U.S., and Baja California. When sardine disappear from the U.S. EEZ under cold-water conditions, anchovy, herring, mackerel, and market squid become more prevalent, filling the void left by sardine both in terms of directed harvests and forage. This pattern was documented for pre-fishery times using sediment samples from the Southern California Bight (Baumgartner et al., 1992). Similarly, the U.S. fishery shows a 70-year pattern of sardine boom, bust, and boom during the 20th century (Norton and Mason, 2003, 2004, 2005).

In response to fluctuations in productivity due to climate variability, the U.S. Pacific sardine fishery is managed with an environmentally-based harvest control rule that determines the annual harvest level (Figure A-4). The harvest control rule is intended to prevent overfishing, sustain consistent yield levels (Herrick et al., 2006), and reduce the exploitation rate if stock biomass decreases or if ocean conditions become cooler and less favorable for the stock (Figure A-4). For sardine, the harvest control rule formula is based on a three-year average of SST observed at the Scripps Institution of Oceanography pier in La Jolla, California (Hill et al., 2011). Including this factor reflects the positive relationship between sardine reproductive success and water temperature; at higher SSTs, a greater fraction of the available biomass can be harvested. However, recent work by McClatchie et al. (2010) finds that the Scripps Institution of Oceanography SST series is no longer valid for predicting sardine reproductive success and should not be used in sardine conservation and management. In its statement on Pacific sardine management for 2012 to the Pacific Fishery Management Council (PFMC) in November 2011, the management team recognized that, although sardine reproductive success is clearly related to environmental conditions, the temperature relationship underlying the harvest control rule needed to be revised. At this time, the PFMC has yet to consider how this finding should be incorporated into its Coastal Pelagic Species Fishery Management Plan.

In general, the motivation for an environmentally-based harvest-control rule is that fishing mortality associated with maximum sustainable yield and the productivity of the sardine stock have both been shown to increase when relatively warm-ocean conditions persist. The use of environmentally-based rules has advantages for responding to climate change.

- Environmentally-based rules can avoid destabilizing feedbacks that can cause a fishery to collapse; an example is the infamous collapse of the California sardine fishery in the 1940s, which has been attributed to a combination of climatic conditions and overfishing (Herrick et al., 2006).

- The use of well-defined harvest control rules reduces uncertainty for fishermen and processors and thus provides direct efficiency gains and promotes incentives for conservation.

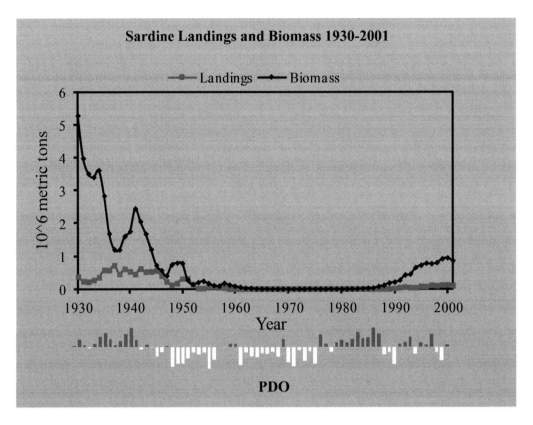

Figure A-2 Pacific sardine yield, abundance, and the Pacific Decadal Oscillation (PDO) through 2001. Sardine productivity is higher during the warm, positive phase of the PDO, and the collapse of the sardine stock that began in the 1940s has been attributed to a regime shift around 1945 (Source: M. Dalton, pers. comm.).

- Environmentally-based harvest control rules can potentially incorporate other ecosystem interactions, including the availability of important forage species like Pacific sardine.

   Given the general lack of existing research and modeling capabilities on socio-economic impacts of climate change on fisheries, the impacts on Pacific sardine fisheries are uncertain. To this end, two scenarios have been considered out the many potential scenarios that could occur in the Eastern Pacific. The first assumes that the increase in ocean temperature due to climate change will be consistent with that experienced during a warm water regime in the CCE, which is favorable to sardine productivity. This results in an increase in the northern sardine stock biomass in the U.S. EEZ off of the West Coast, particularly off of the Pacific Northwest, increased biomass in Canada's EEZ, and, depending on the extent of ocean warming, potentially harvestable biomass in the U.S. EEZ off of Alaska (Figure A-5). This is essentially the status quo scenario, with perhaps a slight increase in U.S. Pacific sardine fishing opportunities followed by a corresponding increase in economic activity and economic value.

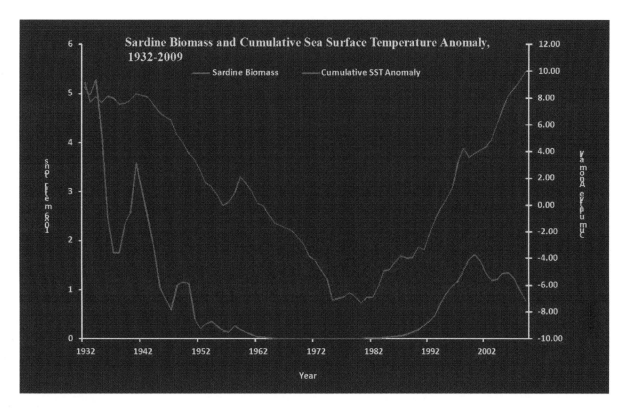

Figure A-3 Pacific sardine abundance and cumulative SST anomalies observed at the Scripps Institution of Oceanography pier in La Jolla, California (Source: S. Herrick, pers. comm.).

The second scenario assumes that an increase in ocean temperature results in a northerly expansion of the entire subtropical marine biota. This would include an increase in abundance and a northerly shift of all Pacific sardine stocks, allowing sardine fisheries, and potentially fisheries of species that prey on sardines, to expand along the entire Northeast Pacific coast. This scenario is suggestive of what has been observed during ENSO events in the eastern Pacific; for example, dring an ENSO event, tropical tunas, which U.S. sardine vessels are capable of harvesting, typically become more available off Southern California. Under these circumstances, fishing opportunities would increase significantly for U.S. vessels that target Pacific sardine as would fishing-related economic activity and economic value.

Northeast

The Northeast region is often divided into New England (Maine to Connecticut) and the Mid-Atlantic (New York to Virginia). In the Northeast, commercial landings in 2009 were 1.3 billion pounds, valued at $1.2 billion. Landings revenue in New England was dominated by American lobster ($298 million) and Atlantic sea scallop ($210 million). These species represented 65 percent of total landings revenue but only 20 percent of total landed pounds (NOAA Fisheries, 2010, page 50). Mid-Atlantic landings revenue

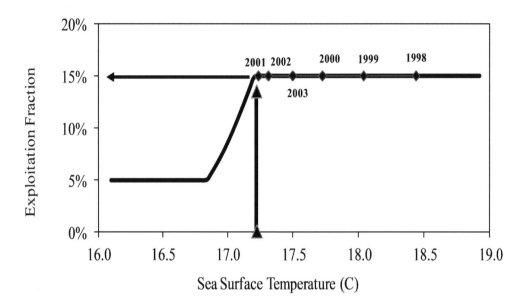

Figure A-4 Pacific sardine harvest-control rule implements a decreasing exploitation fraction in cool years based on a 3-year moving average of sea surface temperatures (SST) at Scripps Pier, San Diego, California. 'Harvest' is the guideline harvest level in metric tons (mt), 'Biomass' is current biomass estimate, 'Cutoff' is the lowest level of estimated biomass at which harvest is allowed (150,000 mt), and 'Fraction (SST)' is the environmentally-based percentage of biomass above the cutoff that can be harvested (Source: M. Dalton, pers. comm.).

came largely from sea scallop ($162 million) and blue crab ($85 million), comprising 57 percent of total landings revenue but only 15 percent of total landed pounds (NOAA Fisheries, 2010, page 74). In terms of overall economic impacts, in 2009, the seafood industry in the Northeast region was responsible for 232 thousand jobs, $25 billion in sales, $6 billion in income, and $16 billion in value added from such activities as processing (NOAA Fisheries, 2010, pages 55 and 78).

In terms of pounds landed, Virginia contributed the most (426 million pounds), followed by Massachusetts (356 million pounds) and Maine (185 million pounds). Massachusetts had the highest landings revenue in the region with $400 million in 2009, followed by Maine ($286 million) and Virginia ($153 million). The Northeast region's seafood industry generated $1.2 billion in sales impacts in Maine, $6.7 billion in Massachusetts, $651 million in New Hampshire, $906 million in Rhode Island, $57 million in Delaware, $1.6 billion in Maryland, $5.8 billion in New Jersey, 5.3 billion in New York, and $1.7 billion in Virginia (NOAA Fisheries, 2010). Massachusetts also generated the largest value-added income and employment impacts ($1.7 billion; $2.6 billion; 77,820 jobs) (NOAA Fisheries, 2010, page 63).

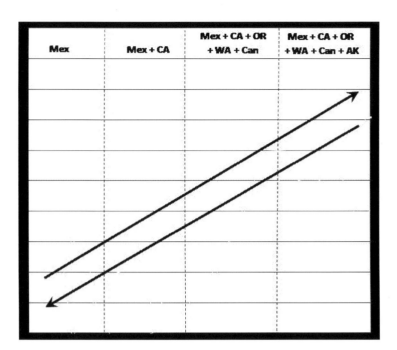

Figure A-5 The expected changes in the harvestable biomass and distribution of the northern Pacific sardine stock due to climate change. The harvestable biomass increases (lower left to upper right) and decreases (upper right to lower left, heavy arrows) depending on the duration of favorable climate conditions. The vertical lines show which fisheries become involved as the harvestable biomass increases and declines: left, at lowest harvestable biomass, is a Mexican (Mex)-only fishery; middle left is Mex and California (CA) fisheries; middle right is Mex, CA, Oregon (OR), Washington (WA) and Canadian (Can) fisheries; and the far right is Mex, CA, OR, WA, Can and Alaskan (AK) fisheries. (Source: Sam Herrick, NOAA).

For commercial fishermen in the Northeast and elsewhere, however, fishing is more than a job or an income; it is a way of life. Studies show that fishermen value the independence and risk-taking afforded by fishing (Apostle et al., 1985; Gatewood and McCay, 1990; Pollnac and Poggie, 1988; Smith and Clay, 2010) and may even subsidize fishing with income from another job (Veltre and Veltre, 1983, pages 185-193) or by using kin as crew, which they often subsidise by providing lesser incomes to individual crewmembers (Doeringer et al., 1986, page 47). In the Northeast, ethnic groups that most commonly use kin as crew include the Sicilians of Gloucester, Massachusetts, the Portuguese of New Bedford and Provincetown, Massachusetts, and small, inshore vessels, especially from rural areas such as Downeast Maine and areas of Maryland and Virginia. Household needs such as debt and cultural norms of equity can also influence both the degree to which fishermen maximize profits (Davis, 1991; Durrenberger, 1997) and the type of fishing, especially day versus trip fishing, that fishermen choose to undertake. Especially while their children are young, or if they desire stronger involvement in community groups and activities, they may choose small boat, inshore fishing over larger, trip vessels, even if this means earning a more modest living (Maurstad, 2000). Festivals celebrating fishing, such as Blessings of the Fleet, are common throughout the Northeast and involve the entire community, not just fishermen and their families (see and NOAA Fisheries, 2009b for examples from the Northeast).

Recreational finfish anglers also value fishing for multiple reasons (Holland and Ditton, 1992), including relaxation and following in the tradition of their parents. Sportfishing is a cultural subset that revolves around tournaments (see NOAA Fisheries, 2009b for examples in the Northeast). Recreational fishing can also mask subsistence fishing, potentially based on ethnicity, gender, or location (Toth and Brown, 1997). In the

Northeast, Steinback et al. (2009) found that 28 percent of presumed recreational anglers in fact fished for reasons other than recreation, including food or income, though fewer than 3 percent stated that they fished all or mostly for food or income. Overall, 54 percent of all interviewed anglers, whatever their motivations for fishing, ate at least some of their catch. Those who fished in part for food or income were also statistically more likely to collect non-finfish marine resources such as shellfish, squid, seaweed, or kelp (Steinback et al., 2009, page 54).

Atlantic cod, primarily found from Maine to Connecticut is likely to experience negative impacts from climate change, while Atlantic croaker, primarily found from New York to Virginia, is likely to experience positive impacts. In 2009 (NOAA Fisheries, 2010), Atlantic cod and haddock (*Melanogrammus aeglefinus*) accounted for $36.6 million and 30.8 million pounds of commercial landings in New England. Of this, 92.1 percent was landed in Massachusetts (Figure A-6), 7.3 percent in New Hampshire, and 0.6 percent in Maine. Total cod plus haddock value was only 4.7 percent of all landed value in New England in 2009 and 4.8 percent of landed pounds, but the groundfish fishery as a whole is the cornerstone of the Northeast region's fishing industry, with the greatest number of vessels of any fishery in the region (1,347 limited access permits and 900 active vessels) in fishing year 2010 (Kitts et al., 2011). Cod is also an important recreational fishing species; in 2009 (NOAA Fisheries, 2010), 483,000 cod were harvested by anglers in New England, with an additional 1.1 million cod captured but then released.

Cod are sensitive to increases in ocean temperature (Fogarty et al., 2008), although the level of impact varies. Some research suggests that certain prey species may not move in synch with cod, creating further difficulties in the Northeast region (Murawski, 1993). If cod move north, likely off of Georges Bank and potentially even completely out of the Gulf of Maine (Fogarty et al., 2008; Nye et al., 2009), then U.S. commercial fishermen will need to substitute other species because they cannot follow the cod north into Canadian waters.

Even where climate change has a positive effect on a species, social and economic impacts can be felt. Atlantic croaker are found from the Gulf of Maine to Argentina (ASMFC, 2011), but in the Northeast, their primary range is from Hudson Canyon off the coasts of New York and New Jersey south to Cape Hatteras, North Carolina (Hare et al., 2010). Croaker is not one of the top commercial species, representing only 0.7 percent of total Mid-Atlantic landings revenue and 2 percent of Mid-Atlantic landed pounds in 2009 (NOAA, 2010), but it is one of the key recreational species, with 15 million fish caught, second only to summer flounder (*Paralichthys dentatus*) in the Mid-Atlantic.

Spawning stock biomass (SSB) estimates for croaker have exhibited a cyclical trend based on environmental conditions (ASMFC, 2007). As of 2010, croaker was coming off a peak and moving into a low biomass status (ASMFC, 2012). Hare et al. (2010, page 452) estimate that as ocean temperatures warm, "[a]t current levels of fishing, the average (2010-2100) spawning biomass of the population is forecast to increase by 60-100 percent. Similarly, the center of the population is forecast to shift 50-100 km northward. A yield analysis, which is used to calculate benchmarks for fishery management, indicates that the maximum sustainable yield will increase by 30-100 percent." These estimates were based on three scenarios of CO_2 emissions into the 21st century, the first being the commitment scenario that makes projections based on a fixed CO_2 concentration of 350

Figure A-6   Gloucester, MA Stern Trawler (Source: http://www.photolib.noaa.gov/htmls/fish0533.htm).

ppm, the second is the B1 scenario that assumes an increase in CO_2 to 550 ppm, and the third is the A1B scenario that assumes an increase in CO_2 to 720 ppm (IPCC, 2007b). With rising ocean temperatures, the fishery is expected to leave the Southeast region but remain in the Mid-Atlantic and expand into southern New England, through Connecticut, Rhode Island, and Massachusetts. Montane and Austin (2005) also found that hurricanes coincide with spikes in juvenile croaker recruitment.

Commercial fishermen in the Mid-Atlantic will likely be able to increase their landings of croaker, while fishermen further north will gain a new species, croaker, with a high biomass, potentially raising the importance of croaker as a commercial species. Further, Mendelsohn and Markowski (1999) and Loomis and Crespi (1999) note that recreational fishing and boating activities may significantly increase with warming. Tournament fishermen in New England (see NOAA Fisheries, 2009a, for examples of New England tournaments) will likely be able to add croaker to their repertoire. Subsistence fishermen in southern New England may add croaker to their landings, providing an additional species at a time when they may be able to collect fewer local shellfish (Steinback et al., 2009) because shellfish are expected to be negatively impacted by increasing ocean acidification.

Pacific Islands

In Hawai'i, commercial landings in 2009 were 27 million pounds of finfish and shellfish, valued at $71 million. Landings of tuna comprised 67 percent ($48 million) of the landings revenue and 54 percent of total landings (15 million pounds; NOAA Fisheries, 2010, page 100). Swordfish ($7.3 million, 3,881 pounds), mahi mahi ($2.9 million, 1,287 pounds), moonfish ($2.4 million, 1,884 pounds), and marlin ($2.1 million, 1,678 pounds) also contributed to the region's overall landings. In terms of overall economic impacts, in 2009, the seafood industry in Hawai'i was responsible for 14 thousand jobs, $1.3 billion in sales, $369 million in income, and $546 million in value added from such activities (NOAA Fisheries, 2010, pages 100 and 120).

Recreational fishing is undertaken throughout the state and adds value in the local economy. In Hawai'i, 2.2 million trips were taken by private anglers in 2009. Recreational fishing in the region was responsible for 4,286 jobs, $460.8 million in sales, $228.6 million in value added from the industry, and $150.9 million in income (NOAA Fisheries, 2010, page 47). The most fished species by anglers in Hawai'i included bigeye, mackerel, and goatfishes.

Southeast

The Southeast region is often divided into the South Atlantic, which is comprised of areas from North Carolina down to the tip of the east coast of Florida, and the Gulf of Mexico, which contains the area from the west coast of Florida to Texas. In the Southeast, commercial landings in 2009 were 1.5 billion pounds, valued at $773.4 million. Landings revenue in the South Atlantic was dominated by blue crab ($35.3 million) and shrimp ($32.7 million). These species represented 47 percent of total landings revenue and 51 percent of total landed pounds (NOAA Fisheries, 2010, page 100). Gulf landings revenue came overwhelmingly from shrimp ($324.7 million); however, although shrimp comprised 51 percent of total landings revenue, it only comprised 17.4 percent of total landed pounds (NOAA Fisheries, 2010, page 120). The next most economically-important species in the Gulf were oysters ($72 million or 11.5 percent) and menhaden ($60.6 million or 9.6 percent). In terms of overall economic impacts, in 2009, the seafood industry in the Southeast region was responsible for 210 thousand jobs, $32.8 billion in sales, $6.5 billion in income, and $11.2 billion in value added from such activities as processing (NOAA Fisheries, 2010, pages 100 and 120). In addition, Adams et al. (2009) reported that the Gulf region supported 20,470 commercial fishing craft, which was approximately one-third of the nation's commercial fleet in 2003, and contains a quarter of the U.S. seafood processing plants and wholesale establishments that generate approximately 25 percent of the total value of domestically-processed fisheries products. Finally, Yoskowitz (2009) estimated that the total productive value of the Gulf of Mexico to the U.S., as defined as the market value of resources extracted from, or value of services generated as a result of, proximity to the Gulf, at nearly $76 billion, assuming crude oil is $28.50 per barrel, including nearly $700 million from fisheries in 2003.

In terms of pounds landed, Louisiana contributed the most by far (1 billion pounds), followed by Mississippi (230 million pounds) and Texas (100 million pounds). Louisiana also had the highest landings revenue in the region with $284 million in 2009, followed by Florida ($157 million) and Texas ($150 million). The Southeast region's seafood industry generated $13 billion in sales impacts in Florida, $1.7 billion in Texas, $1.7 billion in Louisiana, $1 billion in Georgia, $700 million in North Carolina, $400 million in Alabama, $300 million in Mississippi, and $70 million in South Carolina (NOAA Fisheries, 2010).

Florida also generated the largest value-added income and employment impacts ($2.5 billion; $4.3 billion; 64,744 jobs) (NOAA Fisheries, 2010, page 102).

Recreational fishing is also important to the economy of the Southeast. In the South Atlantic, 19.1 million trips were taken by private anglers in 2009. In addition, recreational fishing in the region was responsible for 51,314 jobs, $5.6 billion in sales, $1.8 billion in value added from the industry, and $2.9 billion in income (NOAA Fisheries, 2010, page 101). The most fished species by anglers in the South Atlantic included Atlantic spot and croaker, spotted seatrout, and blue fish. In the Gulf, 22.3 million trips were taken by private anglers in 2009. In addition, recreational fishing in the region was responsible for 92,241 jobs, $9.9 billion in sales, $3.3 billion in value added from the industry, and $5.1 billion in income (NOAA Fisheries, 2010, page 121). In addition, charter and party boat operations in particular have "generated $149.5 million in economic output, $68.5 million in incomes, and 3,487 jobs withing the overall Gulf of Mexico regional economy" (see references cited in Adams et al., 2009, page 39). In 2004, the Gulf of Mexico was estimated to have supported 7.3 million recreational anglers who took 25.6 million fishing trips that represent approximately one-third of U.S. totals. Non-residents took about half of these trips, indicating the importance of these resources to all U.S. residents (Adams et al., 2009). Anglers in the Gulf overwhelmingly target spotted seatrout, although red drum and sand and silver seatrouts are also caught in substantial numbers.

Economic benefits generated from coastal recreation extend beyond the costs associated with traveling. These non-market values, according to Kildow et al. (2009, page 75), "reflect important changes in the net economic value of [Gulf] coastal resources and can result from changes in resource access, availability, or quality." Using a benefits transfer method, Kildow et al. (2009) estimate the non-market economic value of coastal and ocean recreational fishing in Texas, Louisiana, Alabama, and Mississippi to be between $2.2 billion and $2.8 billion in 2004.

In 2009, Kildow et al. (2009, page 47) examined the changing coastal ocean economies of the U.S. Gulf of Mexico and noted that "stresses on the rich natural resources of this special area have been intensifying for many years … and the economies of the Gulf states are inextricably linked to the quality and values of the Gulf's natural resources," which stand to be significantly affected by climate change.

Changes in the Gulf of Mexico marine environment have the potential to change the mix of available species; therefore, fishermen that rely on a portfolio of species will be affected. Perruso et al. (2005) examined optimal species-targeting strategies of the long-line fleet in the Gulf of Mexico and found that the joint production behavior affected targeting strategies as a result of potential area closures. Should climate change affect the distribution of species or development of marine protected areas, these results would mimic those estimated from examining the effect of climate change directly. In addition, with respect to the value generated from commercial fisheries, Yoskowitz (2009, pages 25–26) notes, "The fisheries industry in particular faces several challenges. In the northern Gulf a zone of low oxygen, or hypoxia, is tied to pollution from the Mississippi River [which is likely to be exacerbated due to climate change].… It forces shrimp and certain fish into areas that may be harder and more expensive for fishing boats to reach, and it also has devastating effects on the ecosystem."

A.4 Subsistence Fisheries

For generations, subsistence harvesting of marine resources has been important to rural communities around the U.S. where it makes an important contribution to limited livelihoods. In addition to providing basic nutrition and sustenance, subsistence activities are also recognized by native peoples as supporting spiritual values, community well-being, family structures, and resource conservation.

Based on the numerous physical and biological changes that are occurring in the marine environment due to climate change (see Sections 2 and 3), people who rely on subsistence fisheries are expected to have to deal with a number of impacts. The technology needed to support marine subsistence activities, such as motors, boats, fishing gear motors, snow machines, firearms, ammunition and fuel, is expensive for low-income people to maintain (Callaway et al., 1999). In addition, the distribution of fish stocks such as salmon and marine mammal populations in the North Pacific, which subsistence and other fishermen rely on, have altered or are expected to alter significantly in the coming years. Projected ecosystem shifts are likely to displace or change the resources available for subsistence, requiring communities to change their practices or move (Gerlach et al., 2011).

Perhaps most importantly, subsistence resource users are typically the least mobile constituency of all marine resource users due to their strong, place-based cultural identities and/or limited incomes. Partly because of this, many subsistence resource users fall within categories listed in Executive Order 12898 (59 FR 7629, 1994), "Federal Actions to Address Environmental Justice in Minority Populations and Low-Income Populations." As species distributions shift, subsistence users will have to find new sources of nutrition and are therefore likely to be the most impacted group of marine resource users.

The majority of impacts on subsistence fisheries in the U.S. are expected to occur in Alaska and the Arctic, but impacts are also expected to occur on subsistence fisheries that are not as well documented. The following subsections describe what is known about subsistence fishing in each of the major regions of the U.S. and the known and anticipated effects of climate change on those fisheries.

North Pacific

Rural livelihoods in Alaska are tightly connected to climate, weather, and ecosystems. Northern people have relied for millennia on the landscape for their food through a variety of subsistence activities including hunting, herding, gathering, fishing, and small-scale gardening. The importance of wild fish, whether anadromous species such as salmon or non-anadromous species such as whitefish (*Caulolatilus princeps*) is the notable constant from south to north in rural Alaska. (For more background on subsistence harvest in Alaska, see Wolfe, 2004, Nelson, 1986, Nelson, 1969, and Norris, 2002).

Subsistence fisheries in Alaska are the only officially-documented subsistence fisheries in the U.S. In general, fish, including salmon, halibut, and other species, provide the most substantial contribution to subsistence diets throughout rural Alaska. However, regions vary in important ways; for instance, communities in Western Alaska are much more focused on harvesting fish for subsistence while communities in Northern Alaska are more reliant on harvesting marine mammals.

Reliance on salmon for subsistence harvests is significant throughout the state, both in coastal communities that fish for salmon in the marine environment and inland communities that fish the rivers that flow from Alaska's interior (Figure A-7). The most recent data available about subsistence salmon harvests in Alaska show that, in 2008, 23,780 subsistence salmon-fishing permits were issued and an estimated total of 989,266 salmon, including all five species of salmon, were caught for subsistence use by families in 177 communities around the state (Fall et al., 2011).

Through the Subsistence Halibut Registration Certificate (SHARC) program administered by NOAA Fisheries, residents of rural Alaskan communities have been fishing for halibut for subsistence purposes since the SHARC program was initiated in 2003. As of 2009, residents of 77 Alaskan communities actively fished for halibut and reported having caught 44,989 halibut weighing a total of 851,579 pounds (Fall and Koster, 2011). The majority of subsistence halibut harvests occur in Southeast and Southcentral Alaska, where halibut is important to the diets of residents in small rural communities (Figure A-7).

Between 2002 and 2007, residents of 56 Alaskan communities reported having fished for other non-salmon fish species for subsistence purposes (Figure A-7). During that time frame, a total of 810,757 fish (1,404,217 pounds) were reported to have been caught both in coastal regions and on freshwater rivers in the state. The majority of non-salmon-fish marine harvesting occurs at the mouths of rivers in Western Alaska, on Kodiak Island, and in southeast Alaska. In addition, between 2000 and 2007, residents of 32 Alaskan communities reported having fished for marine invertebrates for subsistence purposes. The Alaska Department of Fish and Game (ADFG) reported that, during this time period, 32,478 invertebrates of multiple marine species totaling 83,033 pounds were caught (ADFG, 2011). The majority of subsistence fishing for marine invertebrates has been documented on Kodiak Island, the southern tip of the Alaska Peninsula, and in Southcentral Alaska.

Harvest of marine mammals throughout Alaska is equally important to Alaska Native communities, especially in Western Alaska (Figure A-8). In 2010, an estimated 13 polar bears (*Ursus maritimus*), 678 walruses (*Odobenidae divergens*) and 969 sea otters (*Enhydra lutris kenyoni*) were taken for subsistence purposes (USFWS, 2011). The most recent data available on beluga whale (*Delphinapterus leucas*) harvests for subsistence show that 222 beluga whales were taken in 2006 (Frost and Suydam, 2010). As seen in Figure 4, the majority of walrus and polar bear harvests occur on St. Lawrence Island, the Seward Peninsula, and in Barrow. Beluga whale harvests mainly occur in Western Alaska and in the community of Point Hope in the Arctic. The majority of sea otter harvests occur in Southeast Alaska, but significant harvests also occur on Kodiak Island and in the communities of Cordova and Valdez.

Other subsistence fisheries

Although anecdotal accounts of subsistence fisheries that occur in other regions of the U.S. exist, NOAA Fisheries and state-level fisheries management agencies do not disaggregate subsistence fishing from recreational fishing. The following provides an overview of what is known about both traditional and non-traditional subsistence fishing in the West Coast, New England, the Pacific Islands, and the Gulf of Mexico.

West coast

Native Americans have harvested marine resources along the West Coast for centuries. Historically, marine species comprised a large portion of the diet of coastal settlements (NOAA Fisheries, 2009a). 50 federally-recognized tribes live in the region (29 in Washington, 10 in Oregon and 11 in California), many of which are culturally and/or economically dependent upon marine resources, specifically Pacific salmon. Of these, 20 have been included under treaties with the U.S. that preserve their rights to harvest marine resources (NOAA Fisheries, 2009a).

As exemplified by coastal Washington State's Makah Nation, coastal Northwest tribes harvest and make use of a diverse range of marine resources. Makah tribal members have identified over 80 species consumed or used in subsistence practices, including primarily marine and anadromous animal resource categories (Sepez, 2001). Moreover, when surveyed, 99 percent of Makah households participated at least annually in subsistence activities (Sepez, 2001). The principal marine resources that Northwest Tribes rely on are coho, chinook, sockeye, and chum salmon; however, tribal communities rely on a vast array of marine resources for subsistence near their reservations or traditional use areas. Tribal communities farther to the north rely more heavily on shellfish harvests such as Dungeness crab (*Cancer magister*), shrimp, geoduck (*Panopea abrupta*), butter clams (*Saxidomus giganteus*), and manila clams (*Corbicula manilensis*). In central Puget Sound, tribes rely more heavily on salmon harvests but also pursue shellfish for subsistence purposes. Tribes in the southern reaches of Puget Sound rely primarily on chum salmon harvests as well as geoduck and crab (Impact Assessment, 2005).

Salmon are at the heart of cultural values within most if not all Northwest Tribes. The cultural context and values of each tribe are intimately tied to salmon harvests and consumption. In addition, members of many tribes collect salmon for ceremonial purposes and for community functions. However, as seen for subsistence harvests elsewhere, little incentive is offered to tribes to report their use of subsistence harvests and mechanisms for reporting are often absent. These realities lend themselves to absent or underreported subsistence harvest data through the states' formal reporting programs. Figure A-10 depicts the total annual salmon harvest for the members of the Northwest Indian Fisheries Commission, including both commercial and subsistence harvests. Based on interviews conducted with tribal members, tribal members are estimated to retain approximately 25 percent of their total catch for personal consumption (Impact Assessment, 2005).

Over the last two decades, a dramatic reduction has been seen in the overall number of salmon taken by Northwest Tribes for subsistence (Figure A-9). Lower species abundance is considered to be one of the main factors in this decline (Impact Assessment, 2005). Salmon abundance is expected to decrease between 20 and 40 percent by 2050 as changes occur in the biological and physical environments off the coast of Washington and Oregon (Battin et al., 2007). Given that so many tribes depend on salmon for their high fat content and large size, as well as the central position of salmon within Northwest tribal cosmology and culture, it is likely that a significant decline in salmon stocks in the Northwest U.S. could dramatically affect the overall health and well-being of tribal communities (Colombi, 2009).

In addition to finfish and shellfish, the Makah have traditionally harvested a variety of marine mammals and seabirds, and the tribe has made a concerted effort over the last

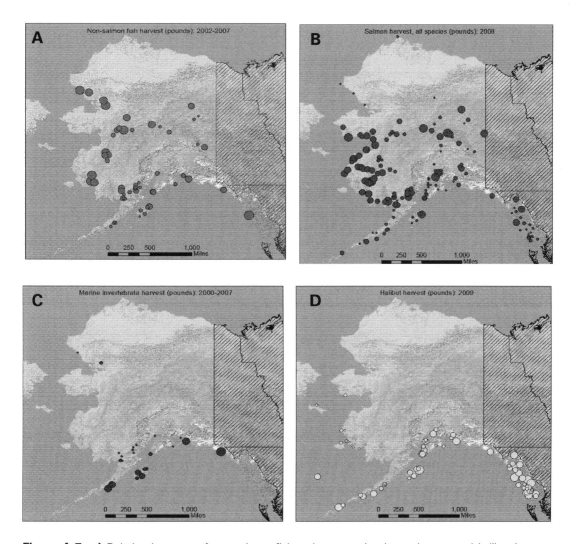

Figure A-7a–d Relative harvests of non-salmon fish, salmon, marine invertebrates, and halibut in subsistence reliance communities around Alaska. Points on the maps represent species group harvests in individual communities. The point size varies based on the relative harvest of each community's residents compared to all other communities harvesting that species group in the state. (Source: Amber Himes-Cornell, NOAA).

decade to formally revive marine harvest treaty rights and the whaling traditions central to its culture (Firestone and Lilley, 2004).

In addition to the subsistence harvests of marine resources by tribal entities, subsistence harvests by non-tribal individuals on urban and rural coastal piers throughout the U.S have been documented , particularly along the West Coast (CIC Research, 2004; Environmental Health Coalition, 2005; Sechena, 2003). Many of these subsistence fishers are a part of low-income, immigrant, and minority populations and therefore fall under the purview of the Environmental Justice Initiative of Executive Order 12898 (Pitchon and Norman, under review). Such subsistence-fishing practices are readily observed in

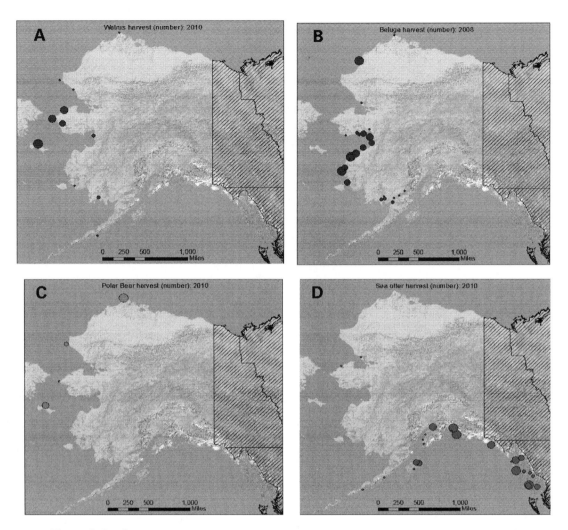

Figure A-8a–d Relative harvests of walrus, beluga whales, polar bears, and sea otters in subsistence reliance communities around Alaska. Points on the maps represent species group harvests in individual communities. The point size varies based on the relative harvest of each community's residents compared to all other communities harvesting that species group in the state. (Source: Amber Himes-Cornell, NOAA).

West Coast urban centers, including cities like Seattle, San Francisco, San Diego, and Los Angeles, but no comprehensive data collection effort at the federal or state level exists to document these activities. Extant data on non-tribal subsistence and personal use fishing comes primarily from independent case studies conducted in urban centers (CIC Research, 2004; Pitchon and Norman, under review). For example, a survey of pier fishers in Los Angeles County revealed that, as opposed to purely recreational fishing, 27 percent of surveyed pier fishers were fishing in order to consume what they caught, and these individuals identified food security issues within their households (Pitchon and Norman, under review).

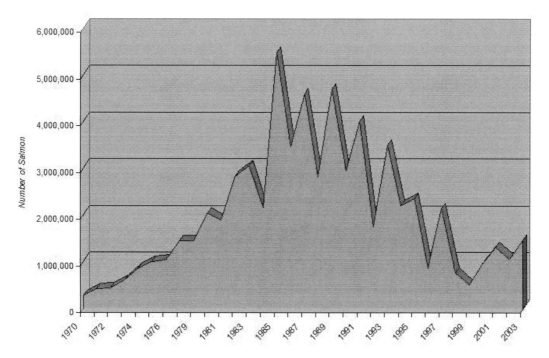

Figure A-9 Total number of salmon taken by Northwest Indian Fisheries Commission tribes (Source: Impact Assessment, 2005).

Northeast

Unlike in the Pacific Northwest, none of the federally-recognized tribes in the Northeast U.S. have fishing rights under treaty. In addition, subsistence use of marine resources is not specifically regulated or systematically documented. Limited research has been conducted regarding the extent of subsistence fishing that occurs in the region. The most recent research highlights the relationship between fishing by saltwater anglers for recreational purposes and behaviors traditionally characterized as subsistence uses such as home consumption and giving away self-caught catch to friends and relatives (Steinback et al., 2009). Steinback et al. (2009) also note that a general lack of knowledge about the size of the angling population who specifically keep fish for personal consumption as compared to those who fish solely for recreation. Given this and the fact that, with few exceptions, selling recreationally-caught fish is illegal, subsistence fishing activities in the Northeast are likely highly under-estimated.

Steinback et al. (2009) estimate that 28 percent of saltwater anglers who live in coastal counties of the Northeast U.S. fish for marine resources for personal consumption or for supplemental income on at least some of their fishing trips. Of these anglers, an estimated 33 percent rely on their catch as a cost-saving food source or as a supplement to household income. In addition, of those anglers who fish primarily for recreation (72 percent), half reported using at least some of their catch for personal consumption.

Steinback et al. (2009) estimated that approximately 783,000 saltwater anglers in the Northeast U.S. relied, at least partially, on self-caught fish for food, 127,000 relied on self-caught fish all or mostly for food, and 105,000 relied on self-caught fish to supplement their income, for a total of 1.22 million anglers who rely to some extent on their recreational catch for food or income. Overall, they estimated that 69.8 percent of those anglers who fish all or mostly for food or income rely to some degree on self-caught catch for personal consumption versus 54.2 percent of all anglers; however, although many anglers report that they personally consume or share their catch with their social network, the results did not show that extensive reliance on self-caught marine resources by anglers in the Northeast. Some anglers, however, do appear to rely more heavily on self-caught marine resources, but the subset who do tended to have lower household incomes, lower education levels, and were more likely to characterize themselves as a minority.

Steinback et al. (2009) also noted that approximately 12 percent of anglers in the Northeast collect non-finfish marine resources such as shellfish, squid, and seaweed, with the percentage being higher (16.5 percent) among those who fished for finfish all or mostly for food or income. The top five species of finfish caught by all anglers in 2005 were estimated to be striped bass (*Morone saxatilis*), summer flounder (*Paralichthys dentatus*), bluefish (*Pomatomus saltatrix*), Atlantic croaker (*Micropogonias undulatus*), and black sea bass (*Centropristis striata*), which together constitute about 75 percent of all consumed species.

Pacific Islands

Nearly all Hawai'i shoreline fishermen are considered subsistence fishermen because they tend to keep their catch for home consumption rather than selling it, but in Hawai'i, as in other places, clearly distinguishing boat-based subsistence fishing from other types of fishing, such as recreational and commercial, is difficult. Many, if not most, small-boat fishermen participate in several of these types of fishing depending on the year, season, or even the trip.

In Hawai'i, recreational fishing is focused on food (Glazier, 2002; Miller, 1996). One reflection of this is that Hawai'i typically has the highest ratio of fish kept to fish released among all states having marine recreational fisheries (NOAA Fisheries, 2011). Fishermen who define themselves as recreational fishermen keep their catch for both personal use and sharing with friends and family, and many will sell a portion of their catch after a good trip to help cover their fishing expenses. In a 2006 study of 400 Hawaiian recreational fishermen (Allen and Duffield, 2011), nearly 40 percent reported that subsistence, defined as fishing for food for one's self, family, friends, or community, was a purpose for all of their fishing trips. An additional 27 percent said it was a motivation for most of their trips, 25 percent for some of their trips, and less than 10 percent said that none of their trips were motivated by subsistence purposes.

Even small-boat commercial fishing has subsistence aspects because some fish, up to one-third of the catch from a trip (Glazier, 2002), are typically taken home or shared with friends and family. Hospital et al. (2011) reported that 38 percent of pelagic fish caught by commercial fishermen is not sold and that 97 percent of fishermen sampled said they participate in fish-sharing networks with friends and relatives. More than 62 percent considered the fish they catch to be an important source of food for their family.

A study conducted by the Molokai Subsistence Task Force (1994) found that 28 percent of respondents' food was acquired through subsistence activities, which increases to 38 percent for Hawaiian families. Nearly everyone believed that subsistence was important to the Molo'kai lifestyle, meeting many other purposes in addition to food procurement, including exercise, recreation, time spent in nature, a sense of environmental kinship, and a means of enhancing family and community cohesion. Subsistence harvest is a basis for sharing and gift-giving as both a process of reciprocity and providing resources to people who cannot obtain them on their own.

On Molokai especially but also on the other Hawaiian islands, subsistence resources continue to provide a source of food for a variety of celebrations and events, further cultivating a sense of communal identity and enhancing social networks. The importance of fishing for food is also strong on Molokai, which, along with Hawai'i, is under the jurisdiction of the Western Pacific Regional Fishery Management Council. The small-boat fleets in these island areas consist primarily of vessels that would meet many definitions of artisanal or subsistence fishing.

Fishermen in the Marianas Islands are also known to keep a significant portion of their catch for personal consumption. A 2011 survey of Marianas archipelago commercial fishermen (Hospital and Beavers, 2012) reported that, for the Commonwealth of the Northern Marianas Islands, a vast majority of the pelagic fishermen (84 percent), the bottomfish fishermen (89 percent), and the reef fishermen (92 percent) said the fish they caught were an important source of food for their families. Consistent with that result, a higher proportion of their catch from the last 12 months was consumed at home (29 percent) or given away to crew, family members, or friends and neighbors (33 percent) than was sold (25 percent) and an additional portion (8 percent) was described as being caught for fiestas or other community and cultural events. Even the highliners, those who catch the largest amount fish, reported that 19 percent of their catch was consumed at home, 24 percent was given away, and 8 percent was caught for fiestas or other community or cultural events.

When asked why they went fishing, Rubinstein's (2001) sample of offshore fishermen on Guam revealed three motivations. The predominant motivation reported by 65 percent of respondents was the personal enjoyment derived from fishing; for some, this enjoyment, as reported especially by the indigenous Chamorros and other Micronesians, included a heightened sense of cultural identity. A second motivation (18 percent) was consumption of fish by the family. The final motivation (16 percent) was income derived from fishing. More than half (51 percent) of the respondents claimed multiple motivations, and, frequently, respondents who indicated that recreation was their primary motivation said they provided fish to family and friends. Nearly all fishermen (96 percent) reported that they share fish regularly, giving fish to family (36 percent), friends (13 percent), or both (47 percent).

Based on creel surveys of Guam fishermen in the mid-1980s, about one-quarter to one-third of the inshore catch was sold while the remainder entered noncommercial channels (Knudson, 1987). Reef fish continue to be important for social obligations such as fiestas and food exchanges with friends and families; one study found a preference for inshore fish species in noncommercial exchanges of food (Amesbury and Hunter-Anderson, 1989). Fishing in Guam continues to be important in terms of contributing to the subsistence needs of the Chamorro population and in preserving their history and identity (Allen and Bartram, 2008). Fishing has assisted Chamorros and immigrant cultures to keep alive what remains of the maritime attributes of their traditional culture and maintaining their connection to the sea and its resources. High value is placed on sharing one's fish catch with relatives and friends, a social obligation that includes part-time and full-time commercial fishermen (Amesbury and Hunter-Anderson, 1989).

Nearshore fishing in American Samoa is largely for subsistence. A 2005 survey of 425 people from 34 villages found that most respondents felt subsistence fishing was an important use of coral reef resources and that 55 percent of respondents fished for subsistence to some degree (Kilarski et al., 2006). Subsistence fishing activity has declined on the reefs over the past two decades (Sabater, 2007) due to societal changes including a shift away from a subsistence economy coupled with a shifting of dietary preferences to canned and frozen fish, poultry, and meat. A trend of decreasing reliance on local fish as a food source is one that might be expected from a society that has been undergoing a shift from a subsistence-oriented economy to a cash economy (Levine and Allen, 2009); however, fishing events such as the annual *atule* and *palolo* harvests continue to organize and mobilize many villages. Brookins and Sabater (2007) reported that the current coral reef fishery has characteristics of a sustainable fishery based on the available data (low effort, stable CPUE, low catch, expanding fish populations).

Southeast

As with the other regions outside of Alaska, although we know that various types of subsistence fishing take place in the Southeast, they cannot be disaggregated from recreational fishing statistics and therefore cannot be quantified. That said, Native American groups in the Southeast are known to have relied on inshore marine resources for subsistence purposes for the last few centuries and continue to maintain strong ties with subsistence livelihoods today (NOAA Fisheries, 2009a). Coastal counties and parishes throughout the Southeast maintain bayou fishing camps and have integrated subsistence fishing into their cultural lifestyles, given the historical abundance of seafood in the region (Gardner, 1989). The central species fished for include red snapper, crawfish, shrimp, and crab species (Gardner, 1989; NOAA Fisheries, 2009b).

References

Aaheim, A. & Sygna, L. 2000. Economic Impacts of climate change on tuna fisheries in Fiji Islands and Kiribati. CICERO Report 2000(4), Oslo, Norway.

Abreu, M.H., Pereira, R., Sousa-Pinto, I. & Yarish, C. 2011. Ecophysiological studies of the non-indigenous species *Gracilaria vermiculophylla* (Rhodophyta) and its abundance patterns in Ria de Aveiro lagoon, Portugal. *European Journal of Phycology* **46**, 453-464.

Acclimatise 2009a. *Understanding the investment implications of adapting to climate change - oil and gas.* Oxford, U.K.: Acclimatise and Climate Risk Management Ltd.

Acclimatise 2009b. *Building Business Resilience to Inevitable Climate Change. Carbon Disclosure Project Report 2008. Global oil and gas.* Oxford, U.K.: Acclimatise and Climate Risk Management Ltd.

Acheson, J. M. 2003. *Capturing the Commons: Devising Institutions to Manage the Maine Lobster Industry.* Hanover and London: University Press of New England.

Adger, N. W., 2003. Social capital, collective action, and adaptation to climate change. *Economic Geography* 79 (4): 387-404.

Agostini V.N., Hendrix, A.N., Hollowed, A.B., Wilson, C.D., Pierce, S. & Francis, R.C. 2007. Climate and Pacific hake: a geostatistical modeling approach. *Journal of Marine Systems* **71**, 237-248.

Agostini, V.N., Francis, R.C., Hollowed, A.B., Pierce, S., Wilson, C., & Hendrix, A.N. 2006. The relationship between hake (*Merluccius productus*) distribution and poleward sub-surface flow in the California Current System. *Canadian Journal of Fisheries and Aquatic Sciences* **63**, 2648-2659.

Alaska Department of Fish and Game (ADFG). 2011. Community Subsistence Information System (CSIS). ADFandG Division of Subsistence. Data compiled by Alaska Fisheries Information Network for Alaska Fisheries Science Center, Seattle. Retrieved February 2011, from http://www.adfg.alaska.gov/sb/CSIS/.

Alekseev, G., Danilov, A., Kattsov, V., Kuz'mina, S. & Ivanov, N., 2009. Changes in the climate and sea ice of the Northern Hemisphere in the 20th and 21st centuries from data of observations and modeling. Izvestiya *Atmospheric and Oceanic Physics* **43**, 675-686.

Alexander, R.B., Smith, R.A., Schwarz, G.E., Boyer, E.W., Nolan, J.V., & Brakebill, J.W. 2008. Differences in phosphorus and nitrogen delivery to the Gulf of Mexico from the Mississippi River Basin, *Environmental Science and Technology* **42**, 822-830.

Allen, S. and P. Bartram. 2008 . Guam as a Fishing Community. 2008. Pacific Islands Fisheries Science Center Administrative Report H-08-01, Feb. 2008.

Allen, S. D. and J. Duffield. 2011. Modeling the behavior of recreational blue marlin anglers in the Western Pacific. Presentation at North American Association of Fisheries Economists, May 13, Honolulu, HI.

Allison, E. H., Badjeck, M.-C. and Meinhold, K. 2011. The Implications of Global Climate Change for Molluscan Aquaculture, in Shellfish Aquaculture and the Environment (ed S. E. Shumway), Wiley-Blackwell, Oxford, UK. doi: 10.1002/9780470960967.ch17

Allison, G.W., Gaines, S.D., Lubchenco, J. & Possingham, H.P. 2003. Ensuring persistence of marine reserves: catastrophes require adopting an insurance factor. *Ecological Applications* **13**, S8-S24.

Altizer, S., Harvell, C.D. & Friedle, E. 2003. Rapid evolutionary dynamics and disease threats to biodiversity. *Trends in Ecology and Evolution* **18**, 589-596.

Alvarez-Filip, L., Dulvy, N.K., Gill, J.A., Côté, I.M. & Watkinson, A.R. 2009. Flattening of Caribbean coral reefs: region-wide declines in architectural complexity. *Proceedings of the Royal Society B* **276**, 3019-3025.

AMAP. 2008. Arctic oil and gas 2007. Oslo: Arctic Monitoring and Assessment Program, 40 pp.

Amesbury, J. R., and R. L. Hunter-Anderson. 1989. Native Fishing Rights and Limited Entry in Guam. Prepared for Western Pacific Regional Fishery Management Council, Honolulu. Micronesian Archaeological Research Services, Guam.

AMSA. 2009. Arctic Marine Shipping Assessment 2009 report. Akureyri, Iceland: Protection of the Arctic Marine Environment.

Anand, G., *Science.* **283**, 2077-2079 (1999). and winners in coral reefs acclimatized to elevated carbon dioxide concentrations. Nature Climate Change 1: 165-169.

Andersen, N., & Malahoff, A. 1977. *The Fate of Fossil Fuel CO_2 in the Oceans.* 749 pp. Plenum, NY.

Anderson, D. 2012. Harmful algae. http://www.whoi.edu/website/redtide/home. Accessed January 10, 2012.

Anderson, D.M. 1989. Toxic algal blooms and red tides. In *Red Tides: Biology, Environmental Science, and Toxicology*, T. Okaichi et al. (eds.). New York: Elsevier, pp. 11-16.

Anderson, D.M. 2009. Approaches to monitoring, control and management of harmful algal blooms (HABs). *Ocean and Coastal Management* **52**, 342-347.

Andrews, T., P. M. Forster, O. Boucher, N. Bellouin, and A. Jones (2010), Precipitation, radiative forcing and global temperature change, Geophys. Res. Lett., 37, L14701, doi:10.1029/2010GL043991.

Anestis, A., Pörtner, H.O. & Michaelidis, B. 2010b. Anaerobic metabolic patterns related to stress responses in hypoxia exposed mussels *Mytilus galloprovincialis. Journal of Experimental Marine Biology and Ecology* **394**, 123-133.

Anestis, A., Pörtner, H.O., Karagiannis, D., Angelidis, P., Staikou, A. & Michaelidis, B. 2010a. Response of *Mytilus galloprovincialis* (L.) to increasing seawater temperature and to marteliosis: metabolic and physiological parameters. *Comparative Biochemistry and Physiology Part A* **156**, 57-66.

Anonymous. n.d. *Color Massachusetts: Teachers Lesson Plan*. Published by Published by William Francis Galvin, Secretary of the Commonwealth, Citizen Information Service, One Ashburton Place, Room 1611, Boston, Massachusetts 02108. Retrieved January 17, 2012, from http://www.sec.state.ma.us/cis/cispdf/color_mass_teachers_guide.pdf.

Anthony, K.R.N., Kline, D.I., Diaz-Pulido, G., Dove, S. & Hoegh-Guldberg, O. 2008. Ocean acidification causes bleaching and productivity loss in coral reef builders. *Proceedings of the National Academy of Sciences USA* **105**, 17442-17446.

Apostle, R., L. Kasdan, and A. Hanson. 1985. Work Satisfaction and Community Attachment among Fishermen in Southwest Nova Scotia. *Canadian Journal of Fisheries and Aquatic Sciences* 42(2): 256-267.

Archer, C.L. & Caldeira, K. 2008. Historical trends in the jet streams. *Geophysical Research Letters* **35**, L08803.

Arctic Council. 2009. Arctic Marine Shipping Assessment Report. Exhibit 31 AEWC and ICAS.

Armstrong, D.A., Wainright, T.C., Jensen, G.C., Dinnel, P.A. and Anderson, H.B. 1993. Taking refuge from bycatch issues: red king crab (*Paralithodes camtschaticus*) and trawl fisheries in the eastern Bering Sea. Canadian Journal of Fisheries and Aquatic Sciences, 50, 1993-2000.

Arzel, O., England, M., de Verdière, A.C. & Huck, T. 2011. Abrupt millennial variability and interdecadal-interstadial oscillations in a global coupled model: sensitivity to the background climate state. *Climate Dynamics* (Online First): 1-17. DOI: 10.1007/s00382-011-1117-y

Atlantic States Marine Fisheries Commission (ASMFC). 2007. Species Profile for Atlantic Croaker. *ASMFC Fisheries Focus*, Vol. 16, Issue 3, April. Retrieved January 17, 2012, from http://www. asmfc.org/atlanticCroaker.htm.

Atlantic States Marine Fisheries Commission (ASMFC). 2011. Atlantic Croaker. Retrieved January 17, 2012, from http://www.asmfc.org/atlanticCroaker.htm.

Atlantic States Marine Fisheries Commission (ASMFC). 2012. Overview of Stock Status: Atlantic Croaker, *Micropogonias undulatus*. Retrieved January 17, 2012, from http://www.asmfc.org/atlanticCroaker.htm.

Aubry, A., Chosidow, O., Caumes, E., Roert, J. and Cambau, E. 2002. Sixty-three cases of Mycobacterium marinum infection. *Archives of Internal Medicine* 162: 1746-1752.

Auster, P.J., & Link, J.S. 2009. Compensation and recovery of feeding guilds in a northwest Atlantic shelf fish community. *Marine Ecology Progress Series* **382**, 163-172.

Austin, D. & A. Sauer. 2002. Changing oil: emerging environmental risks and shareholder value in the oil and gas industry. Washington, DC: World Resources Institute.

Averhoff, F., Young S, Mott J, Fleischauer A, Brady J, Straif-Bourgeois S, Valadez A, Lurie D, Palacio H, Buhner D, Persse D, Guerra F, Morgan J, Zoretic J, Moolenaar R, McNeill KM, Byers P, Kittle TS, Chavez G, Phillips M, Koo D, Groseclose S, Hicks P, Jones NF, Kenneson A, Vranken P, Sergienko E, Bitsko RH, Lorick SA. 2006. Morbidity surveillance after Hurricane Katrina --- Arkansas, Louisiana, Mississippi, and Texas, September 2005. 55(26):727-731.

Badjeck M., Allison E., Halls A., & Dulvy N. 2010. Impacts of climate variability and change on fishery-based livelihoods. *Marine Policy* **34**, 375-383.

Baer, H. & Singer, M. 2009. *Global Warming and the Political Ecology of Health: Emerging Crises and Systemic Solutions*. Walnut Creek, CA: Left Coast Press.

Baer, H., Singer, M. & Susser, I. 2003. *Medical Anthropology and the World System*. 2nd Edition. Westport, CT: Praeger, p. 5.

Baker, J.D., Littnan, C.L. & Johnston, D.W. 2006. Potential effects of sea level rise on the terrestrial habitats of endangered and endemic megafauna in the northwestern Hawaiian Islands. *Endangered Species Research* **2**, 21-30.

Barange, M, & Perry, R.I. 2009. Physical and ecological impacts of climate change relevant to marine and inland capture fisheries and aquaculture. In: K. Cochrane, C. De Young, D. Soto, and T. Bahri (eds). *Climate change implications for fisheries and aquaculture: overview of current scientific knowledge*. FAO Fisheries and Aquaculture Technical Paper No. 530. Rome: Food and Agricultural Organization of the United Nations, 7-106.

Barbieri, L. R., M. E. Chittenden, Jr. and C. M. Jones. 1994. Age, growth, and mortality of Atlantic croaker, *Micropogonias undulatus*, in the Chesapeake Bay region, with a discussion of apparent geographic changes in population dynamics. *Fishery Bulletin* 92:1-12.

Baringer, M. O. & Larsen, J. 2001: Sixteen years of Florida Current transport at 27N. *Geophysical Research Letters*, **28**, 3179-3182.

Barlas, M.E., C.J. Deutsch, M. de Wit, and L. I. Ward-Gieger (editors). 2011. Florida manatee cold-related unusual mortality event, January-April 2010. Final report to USFWS (grant 40181AG037). Florida Fish and Wildlife Conservation Commission, St. Petersburg, Florida. 138 pp.

Barnett, J. 2006. Climate Change, Insecurity, and Injustice. Adger, W.N., Paavola, J., Huq, S. & Mace, M.J. eds. In *Fairness in Adaptation to Climate Change*. Cambridge, MA: MIT Press, 115-129.

Barnosky, A.D., Matzke, N., Tomiya, S., Wogan, G.O.U., Swartz, B. Quental, T.B., Marshall, C., McGuire, J.L., Lindsey, E.L., Maguire, K.C., Mersey, B., and Ferrer, E.A. 2011. Has the Earth's sixth mass extinction already arrived? *Nature* **470**, 51-57.

Barrows, F.T., D. Bellis, A. Krogdahl, J.T. Silverstein, E.M. Herman, W.M. Sealey, M.B. Rust, and D.M. Gatlin III. 2008. Report of the plant products in aquafeed strategic planning workshop: An integrated, interdisciplinary research roadmap for increasing utilization of plant feed-stuffs in diets for carnivorous fish. Reviews in Fisheries Science, 16(4):449-455.

Barry, J.P., Baxter, C.H., Sagarin, R.D. & Gilman, S.E. 1995. Climate-related, long-term faunal changes in a California rocky intertidal community. *Science* **267**, 672-675.

Barth, J.A., Menge, B.A., Lubchenco, J., Chan, F., Bane, J.M., Kirincich, A.R., McManus, M.A., Nielsen, K.J., Pierce, S.D. & Washburn, L. 2007. Delayed upwelling alters nearshore coastal ocean ecosystems in the northern California current *Proceedings of the National Academy of Sciences USA* **104**, 3719-3724.

Barton, A., B. Hales, G.G. Waldbusser, C. Langdon, and R.A. Feely. 2012. The Pacific oyster, *Crassostrea gigas*, shows negative correlations to naturally elevated carbon dioxide levels: implications for near-term ocean acidification impacts. Limnology and Oceanography 57(3): 698-710.

Baskett, M.L., Gaines, S.D. & Nisbet, R.M. 2009. Symbiont diversity may help coral reefs survive moderate climate change. *Ecological Applications* **19**, 3-17.

Battin, J., M.W. Wiley, M.H. Ruckelshaus, R.N. Palmer, E. Korb, K.K. Bartz and H Imaki. 2007. Projected impacts of climate change on salmon habitat restoration. Proceedings of the National Academy of Sciences of the United States of America 104(16): 6720-25.

Baumgartner, T. R., A. Soutar and V. Ferreira-Bartrina. 1992. Reconstruction of the history of Pacific sardine and northern anchovy populations over the past two millennia from sediments of the Santa Barbara Basin, California. California. Calif. Coop. Oceanic Fish. Invest. Rep. 33: 24-40.

Beamish, R.J., Lange, K.L., Riddell, B.E., & Urawa, S. 2010. Climate Impacts on Pacific Salmon: Bibliography. Vancouver, BC: North Pacific Anadromous Fish Commission, Special Publication No. 2.

Beamish, R.J., Mahnken, C. & Neville, C.M. 2004. Evidence That Reduced Early Marine Growth is Associated with Lower Marine Survival of Coho Salmon. Transactions of the American Fisheries Society 133, 26-33

Beamish, R.J., Riddell, B.E., Lange, K.L., Farley Jr., E., Kang, S., Nagasawa, T., Radchenko, V., Temnykh, O. & Urawa, S. 2009. A long-term research and monitoring plan (LRMP) for Pacific salmon (*Oncorhynchus* spp.) in the North Pacific Ocean. Vancouver, BC: North Pacific Anadromous Fish Commission, Special Publication No. 1.

Beaudreau, A. H., P. S. Levin, and K. C. Norman. 2011. Using folk taxonomies to understand stakeholder perceptions for species conservation. *Conservation Letters* 0: 1-13. doi: 10.1111/j.1755-263X.2011.00199.x

Beaufort, L, Probert, I., De Garidel-Thoron, T., Bendif, E.M., Ruiz-Pino, D., Metzl, N., Goyet, C., Buchet, N., Coupel, P., Greland, M., Rost, B., Rickaby, R.E.M. & de Vargas, C. 2011. Sensitivity of coccolithophores to carbonate chemistry and ocean acidification. *Nature* **476**, 80-83.

Beaugrand, G., Brander, K.M., Lindley, J.A., Souissi, S. & Reid, P.C. 2003. Plankton effect on cod recruitment in the North Sea. *Nature* **426**, 661-664.

Beaugrand, G., Reid, P.C., Ibañez, F., Lindley, J.A., & Edwards, M. 2002. Reorganization of North Atlantic marine copepod biodiversity and climate. *Science* **296**, 1692-1694.

Behrenfeld, M.J., O'Malley, R.T., Siegel, D.A., McClain, C.R., Sarmiento, J.L., Feldman, G.C., Milligan, A.J., Falkowski, P.G., Letelier, R.M. & Boss, E.S. 2006. Climate-driven trends in contemporary ocean productivity. *Nature* **444**, 752-755.

Bell, G. & Collins, S. 2008. Adaptation, extinction, and global change. *Evolutionary Applications* **1**, 3-16.

Bell, J. D, C. Reid, M. J. Batty, E. H. Allison, P. Lehodey, L. Rodwell, T. D. Pickering, R. Gillett, J. E. Johnson, A. J. Hobday and A. Demmke. 2011c. Implications of climate change for contributions by fisheries and aquaculture to Pacific Island economies and communities. Ch. 12 in: Bell JD, Johnson JE and Hobday AJ (2011a) Vulnerability of Tropical Pacific Fisheries and Aquaculture to Climate Change. Secretariat of the Pacific Community, Noumea, New Caledonia.

Bell, J. D., Tim J.H. Adams, J.E. Johnson, A. J. Hobday and A. Sen Gupta. 2011b. Pacific communities, fisheries, aquaculture and climate change: An introduction. Chapter 1 in: Bell JD, Johnson JE and Hobday AJ (2011b) Vulnerability of Tropical Pacific Fisheries and Aquaculture to Climate Change. Secretariat of the Pacific Community, Noumea, New Caledonia.

Bell, J.D., Johnson, J.E. & Hobday A.J. 2011. Vulnerability of Tropical Pacific Fisheries and Aquaculture to Climate Change. Secretariat of the Pacific Community, Noumea, New Caledonia.

Beman, J.M., Chow, C.-E., King, A.L., Feng, Y., Fuhrman, J.A., Andersson, A., Bates, N.R., Popp, B.N. & Hutchins, D.A. 2011. Global declines in oceanic nitrification rates as a consequence of ocean acidification. *Proceedings of the National Academy of Sciences USA* **108**, 208-213.

Berger, S.A., Diehl, S., Stibor. H., Trommer, G., Ruhenstroth, M., Wild, A., Weigert, A., Jäger, C.G. & Striebel, M. 2007 Water temperature and mixing depth affect timing and magnitude of events during spring succession of the plankton. *Oecologia,* **150**, 643-654.

Berkman, P.A. & Young, O.R. 2009. Governance and environmental change in the Arctic Ocean. *Science* **324**, 239-240.

Bernard, S., & McGeehin, M. 2004. *Municipal heat wave response plans. American Journal of Public Health* **94**, 1520-1522.

Bertram, D.F., & Kaiser, G.W. 1993. Rhinocerous auklet (*Cerorhinea monocerata*) nestling diet may gauge Pacific sand lance (*Ammodytes hexapterus*) recruitment. *Canadian Journal of Fisheries and Aquatic Sciences* **50**, 1908-1915.

Beukema, J.J., Dekker, R. & Jansen, J.M. 2009. Some like it cold: populations of the tellinid bivalve *Macoma balthica* (L.) suffer in various ways from a warming climate. *Marine Ecology Progress Series* **384**, 135-145.

Biggs, R., Carpenter, S.R., & Brock, W.A. 2009. Turning back from the brink: detecting and impending regime shift in time to avert it. *Proceedings of the National Academy of Sciences USA* **106**, 826-831.

Bindoff, N.L., Willebrand, J., Artale, V., Cazenave, A., Gregory, J.M., Gulev, S., Hanawa, K., Le Quéré, C., Levitus, S., Nojiri, Y., Shum, C.K., Talley, L.D. & Unnikrishnan, A. 2007. Observations: oceanic climate change and sea level. In: *Climate Change 2007: The Physical Science Basis: Contribution of Working Group I to the Fourth Assessment Report of the Intergovernmental Panel on Climate Change*, S. Soloman et al. (eds.). Cambridge, United Kingdom and New York, NY: Cambridge University Press.

Bird, E.C.F. 1994. Physical setting and geomorphology of coastal lagoons. In: Kjerfve, B. (ed) *Coastal Lagoon Processes*. Amsterdam, The Netherlands: Elsevier. 9-40.

Blake, R.E., and J.E. Duffy. 2010. Grazer diversity affects resistance to multiple stressors in an experimental seagrass ecosystem. Oikos 119: 1625-1635.

Bluhm, B.A., & Gradinger, R. 2008. Regional variability in food availability for Arctic marine mammals. *Ecological Applications* **18**, S77-S96.

Blunden, J., Arndt, D.S. & Baringer, M.O. 2011: State of the Climate in 2010. *Bulletin of the American Meteorological Society* **92**, S1-S266.

Boesch DF, Rabalais NN, 1991. Effects of Hypoxia on Continental shelf benthose: comparisons between the New York Bight and the Northern Gulf of Mexico. In "Modern and ancient continental shelf anoxia" (R. V. Tyson and T. H. Pearson, Eds.), Vol. 58, pp, 27-34. Geological Society Publication.

Boesch, D.F., Coles, V.J., Kimmel, D.G., Miller, W.D. 2007. Coastal dead zones and global climate change: ramifications of climate change for the Chesapeake Bay. In *Regional Impacts of Climate Change: Four Case Studies in the United States*. Arlington, VA: Pew Center for Global Climate Change, 54-70.

Bograd, S.J., Castro, C.G., Di Lorenzo, E., Palacios, D.M., Bailey, H., Gilly, W. & Chavez, F.P. 2008. Oxygen declines and the shoaling of the hypoxic boundary in the California Current. *Geophysical Research Letters* **35**, L12607.

Bograd, S.J., Schroeder, I., Sarkar, N., Qiu, X., Sydeman, W.J. & Schwing, F.B. 2009. Phenology of coastal upwelling in the California Current. *Geophysical Research Letters* **36**, L01602.

Boin, A., Hart, P., Stern E., and Sundelius, B. 2007. *The Politics of Crisis Management: Public Leadership Under Pressure*. Cambridge: Cambridge University Press.

Bony, S., Colman, R., Kattsov, V.M., Allan, R.P., Bretherton, C.S., Dufresne, J.-L., Hall, A., Hallegatte, S., Holland, M.M., Ingram, W., Randall, D.A., Soden, B.J., Tselioudis, G. & Webb, M.J. 2006. How well do we understand and evaluate climate change feedback processes? *Journal of Climate* **19**, 3445-3482.

Boos, W. R., J. R. Scott, and K. Emanuel, 2004: Transient diapycnal mixing and the meridional overturning circulation. J. Phys. Oceanogr., 34, 334-341.

Borgerson, S.G. 2008. Arctic Meltdown: The Economic and Security Implications of Global Warming. *Foreign Affairs* **87**, 63-77.

Bossart, G.D., Baden, D.G., Ewing, R.Y., Roberts, B. & Wright, S.D. 1998. Brevetoxicosis in manatees (*Trichechus manatus latirostris*) from the 1996 epizootic: gross, histologic, and immunohistochemical features. *Toxicologic Pathology* **26**, 276-282.

Bossart, G.D., Meisner, R.A., Rommel, S.A., Ghim, S. & Jenson, A.B. 2002. Pathological features of the Florida manatee cold stress syndrom. *Aquatic Mammals* **29**, 9-17.

Boyce, D. G., Lewis, M. R. & Worm, B. 2010. Global phytoplankton decline over the past century. Nature 466, 591-596.

Boyd, P.W., Doney, S.C., Strzepek, R., Dusenberry, J., Lindsay, K., & Fung, I. 2008. Climate-mediated changes to mixed-layer properties in the Southern Ocean: how will key phytoplankton species respond? *Biogeosciences* **5**, 847-864.

Braby, C.E., & Somero, G.N. 2006. Following the heart: temperature and salinity effects on heart rate in native and invasive species of blue mussels (genus *Mytilus*). *Journal of Experimental Biology* **209**, 2554-2566.

Bradley, P.A., Fore, L.S., Fisher, W.S., & Davis, W.S. 2010. *Coral reef biological criteria: using the Clean Water Act to protect a national treasure*. EPA/600/R-10/054, July 2010. Narragansett, RI: U.S. Environmental Protection Agency, Office of Research and Development.

Brander, K. 2010. Impacts of climate change on fisheries. *Journal of Marine Systems* **79**, 389-402.

Brander, L., K. Rehdanz, Tol, R.S.J., van Beukering, P.J.H. 2009. The economic impact of ocean acidification on coral reefs, Papers WP282, Economic and Social Research Institute (ESRI).

Brewer, P.G., & Hester, K. 2009. Ocean acidification and the increasing transparency of the ocean to low-frequency sound. *Oceanography* **22**, 86-93,

Bricker, S., Longstaff, W. Dennison, A. Jones, K. Boicourt, C. Wicks, & Woerner, J. 2007. *Effects of Nutrient Enrichment in the Nation's Estuaries: A Decade of Change. NOAA Coastal Ocean Program Decision Analysis Series No. 26*. Silver Spring, MD: Nation Centers for Coastal Ocean Science.

Brigham, L. W. 2008. Arctic Shipping Scenarios and Coastal State Challenges. WMU Journal of Maritime Affairs 7(2):477-484.

Broecker, W.S. 1991. The great ocean conveyor, *Oceanography* **4**, 79-89.

Broecker, W.S., & Takahashi, T. 1966. Calcium carbonate precipitation on the Bahama Banks. *Journal of Geophysical Research* **71**, 1575-602.

Broitman B.R., & Kinlan, B.P. 2006. Spatial scales of benthic and pelagic producer biomass in a coastal upwelling ecosystem. *Marine Ecology Progress Series* **327**, 15-25.

Brook, B., Sodhi, N. & Bradshaw, C. 2008. Synergies among extinction drivers under global change. *Trends in Ecology & Evolution* **23**, 453-460.

Brooke, C. & Riley, T. 1999. *Erysipelothrix rhusiopathiae*: bacteriology, epidemiology and clinical manifestations of an occupational pathogen. *Journal of Medical Microbiology* **48**, 789-799.

Brookins, K., and M. Sabater. 2007. Coral reef fishery analysis for Tutuila. DMWR, ASG. Presentation to Western Pacific Regional Fisheries Management Council 95th SSC meeting (Item 8.b.2), June 12-14, 2005. Honolulu.

Brown CA, Nelson WG, Boese BL, DeWitt TH, Eldridge PM, Kaldy JE, Lee H II, Power JH, Young DR. 2007. An Approach to Developing Nutrient Criteria for Pacific Northwest Estuaries: A Case Study of Yaquina Estuary, Oregon. U.S.

Brown, K. 1999. Climate Anthropology: Taking Global Warming to the People. *Science* **283**, 1440-1441.

Bruno, J.F., E.R. Selig, K.S. Casey, C.A. Page, B.L. Willis, C.D. Harvell, H. Sweatman, and A.M. Melendy. 2007. Thermal stress and coral cover as drivers of coral disease outbreaks. PLoS Biology 5: e124.

Bryant, M. 2009. "Global climate change and potential effects on Pacific salmonids in freshwater ecosystems of southeast Alaska." Climatic Change 95 (1) (July 1): 169-193. doi:10.1007/s10584-008-9530-x.

Buisson, Y., Marié, J. & Davoust, B. 2008. These infectious diseases imported with food. *Bulletin Société Pathologie Exotique* **101**, 343-347.

Burby, R.J., ed. 1998. Cooperating with Nature: Confronting Natural Hazards with Land-Use Planning for Sustainable Communities. Washington, DC: Joseph Henry Press.

Bureau of Ocean Energy Management (BOEM). 2007. Final Programmatic Environmental Impact Statement for Alternative Energy Development and Production and Alternate Use of Facilities on the Outer Continental Shelf (OCS Report MMS 2007-046; USDOI, MMS, 2007) Retrieved February 2, 2012, from http://ocsenergy.anl.gov/.

Burek, K, Gulland, F. & O'Hara, T. 2008. Effects of climate change on Arctic marine mammal health. *Ecological Applications*. **18**, S126-S134

Burke, L., K. Reytar, M. Spalding, A. Perry, et al. 2011. Reefs at risk revisited. World Resources Institute, Washington, D.C.

Burkett, V., 2011. Global climate change implications for coastal and offshore oil and gas development, Energy Policy, 39, 7719-7725.

Burrows, M.T., Schoeman, D.S., Buckley, L.B., Moore, P., Poloczanska, E.S., Brander, K.M., Brown, C., Bruno, J.F., Duarte, C.M., Halpern, B.S., Holding, J., Kappel, C.V., Kiessling, W., O'Connor, M.I., Pandolfi, J.M., Parmesan, C., Schwing, F.B., Sydeman, W.J. & Richardson, A.J. 2011. The pace of shifting climate in marine and terrestrial ecosystems. *Science* **334**, 652-655.

Bussell, J.A., Gidman, E.A., Causton, D.R., Gwynn-Jones, D., Malham, S.K., Jones, M.L.M., Reynolds, B. & Seed, R. 2008. Changes in the immune response and metabolic fingerprint of the mussel, *Mytilus edulis* (Linnaeus) in response to lowered salinity and physical stress. *Journal of Experimental Marine Biology and Ecology* **358**, 78-85.

Butchart, S.H.M., Walpole, M., Collen, B., van Strien, A., Scharlemann, J.P.W., Almond, R.E.A., Baillie, J.E.M., Bomhard, B., Brown, C., Bruno, J., Carpenter, K.E., Carr, G.M., Chanson, J.,

Chenery, A.M., Csirke, J., Davidson, N.C., Dentener, F., Foster, M., Galli, A., Galloway, J.N., Genovesi, P., Gregory, R.D., Hockings, M., Kapos, V., Lamarque, J.-F., Leverington, F., Loh, J., McGeoch, M.A., McRae, L., Minasyan, A., Hernández Morcillo, M., Oldfield, T.E.E., Pauly, D., Quader, S., Revenga, C., Sauer, J.R., Skolnik, B., Spear, D., Stanwell-Smith, D., Stuart, S.N, Symes, A., Tierney, M., Tyrrell, T.D., Vié, J.-C. & Watson, R. 2010. Global biodiversity: indicators of recent declines. *Science* **328**, 1164-1168.

Button, G. 1992. *Social Conflict and the Formation of Emergent Groups in A Technological Disaster: The Exxon-Valdez Oil Spill and the Response of Residents in the Area of Homer, Alaska.* PhD dissertation, Brandeis University.

Button, G. V. and Kristina Peterson. 2009. Participatory Action Research: Community Partnership with Social and Physical Scientists. In *Anthropology and Climate Change: From Encounters to Actions.* Crate, S.A., & Nuttall, M.,eds. Walnut Creek, CA: Left Coast Press. 327-340.

Button, G.. 2010. Disaster Culture: Knowledge and Uncertainty in the Wake of Human and Environmental Catastrophe. Left Coast Press.

Byrd, V.G., Sydeman, W.J., Renner, H.M. & Minobe, S. 2008. Responses of piscivorous seabirds at the Pribilof Islands to ocean climate. *Deep-Sea Research Part II* **55**, 1856-1867.

Byrne, M. 2011. Impact of ocean warming and ocean acidification on marine invertebrate life history stages: vulnerabilities and potential for persistence in a changing ocean. *Oceanography and Marine Biology: an Annual Review* **49**, 1-42.

Byrne, M., N.A. Soars, M.A. Ho, E. Wong, D. McElroy, P. Selvakumaraswamy, S.A. Dworjanyn, and A.R. Davis. 2010. Fertilization in a suite of coastal marine invertebrates from SE Australia is robust to near-future ocean warming and acidification. Marine Biology 157: 2061-2069.

Byrnes, J.E., Reed, D.C., Cardinale, B.J., Cavanaugh, K.C., Holbrook, S.J., & Schmitt, R.J. 2011. Climate-driven increases in storm frequency simplify kelp forest food webs. *Global Change Biology* **17**, 2513-2524.

Byrnes, J.E., Reynolds, P.L., & Stachowicz, J.J. 2007. Invasions and extinctions reshape coastal marine food webs. *PLoS ONE* **2**, e295.

Caldeira, K., & Wickett , M.E. (2005). Ocean model predictions of chemistry changes from carbon dioxide emissions to the atmosphere and ocean. *Journal of Geophysical Research* **110**, C09S04, 12 pp. doi: 10.1029/2004JC002671.

Caldeira, K., & Wickett, M.E. 2003. Oceanography: Anthropogenic carbon and ocean pH. *Nature*, **425**, 365.

Callaway, D., Eamer, J., Edwardsen, E., Jack, C., Marcy, S., Olrun, A., Patkotak, M., Rexford, D. & Whiting, A. 1999. Effects of climate change on subsistence communities in Alaska, In *Assessing the Consequences of Climate Change in Alaska and the Bering Sea Region,* Weller, G. & Anderson, P.A., eds. Fairbanks, Alaska: Center for Global Change and Arctic System Research, the University of Alaska Fairbanks.

Cameron, C.E., R.L. Zuerner, S. Raverty, K.M. Colegrove, S.A. Norman, D.M. Lambourn, S.J. Jeffries, and F.M. Gulland. 2008. Detection of pathogenic *Leptospira* bacteria in pinniped populations via PCR and identification of a source of transmission for zoonotic leptospirosis in the marine environment. Journal of Clinical Microbiology 46(5): 1728-1733.

Campbell, K.M., Gulledge, J., McNeill, J.R., Podesta, J., Ogden, P., Fuerth, L., Woolsey, R. J., Lennon, A.T.J., Smith, J., Weitz, R. & Mix, D. 2007. *The Age of Consequences: The Foreign Policy and National Security Implications of Global Climate Change.* Washington, DC: Center for Strategic & International Studies and Center for a New American Security.

Canadell, J.G. & Raupach, M.R. 2008. Managing forests for climate change mitigation. *Science* **320**, 1456-57.

Cardinale, B.J., Matulich, K.L., Hooper, D.U., Byrnes, J.E., Duffy, E., Gamfeldt, L., Balvanera, P., O'Connor, M.I. & Gonzalez, A. 2011. The functional role of producer diversity in ecosystems. *American Journal of Botany* **98**, 572-592.

Carlton, J.T. 1996. Pattern, process, and prediction in marine invasion ecology. *Biological Conservation* **78**, 97-106.

Carpenter, K.E., Abrar, M., Aeby, G., Aronson, R.B., Banks, S., Bruckner, A., Chiriboga, A., Cortés, J., Delbeek, J.C., DeVantier, L., Edgar, G.J., Edwards, A.J., Fenner, D., Guzmán, H.M., Hoeksema, B.W., Hodgson, G., Johan, O., Licuanan, W.Y., Livingstone, S.R., Lovell, E.R., Moore, J.A., Obura, D.O., Ochavillo, D., Polidoro, B.A., Precht, W.R., Quibilan, M.C., Reboton, C., Richards, Z.T., Rogers, A.D., Sanciango, J., Sheppard, A., Sheppard, C., Smith, J., Stuart, S., Turak, E., Veron, J.E.N., Wallace, C., Weil, E. & Wood, E. 2008. One-third of reef-building corals face elevated extinction risk from climate change and local impacts. *Science* **321**, 560-563.

Carson, H.S., Lopez-Duarte, P.C., Rasmussen, L., Wang, D. & Levin, L.A. 2010. Reproductive timing alters population connectivity in marine metapopulations. *Current Biology* **20**,1926-1931.

Carter, D.W. and D. Letson. 2009. Structural Vector Error Correction Modeling of Integrated Sportfishery Data. *Marine Resource Economics* 24(1):19-41.

Cazenave, A. & Llovel, W. 2010. Contemporary sea level rise. *Annual Review of Marine Science* **2**, 145-173.

Center for Biological Diversity v. Environmental Protection Agency. 2009. No. 2:09-cv-00670-JCC (W.D. Wash).

Center for Climate and Energy Solutions. 2012. Climate change adaptation: what Federal agencies are doing. February 2012 update. Available for download at: http://www.pewclimate. org/docUploads/federal-agencies-adaptation.pdf

Centers for Disease Control and Prevention (CDC). 1998. Outbreak of *Vibrio parahaemolyticus* infections associated with eating raw oysters–Pacific Northwest, 1997. *Morbidity and Mortality Weekly Report* **47**, 457-462.

Centers for Disease Control and Prevention (CDC). 2005 *Vibrio* illnesses after Hurricane Katrina —multiple states, August-September 2005. *Morbidity and Mortality Weekly*. Report **54**, 928-931

Centers for Disease Control and Prevention (CDC). 2005 *Vibrio* illnesses after Hurricane Katrina—multiple states, August-September 2005. *Morbidity and Mortality Weekly*. Report 54:928-931

Centers for Disease Control and Prevention (CDC). 2006. Morbidity surveillance after Hurricane Katrina --- Arkansas, Louisiana, Mississippi, and Texas, September 2005. *Morbidity and Mortality Weekly Report* **55**, 727-731.

Centers for Disease Control and Prevention (CDC). 2007. Preliminary FoodNet data on the incidence of infection with pathogens transmitted commonly through food—10 states, United States, 2006. *Morbidity and Mortality Weekly Report* 56:336-339.

Centers for Disease Control and Prevention (CDC). 2007. Scombroid fish poisoning associated with tuna steaks—Louisiana and Tennessee, 2006. *Morbidity and Mortality Weekly Report* 56(32):817-819.

Centers for Disease Control and Prevention (CDC). 2009. Preliminary FoodNet data on the incidence of infection with pathogens transmitted commonly through food—10 states, United States, 2009. *Morbidity and Mortality Weekly Report* 59:418-422.

Centers for Disease Control and Prevention (CDC). 2010. Vital Signs: Incidence and Trends of Infection with Pathogens Transmitted Commonly through Food – Foodborne Diseases

Active Surveillance Network, 10 U.S. Sites, 1996-2010. *Morbidity and Mortality Weekly Report* 60:749-755.

Centers for Disease Control and Prevention (CDC). 2011. COVIS Annual Summary, 2009. Atlanta, Georgia: U.S. Department of Health and Human Services. P.3

Centers for Disease Control and Prevention (CDC). 2011. Surveillance for waterborne disease outbreaks and other health events associated with recreational water—United States, 2007-2008. *Morbidity and Mortality Weekly Review* Surveillance Summary **60**, 1-32.

Chan, F., Barth, J.A., Lubchenco, J., Kirincich, A., Weeks, H., Peterson, W.T., Menge, B.A. 2008. Emergence of anoxia in the California current large marine ecosystem. *Science* **319**, 920.

Chapin III, F.S., M.W. Oswood, K. Van Cleve, L.A. Viereck, and D.L. Verbyla. 2006. Alaska's Changing Boreal Forest. Oxford, UK: Oxford University Press.

Charles, A. T. Bio-socio-economic dynamics and multidisciplinary models in small-scale fisheries research, p.603-608 in: *Research and Small-Scale Fisheries*. J.R. Durand, J. Lemoalle and J. Weber (editors), ORSTOM, Paris France (1991).

Charles, A. T. Bio-socio-economic fishery models: labour dynamics and multiobjective management, *Canadian Journal of Fisheries and Aquatic Sciences* 46:1313-1322 (1989).

Charles, A. T. Sustainability assessment and bio-socio-economic analysis: Tools for integrated coastal development, p.115-125 in: *Philippine Coastal Resources Under Stress*. M.A. Juinio-Meñez and G. Newkirk (editors), Coastal Resources Research Network, Halifax, Canada, and Marine Science Institute, Quezon City, Philippines (1995).

Charles, A., C. Burbidge, H. Boyd and A. Lavers, 2009, Fisheries and the Marine Environment in Nova Scotia: Searching for Sustainability and Resilience, Halifax, Nova Scotia: GPI Atlantic. http://www.gpiatlantic.org/pdf/fisheries/fisheries_2008.pdf (accessed 17 March 2010)

Chastel, O., Weimerskirch, H. & Jouventin, P. 1993. High annual variability in reproductive success and survival of an Antarctic seabird, the snow petrel *Pagodroma nivea*: a 27 year study. *Oecologia* **94**, 278-284.

Chavez, F.P., Messié, M. & Pennington, J.T. 2011. Marine primary production in relation to climate variability and change. *Annual Review of Marine Science* **3**, 227-260.

Chavez, F.P., Ryan, J., Lluch-Cota, S.E., & Niquen, M.C. 2003. From anchovies to sardines and back: multidecadal change in the Pacific Ocean. *Science* **299**, 217-221.

Chernook, V.I., & Boltnev, A.I. 2008. Regular instrumental aerial surveys detect a sharp drop in the birth rates of the harp seal in the White Sea. In *Collection of scientific papers from the Marine Mammals of the Holarctic V Conference, Odessa, Ukraine, 14-18 October*, 100-104.

Cheung, W.W.L., Lam, V.W.Y., Sarmiento, J.L., Kearney, K., Watson, R. & Pauly, D. 2009. Projecting global marine biodiversity impacts under climate change scenarios. *Fish and Fisheries* **10**, 235-251.

Cheung, W.W.L., Lam, V.W.Y., Sarmiento, J.L., Kearney, K., Watson, R, Zeller, D. & Pauly, D. 2010. Large-scale redistribution of maximum fisheries catch potential in the global ocean under climate change. *Global Change Biology* **16**, 24-35.

Christensen, J.H., Hewitson, B., Busuioc, A., Chen, A., Gao, X., Held, I., Jones, R., Kolli, R.K., Kwon, W.-T., Laprise, R., Magaña Rueda, V., Mearns, L., Menéndez, C.G., Räisänen, J., Rinke, A., Sarr, A., Whetton, P., University, C., & Press, C.U. 2007. Regional Climate Projections. In *Climate Change 2007: The Physical Science Basis. Contribution of Working Group I to the Fourth Assessment Report of the Intergovernmental Panel on Climate Change*. S. Soloman et al. (eds.), Cambridge, United Kingdom and New York, NY, USA.: Cambridge University Press, 847-940.

CIC Research, Inc. 2004. A Survey on Recreational and Subsistence Fishing in Southern California Coastal Waters. San Diego, CA.

Cinner, J.E., T.R. McClanahan, N.A.J. Graham, T.M. Daw, J. Maina, S.M. Stead, A. Wamukota, K. Brown, and O. Bodin. 2011. Vulnerability of coastal communities to key impacts of climate change on coral reef fisheries. *Global Environmental Change*, In press.

Clark, W., S. Hare, A. Parma, P. Sullivan, and R. Trumble. 1999. Decadal changes in growth and recruitment of Pacific halibut (*Hippoglossus stenolepis*). Canadian Journal of Aquatic and Fisheries Sciences, 56, 242-252.

Clark, W.G. and Hare, S.R. 2002. Effects of climate and stock size on recruitment and growth of Pacific halibut. North American Journal of Fisheries Management, 22, 852-862.

Clay, P. & Olson, J. 2008. Defining 'Fishing Communities': Vulnerability and the Magnuson-Stevens Fisheries Conservation and Management Act. *Human Ecology Review* **15**, 143-160.

Clay, P. M., P. Pinto da Silva and A. Kitts. 2010. Defining Social and Economic Performance Measures For Catch Share Systems in The Northeast U.S. 12 pp. In: Proceedings of the Fifteenth Biennial Conference of the International Institute of Fisheries Economics & Trade, July 13-16, 2010, Montpellier, France: Economics of Fish Resources and Aquatic Ecosystems: Balancing Uses, Balancing Costs. Compiled by Ann L. Shriver. International Institute of Fisheries Economics & Trade, Corvallis, Oregon, USA, 2008. CD ROM. ISBN 0-9763432-6-6

Climate Focus 2011 Blue Carbon: Policy Option Assessment. Report Prepared for the Linden Trust for Conservation by Climate Focus, Washington DC. http://www.climatefocus.com/documents/files/blue_carbon_.pdf

Coastal States Organization (CSO). 2007. The Role of Coastal Zone Management programs in adaptation to climate change: final report of the CSO Climate Change Work Group. Washington, DC: Coastal States Organization.

Cohen, A.N. & Carlton, J.T. 1998. Accelerating invasion rate in a highly invaded estuary. *Science* **279**, 555-558.

Colburn, L. L., P. M. Clay, J. Olson, P. Pinto da Silva, S. L. Smith, A. Westwood, and J. Ekstrom. 2010. Introduction. Community Profiles for Northeast U.S. Marine Fisheries. Available at http://www.nefsc.noaa.gov/read/socialsci/community_profiles/introduction.pdf.

Colburn, L.L. & Jepson, M. Under NEFSC review. Social Indicators of Gentrification Pressure in Fishing Communities: A Context for Social Impact Assessment. Invited paper for special issue of *Coastal Management Journal*.

Colegrove, K.M., Lowenstine, L.J., & Gulland, F.M.D. 2005. Leptospirosis in northern elephant seals (*Mirounga angustirostris*) stranded along the California coast. *Journal of Wildlife Diseases* **41**, 426-430.

Collier, D. 2002. Cutaneous infections from coastal and marine bacteria. *Dermatological Therapy* **15**, 1-9.

Collinge, S.K., Johnson, W.C., Ray, C., Matchett, R., Grentsen, J., Cully, J.F. Jr., Gage, K.L., Kosoy, M.Y., Loye, J.E. & Martin, A.P. 2005. Testing the generality of a trophic-cascade model for plague. *Ecohealth* **2**, 1-11.

Collins, M., An, S.-I., Cai, W., Ganachaud, A., Guilyardi, E., Jin, F.-F., Jochum, M., Lengaigne, M., Power, S., Timmermann, A., Vecchi, G. & Wittenberg, A. The impact of global warming on the tropical Pacific Ocean and El Niño. *Nature Geoscience* **3**, 391-397.

Collins, S., & Bell, G. 2004. Phenotypic consequences of 1000 generations of selection at elevated CO_2 in a green alga. *Nature* **431**, 566-569.

Colombi, B.J. 2009. Salmon nation: Climate change and tribal sovereignty. In *Anthropology and Climate Change: From Encounters to Actions* Crate, S. & Nuttall, M. (eds.). Walnut Creek, CA: Left Coast Press, 186-196.

Comiso, J. & Nishio, F. 2008. Trends in the sea ice cover using enhanced and compatible AMSR-E, SSM/I, and SMMR data. *Journal of Geophysical Research*, **113**, C02S07.

Comiso, J. C., C. L. Parkinson, R. Gersten, and L. Stock (2008), Accelerated decline in the Arctic sea ice cover, Geophys. Res. Lett., 35, L01703, doi:10.1029/2007GL031972.

Committee on Environment and Natural Resources. 2010. Scientific Assessment of Hypoxia in U.S. Coastal Waters. Interagency Working Group on Harmful Algal Blooms, Hypoxia, and Human Health of the Joint Subcommittee on Ocean Science and Technology. Washington, DC.

Connell, S.D., & Russell, B.D. 2010. The direct effects of increasing CO_2 and temperature on non-calcifying organisms: increasing the potential for phase shifts in kelp forests. *Proceedings of the Royal Society of London B* **277,** 1409-1415.

Cook, T., Folli, M., Klinck, J., Ford, S. & Miller, J. 1998. The relationship between increasing sea-surface temperature and the northward spread of *Perkinsus marinus* (Dermo) disease epizootics in oysters. *Estuarine, Coastal and Shelf Science* **46,** 587-597.

Cooley, S.R. & Doney, S.C. 2009. Anticipating ocean acidification's economic consequences for commercial fisheries. *Environmental Research Letters* **4,** 024007.

Cooley, S.R., Kite-Powell, H.L. & Doney, S.C. 2009. Ocean acidification's potential to alter global marine ecosystem services. *Oceanography* **22,** 172-180.

Cooper, L.W., Ashjian, C.J., Smith, S.L., Codispoti, L.A., Grebmeier, J.M., Campbell, R.G. & Sherr, E.B. 2006. Rapid seasonal sea-ice retreat in the Arctic could be affecting Pacific walrus (*Odobenus rosmarus divergens*) recruitment. *Aquatic Mammals* **32,** 98-102.

Cooper, S.R. & Brush, G.S., 1991. Long-term history of Chesapeake Bay anoxia. *Science* **254,** 992-996.

Cooper, T.F., De 'Ath, G., Fabricius, K.E., & Lough, J.M. 2008. Declining coral calcification in massive *Porites* in two nearshore regions of the northern Great Barrier Reef. *Global Change Biology* **14,** 529-538.

Coulthard, S. 2008. Adapting to environmental change in artisanal fisheries—insights from a South Indian Lagoon. *Global Environmental Change* **18,** 479-89.

Coulthard, S. 2009. Adaptation and conflict within fisheries: insights for living with climate change. In Adger, W. N., Lorenzoni, I. and O'Brien, K. (eds) (2009 in press) Adapting to Climate Change: Thresholds, Values, Governance. Cambridge University Press: Cambridge.

Council on Environmental Quality (CEQ). 2010. Memorandum from on Draft NEPA Guidance on Consideration of the Effects of Climate Change and Greenhouse Gas Emissions, from Nancy H. Sutley, Chair, to Heads of Federal Departments and Agencies. Feb. 18, 2010.

Coyle, K., A. Pinkchuk, L. Eisner, and J. Napp. 2008. Zooplankton species composition, abundance and biomass on the southeastern Bering Sea shelf during summer: the potential role of water column stability in structuring the zooplankton community and influencing the survival of planktivorous fishes. Abstract of NPAFC International Symposium on Bering-Aleutian Salmon International Survey (BASIS), November 23 - 25, 2008, Seattle: 26. (Available at www.npafc.org).

Craig, P., B. Ponwith, F. Aitaoto, and D. Hamm. 1993. The commercial, subsistence, and recreational fisheries of American Samoa – Fisheries of Hawaii and U.S. – associated pacific Islands. Marine Fisheries Review, Spring.

Craig, R.K. 2009. The Clean Water Act on the cutting edge: climate change regulation and water-quality regulation. *Natural Resources and Environment* **24,** 14-18.

Craig, R.K. 2010. Stationarity is dead – long live transformation: five principles for climate change adaptation law. *Harvard Environmental Law Review* **34,** 9-75.

Crain, C.M., Kroeker, K. & Halpern, B.S. 2008. Interactive and cumulative effects of multiple human stressors in marine systems. *Ecology Letters* **11,** 1304-1315.

Crate, Susan A., and Mark Nuttall, eds. 2009. *Anthropology and Climate Change: From Encounters to Actions.* Walnut Creek, CA: Left Coast Press.

Cravatte, S., Delcroix, T., Zhang, D., McPhaden, M., Leloup, J. 2009. Observed freshening and warming of the western Pacific Warm Pool. *Climate Dynamics* **33**, 565-589.

Cressey, D. 2007. Arctic melt opens Northwest passage. *Nature* **449**, 267.

Crook, E.D., Potts, D., Rebolledo-Vieyra, M., Hernandez, L., & Paytan, A. 2011. Calcifying coral abundance near low-pH springs: implications for future ocean acidification. *Coral Reefs* **31**, 239-245.

Crooks, S., Findsen, J., Igusky, K., Orr, M.K., and Brew, D., 2009. Greenhouse Gas Mitigation Typology Issues Paper Tidal Wetlands Restoration (2009). Report prepared for California Climate Action Registry. http://www.climateactionreserve.org/wp-content/uploads/2009/03/future-protocol-development_tidal-wetlands.pdf

Crooks, S., Herr, D., Tamelander, J., Laffoley, D. & Vandever, J. 2011. *Mitigating Climate Change through Restoration and Management of Coastal Wetlands and Near-shore Marine Ecosystems: Challenges and Opportunities.* Environment Department Paper 121, Washington, DC: World Bank,.

Cubillos, J.C., Wright, S.W., Nash, G., de Salas, M.F., Griffiths, B., Tilbrook, B., Poisson, A. & Hallegraeff, G.M. 2007. Calcification morphotypes of the coccolithophorid *Emiliania huxleyi* in the Southern Ocean: changes in 2001 to 2006 compared to historical data. *Marine Ecology Progress Series* **348**, 47-54.

Cushing, D.H. 1996. *Towards a science of recruitment in fish populations.* Oldendorf/Luhe, Germany: Ecology Institute.

Cutter, S.L., L. Barnes, M. Berry, C. Burton, E. Evans, E. Tate, and J. Webb. 2008. A place-based model for understanding community resilience to natural disasters. *Global Environmental Change* 18:598-606.

Czaja, A., Robertson, A. & T. Huck, 2003. The role of coupled processes in producing NAO variability. *Geophysical Monograph 134, American Geophysical Union*, 147-172.

Dadisman, T., Nelson, R., Molenda, J. and Carber, H. 1972. *Vibrio* parahaemolyticus gastroenteritis in Maryland. I. Clinical and epidemiologic aspects. *American Journal of Epidemiology* 96:414-426.

Dahlhoff, E.P. 2004. Biochemical indicators of stress and metabolism: applications for marine ecological studies. *Annual Review of Physiology* **66**, 183-207.

Dalton, M. 2001. El Niño, expectations, and fishing effort in Monterey Bay, California. *Journal of Environmental Economics and Management*, **42**, 336-359.

Danaei G, Ding EL, Mozaffarian D, Taylor B, Rehm J, et al. (2009) The Preventable Causes of Death in the United States: Comparative Risk Assessment of Dietary, Lifestyle, and Metabolic Risk Factors. PLoS Med 6(4): e1000058. doi:10.1371/journal.pmed.1000058

Daniels, N., and Shafaie, A. 2000. A Review of Pathogenic *Vibrio* Infections for Clinicians. *Infectious Medicine* 17(10):665-685.

Daniels, N., Ray, B., Easton, A., Marano, N., Kahn, E., McShan, A., Del Rosario, L., Baldwin, T., Kingley, M., Puhr, N., Wells, J. and Anqulo, F. 2000. Emergence of a new *Vibrio* parahaemolyticus serotype in raw oysters: A prevention quandary. *JAMA* 284(12):1541-1545.

Davidson, E. 2009. The contribution of manure and fertilizer nitrogen to atmospheric nitrous oxide since 1860. *Nature Geoscience* **2**, 659-662

Davis, A. 1991. Insidious Rationalities: The Institutionalization of Small Boat Fishing and the Rise of the Rapacious Fisher. *Maritime Anthropological Studies (MAST)* 4(1): 13-31.

Day, J., D. Boesch, E. Clairain, P. Kemp, S. Laska, W. Mitsch, K. Orth, H. Mashriqui, D. Reed, L. Shabman, C. Simenstad, B. Streever, R. Twilley, C. Watson, J. Wells, and D. Whigham (2007).

Restoration of the Mississippi delta: lessons from Hurricanes Katrina and Rita. Science 315: 1679-1684.

Dayton, P.K. 1985. Ecology of kelp communities. *Annual Review of Ecology and Systematics* **16**, 215-245.

Dayton, P.K., Tegner, M.J., Edwards, P.B. & Riser, K.L.. 1999. Temporal and spatial scales of kelp demography: the role of oceanographic climate. *Ecological Monographs* **69**, 219-250.

de Rivera, C.E., Steves, B.P., Fofonoff, P.W., Hines, A.H. & Ruiz, G.M. 2011. Potential for high-latitude marine invasions along western North America. *Diversity and Distributions* **17**, 1198-1209.

De Silva, S.S. & Soto, D. 2009. Climate change and aquaculture: potential impacts, adaptation and mitigation. FAO Fisheries and Aquaculture Technical Paper 537, Rome, Italy: Food and Agricultural Organization of the United Nations, pp. 137-215.

Deacutis, C.F., Murray, D.P., Prell, W., Saarman, E. & Korhun, L. 2006. Hypoxia in the upper half of Narragansett Bay, RI, during August 2001 and 2002. *Northeastern Naturalist* 13, 173-198.

Denman, K., Christian, J. R., Steiner, N., Pörtner, H.-O. & Nojiri, Y. 2011. Potential impacts of future ocean acidification on marine ecosystems and fisheries: current knowledge and recommendations for future research. *ICES Journal of Marine Science* **68**, 1019-1029.

Denny, M.W., Hunt, L.J.H., Miller, L.P. & Harley, C.D.G. 2009. On the prediction of extreme ecological events. *Ecological Monographs* **79**, 397-421.

Deser, C. & Teng, H., 2008. Evolution of Arctic sea ice concentration trends and the role of atmospheric circulation forcing, 1979-2007. *Geophysical Research Letters* **35**, L02504.

Dessler, A.E. 2011. Cloud variations and the Earth's energy budget. *Geophysical Research Letters* **38**, L19701.

Deverel, S.J. & Leighton, D.A. 2010. Historic, Recent and Future Subsidence, Sacramento- San Joaquin Delta, California, USA. San Francisco Estuary and Watershed Science, 8-1-23. http://escholarship.org/uc/item/7xd4x0xw

Di Lorenzo, E., Miller, A.J., Scheider, N. & McWilliams, J.C. 2005. The warming of the California Current System: Dynamics and ecosystem implications *Journal of Physical Oceanography* **35**, 336-362.

Di Nezio, P.N., L.J. Gramer, W.E. Johns, C.S. Meinen, and M.O. Baringer, 2009: Observed inter-annual variability of the Florida Current: Wind forcing and the North Atlantic oscillation. *Journal of Physical Oceanography*, 39(3):721-736.

Diaz R.J. & Breitburg D.L. 2009. The hypoxic environment. In: Richards JG, Farrell AP, Brauner CJ (eds.), Fish Physiology, Hypoxia. Vol. 27. Academic Press, Burlington, VT. pp.1-23.

Diaz, R.J. & Rosenberg, R. 2008. Spreading dead zones and consequences for marine ecosystems. *Science* **321**, 926-929.

Diaz, R.J., & Rosenberg, R. 1995. Marine benthic hypoxia: a review of its ecological effects and the behavioural responses of benthic macrofauna. *Oceanography and Marine Biology: an Annual Review*, **33**, 245-303.

Diaz-Pulido, G., Anthony, K.R.N., Kline, D.I., Dove, S. & Hoegh-Guldberg, O. 2012. Interactions between ocean acidification and warming on the mortality and dissolution of coralline algae. *Journal of Phycology* **48**, 32-39.

Diaz-Pulido, G., Gouezo, M., Tilbrook, B., Dove, S. & Anthony, K.R. 2011. High CO_2 enhances the competitive strength of seaweeds over corals. *Ecology Letters* **14**, 156-162.

Diaz-Pulido, G., McCook, L.J., Dove, S., Berkelmans, R., Roff, G., Kline, D.I., Weeks, S., Evans, R.D., Williamson, D.H. & Hoegh-Guldberg, O. 2009. Doom and boom on a resilient reef: climate change, algal overgrowth and coral recovery. *PLoS ONE* **4**, e5239.

Dijkstra, J.A., Westerman, E.L.,& Harris, L.G. 2011. The effects of climate change on species composition, succession and phenology: a case study. *Global Change Biology*.17, 2360-2369.

Dinsdale, E.A., O. Pantos, S. Smriga, R.A. Edwards, F. Angly, et al. 2008. Microbial ecology of four atolls in the Northern Line Islands. PLoS ONE 3: e1584.

DIPNET. 2007. Review of disease interactions and pathogen exchange between farmed and wild finfish and shellfish in Europe (DIPNET). Editors: Rob Raynard, Thomas Wahli, Ioannis Vatsos, Stein Mortensen. Oslo: VESO Publisher.

Dirzo, R., & Raven, P.H. 2003. Global State of Biodiversity and Loss. *Annual Review of Environment and Resources* **28**, 137-167.

Dobos, K., Quinn, F., Ashford, D., Horsburgh, C. & King, C. 1999. Emergence of a unique group of necrotizing mycobacterial diseases. *Emerging Infectious Disease* **5**, 367-378.

Doeringer, P. B., P. I. Moss and D. G. Terkla. 1986. The New England Fishing Economy: Jobs, Income and Kinship. Amherst, MA: University of Massachusetts Press.

Donato D.C., Kauffman J.B., Murdiyarso D., Kurnianto S., Stidman M., Kanninen M. (2011) Mangroves among the most carbon-rich forests in the tropics. Nature Geoscience 4: 293-297.

Doney S.C., Mahowald N, Lima I et al. (2007). Impact of anthropogenic atmospheric nitrogen and sulfur deposition on ocean acidification and the inorganic carbon system. PNAS, 104, 14580-5.

Doney, S.C. 2010. The growing human footprint on coastal and open-ocean biogeochemistry. Science 328:1512-1516.

Doney, S.C., Fabry, V.J., Feely, R.A. & Kleypas, J.A. 2009. Ocean acidification: the other CO_2 problem. *Annual Review of Marine Science* **1**, 169-192.

Doney, S.C., Ruckelshaus, M., Duffy, J.E., Barry, J.P., Chan, F., English, C.A., Galindo, H.M., Grebmeier, J.M., Hollowed, A.B., Knowlton, N., Polovina, J., Rabalais, N.N., Sydeman, W.J. & Talley, L.D. 2012. Climate change impacts on marine ecosystems. *Annual Review of Marine Science* **4**, 11-37.

Dong B, Gregory J M and Sutton R T 2009 Understanding land–sea warming contrast in response to increasing greenhouse gases. Part I: transient adjustment J. Clim. 22 3079-97

Dowty, R.A., & Allen, B.L. 2011. Afterward. In *Dynamics of Disaster: Lessons on Risk, Response and Recovery*. Dowty, R.A., & Allen, B.L. (eds.) London: Earthscan, 203-207.

Dreier, N., C. Schlamkow, and P. Frohle (2011) Assessment of future wave climate on basis of wind-wave-correlations and climate change scenarios. Journal of Coastal Research, Special Issue, 64, pp. 210-214.

Drinkwater, K.F., A. Belgrano, A. Borja, A. Conversi, M. Edwards, et al. 2003. The response of marine ecosystems to climate variability associated with the North Atlantic Oscillation. pp. 211-233 In: J.W. Hurrell, Y. Kushnir, G. Ottersen, and M. Visbeck, eds. The North Atlantic Oscillation climate significance and environmental impact. American Geophysical Union, Washington, DC.

Dudgeon, S.R., Aronson, R.B., Bruno, J.F. & Precht, W.F. 2010. Phase shifts and stable states on coral reefs. *Marine Ecology Progress Series* **413**, 201-216.

Durant, J.M., Hjermann, D.O., Ottersen, G. & Stenseth, N.C. 2007. Climate and the match or mismatch between predator requirements and resource availability. *Climate Research* **33**, 271-283.

Durrenberger, E. P. 1997. Fisheries Management Models: Assumptions and Realities or, Why Shrimpers in Mississippi Are Not Firms. *Human Organization* 56(2):158-166.

Eakin, H. & Luers, A.L. 2006. Assessing the vulnerability of socio-environmental systems. *Annual Review of Environment and Resources* **31**, 365-394.

Ebeling, A.W., Laur, D.R. & Rowley, R.J. 1985. Severe storm disturbances and reversal of community structure in a southern California kelp forest. *Marine Biology* **84**, 287-294.

Ecosystem Assessment Program (EAP). 2009. *Ecosystem assessment report for the Northeast U.S. Continental Shelf Large Marine Ecosystem.* Woods Hole, MA: U.S. Department of Commerce, Northeast Fisheries Science Center Reference Document 09-11.

Ecosystem Assessment Program (EAP). 2012. Ecosystem Status Report for the Northeast U.S. Continental Shelf Large Marine Ecosystem. Woods Hole, MA: U.S. Department of Commerce, Northeast Fisheries Science Center, in press.

Edwards, M., & Richardson, A.J. 2004. Impact of climate change on marine pelagic phenology and trophic mismatch. *Nature* **430**, 881-884.

Ehler, C. & Douvere, F. 2009. *Marine spatial planning: A step-by-step approach toward ecosystem-based management.* IOC Manual and Guides No. 53, IOCAM Dosier No. 6. Paris, France: Intergovernmental Oceanographic Commission and Man and the Biosphere Programme.

Eisen, R.J., Glass, G.E., Eisen, L., Cheek, J., Enscore, R.E., Ettestad, P., & Gage, K.L. 2007. A spatial model of shared risk for plague and Hantavirus pulmonary syndrome in the southwestern United States. *American Journal of Tropical Medicine and Hygiene* **77**, 999-1004.

Emmett, R. L., Brodeur, R. D., Miller, T. W., Pool, S. S., Krutzikowsky, G. K., Bentley, P. J., and McCrae, J. 2005. Pacific sardine (Sardinops sagax) abundance, distribution, and ecological relationships in the Pacific Northwest. California. Calif. Coop. Oceanic Fish. Invest. Rep. 46:122.

Enscore, R.E., Biggerstaff, B.J., Brown, T.L., Fulgham, R.F., Reynolds, P.J., Engelthaler, D.M., Levy, C.E., Parmenter, R.R., Montenieri, J.A., Cheek, J.E., Grinnell, R.K., Ettestad, P.J., & Gage, K.L. 2002. Modeling relationships between climate and the frequency of human plague cases in the southwestern United States, 1960-1997. *American Journal of Tropical Medicine and Hygiene* **66**, 186-196.

Environenmental Health Coalition. 2005. Survey of Fishers on Piers in San Diego Bay: Results and Conclusions. National City, CA.

Erlandsson, J. & McQuaid, C.D. 2004. Spatial structure of recruitment in the mussel *Perna perna* at local scales: effects of adults, algae and recruit size. *Marine Ecology Progress Series* **267**, 173-185.

Euskirchen, E. S., A. D. McGuire, and F.S. Chapin III. 2007. "Energy feedbacks of northern high-latitude ecosystems to the climate system due to reduced snow cover during 20th century warming." Global Change Biology 13 (11): 2425-2438. doi:10.1111/j.1365-2486.2007.01450.x.

Fabricius, K.A., C. Langdon, S. Uthicke, C. Humphrey, S. Noonan, et al. 2011. Losers and winners in coral reefs acclimatized to elevated carbon dioxide concentrations. Nature Climate Change 1: 165-169.

Fabry, V.J. 2008. Marine calcifiers in a high-CO_2 ocean. *Science* **320**, 1020-1022.

Fabry, V.J., McClintock, J.B., Mathis, J.T. & Grebmeier, J.M. 2009. Ocean acidification at high latitudes: the bellwether. *Oceanography* **22**, 160-171.

Fabry, V.J., Seibel, B.A., Feely, R.A. & Orr, J.C. 2008. Impacts of ocean acidification on marine fauna and ecosystem processes. *ICES Journal of Marine Science* **65**, 414-432.

Falk, J. M., A. R. Graefe and R. B. Ditton. 1989. Patterns of Participation and Motivation among Saltwater Tournament Anglers. *Fisheries* 14:4, 10-17.

Fall, J.A. and D. Koster. 2011. Subsistence harvests of Pacific halibut in Alaska, 2009. Alaska Department of Fish and Game Division of Subsistence, Technical Paper No. 357, Anchorage. Data compiled by Alaska Fisheries Information Network for Alaska Fisheries Science Center, Seattle.

Fall, J.A., C. Brown, N. Braem, J.J. Simon, W.E. Simeone, D.L. Holen, L. Naves, L. Hutchinson-Scarborough, T. Lemons, and T.M. Krieg. 2011, revised. Alaska subsistence salmon fisheries 2008 annual report. Alaska Department of Fish and Game Division of Subsistence, Technical Paper No. 359, Anchorage. Data compiled by Alaska Fisheries Information Network for Alaska Fisheries Science Center, Seattle.

FAO. 2010. World aquaculture 2010. FAO Fisheries and Aquaculture Technical Paper No. 500/1, Rome, Italy: 120 p.

FAO/WHO. 2011. Report of the Joint FAO/WHO Expert Consultation on the Risks and Benefits of Fish Consumption. Rome, Food and Agriculture Organization of the United Nations; Geneva, World Health Organization, 50 pp.

Farmer, P. 1995. Social Inequalities and Emerging Infectious Diseases. *Emerging Infectious Disease* 2(4): 259-269.

Fautin, D., Dalton, P., Incze, L.S., Leong, J.A.C., Pautzke, C., Rosenberg, A., Sandifer, P., Sedberry, G., Tunnell, Jr., J.W., Abbott, I., Brainard, R.E., Brodeur, M., Eldredge, L.G., Feldman, M., Moretzsohn, F., Vroom, P.S., Wainstein, M. & Wolff, N. 2010. An overview of marine biodiversity in United States waters. *PLoS ONE* **5**, e11914.

Fauver, S. 2008. *Climate-related needs assessment synthesis for coastal management.* Charleston, S.C.: National Oceanic and Atmospheric Administration, Coastal Services Center.

Fazzino, D., and P.A. Loring. 2009. "From crisis to cumulative effects: food security challenges in Alaska." NAPA Bulletin 32: 152-177.

Federal Register. 2008. *Department of the Interior: endangered and threatened wildlife and plants; determination of threatened status for the polar bear (*Ursus maritimus*) throughout its range.* 73 Fed. Reg. 28212, May 15, 2008.

Federal Register. 2010a. *Environmental Protection Agency: Clean Water Act Section 303(d): notice of call for public comment on 303(d) program and ocean acidification.* 75 Fed. Reg. 13537, Mar. 22, 2010.

Federal Register. 2010b. *National Oceanic and Atmospheric Administration: endangered and threatened wildlife and plants; threatened status for the southern distinct population segment of the spotted seal.* 75 Fed. Reg. 65,239, Oct. 22, 2010.

Federal Register. 2010c. *National Oceanic and Atmospheric Administration: endangered and threatened wildlife; notice of 90-day finding on a petition to list 83 species of corals as threatened or endangered under the Endangered Species Act (ESA).* 75 Fed. Reg. 6,616, Feb. 10, 2010.

Federal Register. 2011. *Department of the Interior: endangered and threatened wildlife and plants; 12-month finding on a petition to list the Pacific walrus as endangered or threatened.* 76 Fed. Reg. 7634, Feb. 10, 2011.

Fedler, A.J. & Ditton, R.B. 1994. Understanding angler motivations in fisheries management. *Fisheries* **19**, 6-13.

Feely, R. A., Doney, S.C. & Cooley, S.R. 2009. Ocean Acidification: Present Conditions and Future Changes in a High-CO2 World. *Oceanography*, **22**, 36-47.

Feely, R.A., Fabry, V.J., Dickson, A., Gattuso, J.-P., Bijma, J., Riebesell, U., Doney, S., Turley, C., Saino, T., Lee, K., Anthony, K. & Kelypas, J. 2010. An international observational network for ocean acidification. In *Proceedings of OceanObs ' 09: Sustained Ocean Observations and Information for Society (Vol. 2), Venice, Italy, 21 - 25 September 2009.* Hall, J., Harrison, D.E. & Stammer, D. (eds.). ESA Publication WPP-306. doi:10.5270/OceanObs09.

Feely, R.A., Sabine, C.L., Hernandez-Ayon, J.M., Ianson, D. & Hales, B. 2008. Evidence for upwelling of corrosive "acidified" water along the continental shelf. *Science* **320**,1490-1492.

Feely, R.A., Sabine, C.L., Lee, K., Berelson, W., Kleypas, J., Fabry, V.J. & Millero, F.J. 2004. Impact of anthropogenic CO_2 on the $CaCO_3$ system in the oceans. *Science* **305**, 362-366.

Felthoven, R.G., Paul, C. J. M. and Torres, M. 2009. Measuring Productivity and Its Components for Fisheries: The Case of the Alaskan Pollock Fishery, 1994-2003. *Natural Resource Modeling* 22(1):105-136.

Feng, Y., Warner, M.E., Zhang, Y., Sun, J., Fu, F.-X., Rose, J.M. & Hutchins, D.A. 2008. Interactive effects of increased pCO_2, temperature and irradiance on the marine coccolithophore *Emiliania huxleyi* (Prymnesiophyceae). *European Journal of Phycology* **43**, 87 - 98.

Ferguson, S.H., Stirling, I. & McLoughlin, P. 2005. Climate change and ringed seal (*Phoca hispida*) recruitment in western Hudson Bay. *Marine Mammal Science* **21**, 121-135.

Findlay, H.S., Burrows, M.T., Kendall, M.A., Spicer, J.I. & Widdicombe, S. 2010. Can ocean acidification affect population dynamics of the barnacle *Semibalanus balanoides* at its southern range edge? *Ecology* **91**, 2931-2940.

Fire, S.E., Fauquier, D., Flewelling, L.J., Henry, M., Naar, J., Pierce, R. & Wells, R.S. 2007. Brevetoxin exposure in bottlenose dolphins (*Tursiops truncatus*) associated with *Karenia brevis* blooms in Sarasota Bay, Florida. *Marine Biology* **152**(4), 827-834.

Firestone, J. and Lilley, J. 2004. An endangered species: Aboriginal whaling and the right to self-determination and cultural heritage in a national and international context. *Environmental Law Reporter News* and *Analysis*, 34, 10763-10787.

Firth, L.B., Knights, A.M. & Bell, S.S. 2011. Air temperature and winter mortality: Implications for the persistence of the invasive mussel, *Perna viridis* in the intertidal zone of the southeastern United States. *Journal of Experimental Marine Biology and Ecology* **400**, 250-256.

Fischbach, A.S., Amstrup, S.C. & Douglas, D.C. 2007. Landward and eastward shift of Alaskan polar bear denning associated with recent sea ice changes. *Polar Biology* **30**, 1395-1405.

Fischer, F. 2000. *Citizens, experts, and the environment: The politics of local knowledge.* Durham, NC: Duke University Press.

Fletcher, C. 2010. *Hawai'i's Changing Climate. Briefing Sheet, 2010 UH Sea Grant Program.* Honolulu, HI: Center for Island Climate Adaptation and Policy (ICAP).

Flewelling, L.J., Naar, J.P., Abbott, J.P., Baden, D.G., Barros, N.B., Bossart, G.D., Dottein, M.-Y.D., Hammond, D.G., Haubold, E.M., Heil, C.A., Henry, M.S., Jacocks, H.M., Leighfield, T.A., Pierce, R.H., Pithford, T.D., Rommel, S.A., Scott, P.S., Steiginger, K.A., Truby, E.W., Van Dolah, F.M. & Landsberg, J.H. 2005. Brevetoxicosis: red tides and marine mammal mortalities. *Nature* **435**, 755-756.

Florida Atalantic University. 2011. FAU's Southeast National Marine Renewable Energy Center Achieves Major Milestone. Press release issued April 18, 2011. Accessed January 9, 2012 from http://www.fau.edu/communications/mediarelations/Releases0411/041119.php.

Fodrie, F.J., Heck, K.L., Powers, S.P., Graham, W.M. & Robinson, K.L. 2009. Climate-related, decadal-scale assemblage changes of seagrass-associated fishes in the northern Gulf of Mexico. *Global Change Biology* **16**, 48-59.

Fogarty, M. J., Incze, L., Hayhoe, K., Mountain, D.G. & Manning, J. 2008. Potential climate change impacts on Atlantic cod (*Gadus morhua*) off the northeastern USA. *Mitigation and Adaptation Strategies for Global Change* **13**, 453-466.

Foley, A.M., Singel, K.E., Dutton, P.H., Summers, T.M., Redlow, A.E. & Lessman, J. 2007. Characteristics of a green turtle (*Chelonia mydas*) assemblage in northwestern Florida determined during a hypothermic stunning event. *Gulf of Mexico Science* **25**, 131-143.

Folke, C., Carpenter, S., Walker, B., Scheffer, M., Elmqvist, T., Gunderson, L. & Holling, C.S. 2004. Regime shifts, resilience, and biodiversity in ecosystem management. *Annual Review of Ecology, Evolution, and Systematics* **35**, 557-581.

Food Agric. Org. U.N. 2007. The world's mangroves 1980-2005. *FAO For. Pap. 153*, Food Agric. Org. U.N.,Rome

Food Agric. Org. U.N. 2010. Fishery and aquaculture statistics. *FAO Yearb.*, Food Agric. Org. U.N., Rome

Food and Agriculture Organization of the United Nations & World Health Organization (WHO). 2011. *Report of the Joint FAO/WHO Expert Consultation on the Risks and Benefits of Fish Consumption.* Rome: Food and Agriculture Organization of the United Nations; Geneva: World Health Organization.

Ford, J.D. 2009. Vulnerability of Inuit food systems to food insecurity as a consequence of climate change: a case study from Igloolik, Nunavut. *Regional Environmental Change* 9:83-100.

Ford, J.D., T. Pearce, F. Duerden, C. Furgal, and B. Smit. 2010. Climate change policy responses for Canada's Inuit population: The importance of and opportunities for adaptation. *Global Environmental Change* 20(1):177-191.

Ford, S.E. 1996. Range extension by the oyster parasite *Perkinsus marinus* into the northeastern United States: response to climate change? *Journal of Shellfish Research* **15**, 45-56.

Fowler, K. and H. Etchegary. 2008. Economic crisis and social capital: The story of two rural fishing communities. *Journal of Occupational and Organizational Psychology* 81(2):319-341.

Friedland, K. D., and J. A. Hare. 2007. Long term trends and regime shifts in sea surface temperature on the continental shelf of the northeastern United States. *Continental Shelf Research* 27: 2313-2328.

Friedlingstein, P., Bopp, L., Ciais, P. 2001. Positive feedback between future climate change and the carbon cycle. *Geophysical Research Letters*, **28**, 1543-1546.

Friedman, M., Deputy Commissioner for Operations, Food and Drug Administration. 1996. Testimony Before the Subcommittee on Livestock, Dairy and Poultry. Committee on Agriculture: U.S. House of Representatives. May 22.

Frost, K. J., and Suydam, Robert S. 2010. Subsistence harvest of beluga or white whales (Delphinapterusleucas) in northern and western Alaska, 1987-2006. Journal of Cetacean Research and Management 11(3): 293-299. Data compiled by Alaska Fisheries Information Network for Alaska Fisheries Science Center, Seattle.

Frumhoff, P.C., McCarthy, J.J., Melillo, J.M., Moser, S.C. & Wuebbles, D.J. 2007. *Confronting Climate Change in the U.S. Northeast: Science, Impacts, and Solutions. Synthesis report of the Northeast Climate Impacts Assessment (NECIA).* Cambridge, MA: Union of Concerned Scientists (UCS).

Fu, F.-X., Mulholland, M.R., Garcia, N., Beck, A., Bernhardt, P.W., Warner, M.E., Sañudo-Wilhelmy, S.A. & Hutchins, D.A. 2008. Interactions between changing pCO_2, N_2 fixation, and Fe limitation in the marine unicellular cyanobacterium *Crocosphaera*. *Limnology and Oceanography* **53**, 2472-2484.

Fu, F.-X., Place, A.R., Garcia, N.S. & Hutchins, D.A. 2010. Effects of changing pCO_2 and phosphate availability on toxin production and physiology of the harmful bloom dinoflagellate *Karlodinium veneficum*. *Aquatic Microbial Ecology* **59**, 55-65.

Gage, K., Burkot, T., Eisen, R. & Hayes, E. 2008. Climate and Vectorborne Diseases. *American Journal of Preventive Medicine* **35**, 436-450.

Gaines, S. D., Gaylord, B. & Largier, J. L. 2003. Avoiding current oversights in marine reserve design. Ecol. Appl. 13, 32:46

Gale, R.P. 1991. Gentrification of America's coasts: Impacts of the growth machine on commercial fishermen. *Society and Natural Resources* **4**, 101-121.

Galloway, J.N., Townsend, A.R., Erisman, J.W., Bekunda, M., Cai, Z., Freney, J.R., Martinelli, L.A., Seitzinger, S.P. & Sutton, M.A. 2008. Transformation of the Nitrogen Cycle: Recent Trends, Questions, and Potential Solutions. *Science* **320**, 889-892.

Gao, K., Helbling, E.W., Häder, D.-P. & Hutchins, D.A. In press. Ocean acidification and marine primary producers under the sun: a review of interactions between CO_2, warming, and solar radiation. *Marine Ecology Progress Series*.

Garber-Yonts, B. and J. Lee. 2011. Crab SAFE 2011 Economic Status Report. Document presented to the 9-11 September 2011 meeting in the NPFMC crab plan team. http://www.afsc.noaa.gov/REFM/Socioeconomics/PDFs/crabsafe2011_draft.pdf

García-Quijano, C. G. 2007. Fishers' Knowledge of Marine Species Assemblages: Bridging between Scientific and Local Ecological Knowledge in Southeastern Puerto Rico. *American Anthropologist* 109(3):529-536.

Garcia-Reyes, M. & Largier, J.L. 2010. Observations of increased wind-driven coastal upwelling off central California. *Journal of Geophysical Research* **115**, CO4011.

Gardner, J. 1989. Folklife in the Florida Parishes. Retrieved January 4, 2012 from http://www.louisianafolklife.org/LT/Virtual_Books/Fla_Parishes/book_florida_overview.html. Louisiana Folklife Program, Division of the Arts and the Center for Regional Studies, Southeastern Louisiana University: Baton Rouge, LA.

Gardner, T.A., Cote, I.M., Gill, J.A., Grant, A. & Watkinson, A.R. 2003. Long-term region-wide declines in Caribbean corals. *Science* **301**, 958-960.

Garlich-Miller, J., MacCracken, J.G., Snyder, J., Meehan, R., Myers, M., Wilder, J. M., Lance, E. & Matz, A. 2011. *2011: Status review of the Pacific walrus (Odobenus rosmarus divergens)*. Anchorage, AK: U.S. Fish and Wildlife Service, Marine Mammals Management.

Gasalla, M. A. and A.C.S. Diegues. 2011. "Ethno-oceanography" as an Interdisciplinary Means to Approach Marine Ecosystem Change, pp. 120-136. In Rosemary Ommer, Ian Perry, Kevern L. Cochrane, and Philippe Cury, eds. *World Fisheries: A Social-Ecological Analysis*. Hoboken, NJ and Oxford, U.K.: Wiley-Blackwell.

Gascard, J.C., Festy, J., Ie Gogg, H., Weber, M., Bruemmer, B., Offermann, M., Doble, M., Wadhams, P., Forsberg, R., Hanson, S., Skourup, H., Gerland, S., Nicolaus, M., Metaxin, J.P., Grangeon, J., Haapala, J., Rinne, E., Haas, C., Heygster, G., Jakobson, E., Palo, T., Wilkinson, J., Kaleschke, L., Claffey, K., Elder, B. & Bottenheim, J. 2008. Exploring Arctic transpolar drift during dramatic sea ice retreat. Eos, **89**(3), 21-28.

Gaston, A.J., and K. Woo. 2008. Razorbills (*Alca torda*) follow subarctic prey into the Canadian Arctic: colonization results from climate change? *The Auk* **125**, 939-942.

Gaston, A.J., Gilchrist, H.G., Mallory, M.L. & Smith, P.A. 2009. Changes in seasonal events, peak food availability, and consequent breeding adjustment in a marine bird: a case of progressive mismatching. *The Condor* **111**, 111-119.

Gatewood, J. B. and B. J. McCay. 1990. Comparison of Job Satisfaction in Six New Jersey Fisheries: Implications for Management. *Human Organization* 49(1): 14-25.

Gatlin, D. M., F. T. Barrows, D. Bellis, P. Brown, K. Daborwski, T. G. Gaylord. R. W. Hardy, E. M. Herman, G. Hu, A. Krogdahl, R. Nelson, K. Overturf, M. B. Rust, W. M. Sealey, D. Skonberg, E. J. Souza, D. Stone, R. Wilson and E Wurtele. 2007. Expanding the utilization of sustainable plant products in aquafeeds: a review. Aquaculture Research, 38: 551-579.

Gattuso, J.P. & Hansson, L. (eds.) 2011. *Ocean acidification*. Oxford, U.K.: Oxford University Press.

Gaynor, K., A.R. Katz, S.Y. Park, M. Nakata, T.A. Clark, and P.V. Effler. 2007. Leptospirosis on Oahu: an outbreak associated with flooding of a university campus. American Journal of Tropical Medicine and Hygiene 76: 882-885.

Gazeau, F., Gattuso, J.P., Dawber, C., Pronker, A.E., Peene, F., Peene, J., Heip, C.H., & Middel-burg, J.J. 2010. Effect of ocean acidification on the early life stages of the blue mussel (*Mytilus edulis*). *Biogeosciences* **7**, 2051-2060.

Gazeau, F., Quiblier, C., Jansen, J.M., Gattuso, J.-P., Middelburg, J.J. & Heip, C.H.R. 2007. Impact of elevated CO_2 on shellfish calcification. *Geophysical Research Letters* **34**, LO7603.

Gearhead, S., Matumeak, W., Angutikjuaq, I., Maslanik, J., Huntington, H.P., Leavitt, J., Kagak, D.M., Tigullaraq, G. & Barry, R.G. 2006. 'It's not that simple: A collaborative comparison of sea ice environments, their uses, observed changes, and adaptations in Barrow, Alaska, USA, and Clyde River, Nunavut, Canada. *Ambio* **35**, 203-211.

Gedan, K.B., & Bertness, M.D. 2010. How will warming affect the salt marsh foundation species *Spartina patens* and its ecological role? *Oecologia* **164**, 479-487.

Gerlach, S.C., P.A. Loring, and A.M. Turner. 2011. Food Systems, Climate Change, and Commu-nity Needs. In North by 2020, ed. Hajo Eicken and A.L. Lovecraft. Fairbanks, AK: University of Alaska Press.

Giles, K., Laxon, S. & Ridout, A. 2008. Circumpolar thinning of Arctic sea ice following the 2007 record ice extent minimum. *Geophyical. Research Letters.* **35**, L22502.

Giri, C., Ochieng, E., Tieszen, L, Zhu, Z., Singh, A, Loveland, T., Masek, J., Duke, N. 2010. Status and distribution of mangrove forests of the world using earth observation satellite data. *Global Ecology and Biogeography* **20**(1), 154-159.

Gjøen, H.M & Bentsen, H.B. 1997. Past, present, and future of genetic improvement in salmon aquaculture ICES *Journal of Marine Science* **54**, 1009-1014.

Glass, G.E., Cheek, J.E., Patz, J.A., Shields, T.M., Doyle, T.J., Thoroughman, D.A., Hunt, D.K., Enscore, R.E., Gage, K.L., Irland, C., & Bryan, R. 2000. Anticipating risk areas for Hantavirus pulmonary syndrome with remotely sensed data: re-examination of the 1993 outbreak. *Emerging Infectious Diseases* **6**, 238-247.

Gleeson, M.W., & Strong, A.E. 1995. Applying MCSST to coral reef bleaching. *Advances in Space Research* **16**, 151-154.

Glenn, R.P., & Pugh, T.L. 2006. Epizootic shell disease in American lobster (*Homarus americanus*) in Massachusetts coastal waters: interactions of temperature, maturity, and intermolt dura-tion. *Journal of Crustacean Biology* **26**, 639-645.

Glenn, S., Arnone, R., Bergmann, T., Bissett, W.P., Crowley, M., Cullen, J., Gryzmski, J., Haid-vogel, D., Kohut, J., Moline, M., Oliver, M., Orrico, C., Sherrell, R., Song, T., Weidemann, A., Chant, R., & Schofield, O. 2004. Biogeochemical impact of summertime coastal upwelling on the New Jersey shelf. *Journal of Geophysical Research* **109**, C12S02.

Glenn, S.M., Crowley, M.F., Haidvogel, D.B., & Song, Y.T. 1996. Underwater observatory captures coastal upwelling off New Jersey, *EOS*, **77**, 233-236.

Glick, P., Staudt, A. & Stein, B. 2009. *A new era for conservation: review of climate change adaptation literature*. Washington, DC: National Wildlife Federation.

Glick, P., Stein, B.A. & Edelson, N.A. (eds). 2011. Scanning the conservation horizon: a guide to climate change vulnerability assessment. National Wildlife Federation, Washington, DC. http://www.nwf.org/News-and-Magazines/Media-Center/Reports/Archive/2011/Scanning-the-Horizon.aspx

Global Climate Change Impacts in the United States, Thomas R. Karl, Jerry M. Melillo, and Thomas C. Peterson, (eds.). Cambridge University Press, 2009.

Godduhn, A., Duffy, L.K., 2003. Multi-generation health risks of persistent organic pollution in the far north: use of the precautionary approach in the Stockholm convention. Environ-mental Science and Policy 6: 349-391.

Gold, W. and Salit, I. 1993. Aeromonas hydrophila infections of skin and soft tissue: report of 11 cases and review. *Clinical Infectious Disease* 16: 69-74.

Golden J, Hartz D, Brazel A, Luber G, and Phelan P. In press. *A biometeorology study of climate and heat-related morbidity in Phoenix from 2001 to 2006. International Journal of Biometeorology.*

Goldewijk, K.K. 2005. Three centuries of global population growth: a spatial referenced population (density)cdatabase for 1700-2000. *Population and Environment* **26**, 343-367.

Goldstein, T., J.A.K. Mazet, T.S. Zabka, G. Langlois, K.M. Colegrove, et al. 2008. Novel symptomatology and changing epidemiology of domoic acid toxicosis in California sea lions (*Zalophus californianus*): an increasing risk to marine mammal health. Proceedings of the Royal Society B 275(1632): 267-276.

Gooding, B., C.D.G. Harley, and E. Tang. 2009. Multiple climate variables increase growth and feeding rates of a keystone predator. Proceedings of the National Academy of Sciences USA 106: 9316-9321.

Gooding, B., Harley, C.D.G. & Tang, E. 2009. Multiple climate variables increase growth and feeding rates of a keystone predator. *Proceedings of the National Academy of Sciences USA* **106**, 9316-9321.

Goodkin, N.F., Hughen, K.A., Doney, S. & Curry, W.B. 2008. Increased multidecadal variability of the North Atlantic Oscillation since 1781. *Nature Geoscience* **1**, 844-848.

Gouhier, T.C., Guichard, F. & Menge, B.A. 2010. Ecological processes can synchronize marine population dynamics over continental scales. *Proceedings of the National Academy of Sciences USA* **107**, 8281-8286.

Grafton, R.Q. 2010. Adaptation to climate change in marine capture fisheries. *Marine Policy* **34**, 606-615.

Graham, M.H. 2004. Effects of local deforestation on the diversity and structure of Southern California giant kelp forest food webs. *Ecosystems* **7**, 341-357.

Graham, M.H., Halpern, B.S. & Carr, M.H. 2008. Diversity and dynamics of Californian subtidal kelp forests. In *Food webs and the dynamics of marine benthic ecosystems.* McClanahan, T.R. & Branch, G.R. (eds.), Oxford, U.K: Oxford University Press, 103-134.

Graham, M.H., Vásquez, J.A. & Buschmann, A.H. 2007. Global ecology of the giant kelp *Macrocystis*: from ecotypes to ecosystems. *Oceanography and Marine Biology: an Annual Review* **45**, 39-88.

Graham, N.A.J., Wilson, S.K., Jennings, S., Polunin, N.V.C., Bijoux, J.P. & Robinson, J. 2006. Dynamic fragility of oceanic coral reef ecosystems. *Proceedings of the National Academy of Sciences USA* **103**, 8425-8429.

Grantham, B.A., Chan, F., Nielsen, K.J., Fox, D.S., Barth, J.A., Huyer, A., Lubchenco, J. & Menge, B.A. 2004. Upwelling-driven nearshore hypoxia signals ecosystem and oceanographic changes in the northeast Pacific. *Nature* **429**, 749-754.

Grebmeier, J. M., J. E. Overland, S. E. Moore, E. V. Farley, E. C. Carmack, L. W. Cooper, K. E. Frey, J. H. Helle, F. A. McLaughlin, and S. L. McNutt. 2006. A Major Ecosystem Shift in the Northern Bering Sea. Science 311 (5766) (March 10): 1461-1464. doi:10.1126/science.1121365.

Greene, C.H., Pershing, A.J., Cronin, T.M. & Ceci, N. 2008. Arctic climate change and its impacts on the ecology of the North Atlantic. *Ecology* **89**, S24-S38.

Greene, R.M., Lehrter, J.C., Hagy, J.D. III. 2009. Multiple regression models for hindcasting and forecasting midsummer hypoxia in the Gulf of Mexico. *Ecological Applications* **19**, 1161-1175.

Greenough, G., McGeehin, M., Bernard, S., Trtanj, J. 2001. *The potential impacts of climate change on health impacts of extreme weather events in the United States. Environmental Health Perspectives* **109**,185-189.

Greer, A., Ng, V. & Fisman, D. 2008. Climate change and infectious diseases in North America: the road ahead. *Canadian Medical Association Journal* **178**, 715-722.

Gregg, R.M., Hansen, L.J., Feifel, K.M., Hitt, J.L., Kershner, J.M., Score, A. & Hoffman, J.R. 2011. *The State of Marine and Coastal Adaptation in North America: A Synthesis of Emerging Ideas.* Bainbridge Island, WA: EcoAdapt.

Gregory J M and Webb M 2008 Tropospheric adjustment induces a cloud component in CO_2 forcing J. Clim. 21 58-71

Gregory, J. M., K. W. Dixon, et al. (2005). "A model intercomparison of changes in the Atlantic thermohaline circulation in response to increasing atmospheric CO_2 concentration." Geophys. Res. Lett. **32**(12): L12703.

Grémillet, D., & Boulinier, T. 2009. Spatial ecology and conservation of seabirds facing global climate change: a review. *Marine Ecology Progress Series* **391**, 121-137.

Grinsted, A., J. Moore, et al. (2010). "Reconstructing sea level from paleo and projected temperatures 200 to 2100." Climate Dynamics **34**(4): 461-472.

Guerra, M.A. 2009. Leptospirosis. Journal of the American Veterinary Medical Association 234 (4): 472-478.

Gulland, F.M.D., Koski, M., Lowenstine, L.J., Colagross, A., Morgan, L. & Spraker, T. 1996. Leptospirosis in California sea lions (*Zalophus californianus*) stranded along the central California coast, 1981-1994. *Journal of Wildlife Diseases* **32**, 572-580.

Gunter, G., Williams, R.H., Davis, C.C. & Smith, F.G.W. 1948. Catastrophic mass mortality of marine animals and coincident phytoplankton bloom on the west coast of Florida, November 1946 to August 1947. *Ecological Monographs* **18**, 309-324.

Hagen, P. and T. Quinn. 1991. Long-term growth dynamics of young Pacific halibut: Evidence of temperature-induced variation. Fisheries Research, 11, 283-306.

Hagy III, J.D. & Murrell, M.C. 2007. Susceptibility of a northern Gulf of Mexico estuary to hypoxia: An analysis using box models. *Estuarine, Coastal and Shelf Science* **74**, 239-253.

Hall, S. J., A. Delaporte, M. J. Phillips, M. Beveridge and M. O'Keefe. 2011. Blue Frontiers: Managing the Environmental Costs of Aquaculture. The WorldFish Center, Penang, Malaysia.

Hallegraeff, G.M. 1993. A review of harmful algal blooms and their apparent global increase. *Phycologia* **32**, 79-99.

Hallegraeff, G.M. 2010. Ocean climate change, phytoplankton community responses, and harmful algal blooms: a formidable predictive challenge. *Journal of Phycology* **46**, 220-235.

Hall-Spencer, J.M., Rodolfo-Metalpa, R., Martin, S., Ransome, E., Fine, M., Turner, S.M., Rowley, S.J., Tedesco, D. & Buia, M.-C.. 2008. Volcanic carbon dioxide vents show ecosystem effects of ocean acidification. *Nature* **454**, 96-99.

Halpern, B.S., C. Longo, D. Hardy, K.L. McLeod, J.F. Samhouri, et al. In review. An ocean health index: quantifying and mapping the health of global marine ecosystems. Nature.

Halpern, B.S., C.V. Kappel, K.A. Selkoe, F. Micheli, C. Ebert, C. Kontgis, C.M. Crain, R. Martone, C. Shearer, and S.J.Teck. 2009b. Mapping cumulative human impacts to California Current marine ecosystems. Conservation Letters 2: 138-148.

Halpern, B.S., Ebert, C.M., Kappel, C.V., Madin, E.M.P., Micheli, F., Perry, M., Selkoe, K.A. & Walbridge, S. 2009a. Global priority areas for incorporating land-sea connections in marine conservation. *Conservation Letters* **2**, 189-196.

Halpern, B.S., Lester, S.E. & McLeod, K.L. 2010. Placing marine protected areas onto the ecosystem-based management seascape. *Proceedings of the National Academy of Sciences USA* **107**, 18312-18317.

Halpern, B.S., McLeod, K.L., Rosenberg, A.A. & Crowder, L.B. 2008a. Managing for cumulative impacts in ecosystem-based management through ocean zoning. *Ocean & Coastal Management* **51**, 203-211.

Halpern, B.S., Walbridge, S., Selkoe, K.A., Kappel, C.V., Micheli, F., D'Agrosa, C., Bruno, J., Casey, K.S., Ebert, C., Fox, H.E., Fujita, R., Heinemann, D., Lenihan, H.S., Madin, E.M.P., Perry, M., Selig, E., Spalding, M., Steneck, R. & Watson, R. 2008b. A global map of human impact on marine ecosystems. *Science* **319**, 948-952.

Hamilton, L.C., Brown, B.C. and Rasmussen, R.O., 2003. Greenland's Cod-to-Shrimp Transition: Local Dimensions of Climatic Change. Arctic, 56(3), pp.271-282.

Hannah, C.G., F. Dupont, and M. Dunphy. 2009. "Polynyas and tidal currents in the Canadian Arctic Archipelago." Arctic 62 (1): 83-95.

Hannesson, R. 2007. Introduction to Special Issue: Economic Effects of Climate Change on Fisheries, Natural Resource Modeling. 20: 157-162.

Hansen, I.S., Keul, N., Sorensen, J.T., Erichsen, A., & Anderson, J.H.. 2007. Oxygen map for the Baltic Sea. BALANCE Interim Report No. 17

Hansen, J., Sato, M., Kharecha, P., Russell, G., Lea, D.W. & Siddall, M. 2007. Climate change and trace gases. *Philosophical Transactions of the Royal Society A: Mathematical, Physical and Engineering Sciences* **365**, 1925-1954.

Hansen, L., Hoffman, J., Drews, C. & Mielbrecht, E. 2010. Designing climate-smart conservation: guidance and case studies. *Conservation Biology* **24**, 63-69.

Hare, C.E., Leblanc, K., Ditullio, G.R., Kudela, R.M., Zhang, Y., Lee, P.A., Riseman, S. & Hutchins, D.A. 2007. Consequences of increased temperature and CO_2 for phytoplankton community structure in the Bering Sea. *Marine Ecology Progress Series* **352**, 9-16.

Hare, J.A. & Able, K.W. 2007. Mechanistic links between climate and fisheries along the east coast of the United States: explaining population outbursts of Atlantic croaker (*Micropogonias undulatus*). *Fisheries Oceanography* **16**, 31-45.

Hare, J.A., & Cowen, R.K. 1997. Size, growth, development, and survival of the planktonic larvae of *Pomatomus saltatrix* (Pisces: Pomatomidae). *Ecology* **78**, 2415-2431.

Hare, J.A., Alexander, M.A., Fogarty, M.J., Williams, E.H. & Scott, J.D. 2010. Forecasting the dynamics of a coastal fishery species using a coupled climate-population model. *Ecological Applications* **20**, 452-464.

Hare, S.R., & Mantua, N.J. 2000. Empirical evidence for North Pacific regime shifts in 1977 and 1989. *Progress in Oceanography* **47**, 103-145.

Harley, C.D.G. 2003. Abiotic stress and herbivory interact to set range limits across a two-dimensional stress gradient. *Ecology* **84**, 1477-1488.

Harley, C.D.G. 2008. Tidal dynamics, topographic orientation, and temperature-mediated mass mortalities on rocky shores. *Marine Ecology Progress Series* **371**, 37-46.

Harley, C.D.G. 2011. Climate change, keystone predation, and biodiversity loss. *Science*   **334**, 1124-1127.

Harley, C.D.G., & Paine, R.T. 2009. Contingencies and compounded rare perturbations dictate sudden distributional shifts during periods of gradual climate change. *Proceedings of the National Academy of Sciences USA* **106**, 11172-11176.

Harley, C.D.G., Hughes, A.R., Hultgren, K.M., Miner, B.G., Sorte, C.J.B., Thornber, C.S., Rodriguez, L.F., Tomanek, L. & Williams, S.L. 2006. The impacts of climate change in coastal marine systems. *Ecology Letters* **9**, 228-241.

Harrington, C.R. 2008. The evolution of Arctic marine mammals. Ecological Applications 18: S23-S40.

Harrison D.E., D. Stammer). ESA Publication WPP - 306.

Harrison, G.G., M. Sharp, G. Manalo-LeClair, A. Ramirez, and N. McGarvey. 2007. Food security among California's low-income adults improves, but most severely affected do not share in improvement. *Policy Brief (UCLA Center for Health Policy Research)*, (PB2007-6), 1-11.

Harrison, G.G., Sharp, M., Manalo-LeClair, G., Ramirez, A. & McGarvey, N. 2007. Food security among California's low-income adults improves, but most severely affected do not share in improvement. *Policy Brief (UCLA Center for Health Policy Research)*, (PB2007-6), 1-11.

Harrold, C., & Reed, D.C. 1985. Food availability, sea urchin grazing, and kelp forest community structure. *Ecology* **66**, 1160-1169.

Harvell, C.D., Kim, K., Burkholder, J.M., Colwell, R.R., Epstein, P.R., Grimes, D.J., Hofmann, E.E., Lipp, E.K., Osterhaus, A.D.M.E., Overstreet, R.M., Porter, J.W., Smith, G.W. & Vasta, G.R. 1999. Emerging marine diseases - climate links and anthropogenic factors. *Science* **285**, 1505-1510.

Harvell, D., Altizer, S., Cattadori, I.M., Harrington, L. & Weil, E. 2009. Climate change and wild-life diseases: when does the host matter the most? *Ecology* **90**, 912-920.

Hatala, J.A., Detto, M., Sonnentag, O., Deverel, S.J., Verfaillie, J. & Baldocchi, D.D. 2012. Green-house gas (CO_2, CH_4, H_2O) fluxes from drained and flooded agricultural peatlands in the Sacramento San Joaquin Delta. *Agriculture, Ecosystems and Environment* **150**, 1-18.

Hawkes, L.A., Broderick, A.C., Coyne, M.S., & Godfrey, M.H. 2006. Phenotypically linked dichotomy in sea turtle foraging requires multiple conservation approaches. *Current Biology* **16**, 990-995.

Hawkes, L.A., Broderick, A.C., Godfrey, M.H. & Godley, B.J. 2007. Investigating the potential impacts of climate change on a marine turtle population. *Global Change Biology* **13**, 923-932.

Hawkes, L.A., Broderick, A.C., Godfrey, M.H. & Godley, B.J. 2009. Climate change and marine turtles. *Endangered Species Research* **7**, 137-154.

Haynie, A. C., and Pfeiffer, L. 2012a. Why Economics Matters for Understanding the Effects of Climate Change on Fisheries. ICES Journal of Marine Science, in press.

Haynie, A. C., and Pfeiffer, L. 2012b. The climatic and economic drivers of the Bering Sea pollock fishery: Implications for the future . In review.

Hays, G.C., Richardson, A.J. & Robinson, C. 2005. Climate change and marine plankton. *Trends in Ecology and Evolution* **20**, 337-344.

Heide-Jørgensen, M.P., Laidre, K.L., Quackenbush, L.T. & Citta, J.J. 2011. The Northwest Passage opens for bowhead whales. *Biology Letters* **8**, 270-273.

Heinitz, M., Ramona, R., Wagner, D. and Tatini, S. 2000. Incidence of *Salmonella* in Fish and Seafood. *Journal of Food Protection* 63(5):579-592.

Heinz Center. 2008. *Strategies for managing the effects of climate change on wildlife and ecosystems.* Washington, DC: The Heinz Center.

Heisler, J., Gilbert, P.M., Burkholder, J.M., Anderson, D.M., Cochlan, W., Dennison, W.C., Dortch, Q., Gobler, C.J., Heil, C.A., Humphries, E., Lewitus, A., Magnien, R., Marshall, H.G., Sellner, K., Stockwell, D.A., Stoecker, D.K. & Suddleson, M. 2008. Eutrophication and harmful algal blooms: a scientific consensus. *Harmful Algae* **8**, 3-13.

Held, I.M. & Soden, B.J. 2006, Robust responses of the hydrological cycle to global warming *Journal of Climate* **19**, 5686-5699.

Heller, N.E. and E.S. Zavaleta. 2009. Biodiversity management in the face of climate change: A review of 22 years of recommendations. *Biological Conservation* 142:14-32.

Hellmann, J.J., Byers, J.E., Bierwagen, B.G. & Dukes, J.S. 2008. Five potential consequences of climate change for invasive species. *Conservation Biology* **22**, 534-543.

Helly, J.J. & Levin, L.A. 2004. Global distribution of naturally occurring marine hypoxia on continental margins. *Deep Sea Research Part I: Oceanographic Research Papers* **51**, 1159-1168.

Helm, K.P., Bindoff, N.L., & Church, J.A., 2010. Changes in the global hydrological-cycle inferred from ocean salinity. Geophysical Research Letters **37**, L18701.

Helmuth, B. 2009. From cells to coastlines: how can we use physiology to forecast the impacts of climate change? *The Journal of Experimental Biology* **212**, 753-760.

Helmuth, B., Broitman, B.R., Blanchette, C.A., Gilman, S., Halpin, P., Harley, C.D.G., O'Donnell, M.J., Hofmann, G.E., Menge, B. & Strickland, D. 2006a. Mosaic patterns of thermal stress in the rocky intertidal zone: implications for climate change. *Ecological Monographs* **76**, 461-479.

Helmuth, B., Mieszkowska, N., Moore, P. & Hawkins, S.J. 2006b. Living on the edge of two changing worlds: forecasting the responses of rocky intertidal ecosystems to climate change. *Annual Review of Ecology, Evolution, and Systematics* **37**, 373-404.

Helser, T.E. & Alade, L. 2012. A retrospective of the hake stocks off the Atlantic and Pacific coasts of the United States: Uncertainties and challenges facing assessment andmanagement in a complex environment. *Fisheries Research* **114**, 2-18.

Herbert, R.J.H., Southward, A.J., Clarke, R.T., Sheader, M. & Hawkins, S.J. 2009. Persistent border: an analysis of the geographic boundary of an intertidal species. *Marine Ecology Progress Series* **379**, 135-150.

Herman, M., et al., 2000. *Contaminants in Alaska: Is America's Arctic at Risk?* U.S. Department of the Interior, Department of Environmental Conservation.

Herrick, Jr. S. F., Norton, J.G., Mason, J.E. & Bessey, C. 2007. Management application of an empirical model of sardine-climate regime shifts. *Marine Policy* **31**, 71-80.

Herrick, S.F., K. Hill and C. Reiss. 2006. An optimal harvest policy for the recently renewed United States Pacific sardine fishery. In Hannesson R, Barange M, Herrick S, editors. Climate Change and the Economics of the World's Fisheries: Examples of Small Pelagic Stocks. Glos, UK: Edward Elgar; pp. 126-150.

Hester, K.C., Peltzer, E.T., Kirkwood, W.J. & Brewer, P.G. 2008. Unanticipated consequences of ocean acidification: A noisier ocean at lower pH. *Geophysical Research Letters* **35**, L19601.

Hiatt, T., Dalton, M., Felthoven, R., Fissel, B., Garber-Yonts, B., Haynie, A., Himes-Cornell, A., Kasperski, S., Lee, J., Lew, D., Pfeiffer, L., Sepez, J. & Seung, C. 2010. Stock Assessment and Fishery Evaluation Report for the Groundfish Fisheries of the Gulf of Alaska and Bering Sea/Aleutian Islands Area: Economic Status of the Groundfish Fisheries Off Alaska. Anchorage, AK: North Pacific Fishery Management Council.

Hiatt, T., M. Dalton, R. Felthoven, B. Fissel, B. Garber-Yonts, A. Haynie, A. Himes-Cornell, S. Kasperski, J. Lee, D. Lew, L. Pfeiffer, J. Sepez and C. Seung (2010). Stock Assessment and Fishery Evaluation Report for the Groundfish Fisheries of the Gulf of Alaska and Bering Sea/Aleutian Islands Area: Economic Status of the Groundfish Fisheries Off Alaska. Seattle, Washington.

Higdon, J.W. & Ferguson, S.H. 2009. Loss of Arctic sea ice causing punctuated change in sightings of killer whales (*Orcinus orca*) over the past century. *Ecological Applications* **19**, 1365-1375

Hilborn, R., Quinn, T.P., Schindler, D.E. & Rogers, D.E. 2003. Biocomplexity and fisheries sustainability. *Proceedings of the National Academy of Sciences USA* **100**, 6564-6568.

Hill, K., N. Lo, B. Macewicz, P. Crone, R. Felix-Uraga. Assessment of the Pacific Sardine Resource in 2010 for U.S. Management in 2011. NOAA Technical Memorandum NOAA-TM-NOAA Fisheries-SWFSC-469, U.S. Department of Commerce.

Hlavsa, M.C., Roberts, V.A., Anderson, A.R., Hill, V.R., Kahler, A.M., Orr, M., Garrison, L.E., Hicks, L.A., Newton, A., Hilborn, E.D., Wade, T.J., Beach, M.J. & Yoder, J.S. 2011.

Surveillance for waterborne disease outbreaks and other health events asociated with recreational water --- United States, 2007--2008. *Morbidity and Mortality Weekly Report* **60**, 1-32.

Hoagland, P., Anderson, D.M., Kaoru, Y. & White, A.W. 2002. The economic effects of harmful algal blooms in the United States: Estimates, assessment issues, and information needs. *Estuaries* **25**, 819-837.

Hobbs, R.J., Arico, S., Aronson, J., Baron, J.S., Bridgewater, P., Cramer, V.A., Epstein, P.R., Ewel, J.J., Klink, C.A., Lugo, A.E., Norton, D., Ojima, D., Richardson, D.M., Sanderson, E.W., Valladares, F. Vilà, M., Zamora, R. & Zobel, M. 2006. Novel ecosystems: theoretical and management aspects of the new ecological world order. *Global Ecology and Biogeography* **15**, 1-7.

Hochachka, P.W., & Somero, G.N. 2002. *Biochemical adaptation: mechanism and process in physiological evolution.* New York: Oxford University Press.

Hoegh-Guldberg, O., and J.F. Bruno. 2010. The impact of climate change on the world's marine ecosystems. Science 328: 1523-1528.

Hoegh-Guldberg, O., Mumby, P.J., Hooten, A.J., Steneck, R.S., Greenfield, P., Gomez, E., Harvell, C.D., Sale, P.J., Edwards, A.J., Caldeira, K., Knowlton, N., Eakin, C.M., Iglesias-Prieto, R., Muthiga, N., Bradbury, R.H., Dubi, A. & Hatziolos, M.E. 2007. Coral reefs under rapid climate change and ocean acidification. *Science* **318**, 1737-1742.

Hoegh-Guldberg, O., Ortiz, J.C. & Dove, S. 2011. The future of coral reefs. *Science* **334**, 1494-1495.

Hoffman, S.M., & Oliver-Smith, A. 2002. *Catastrophe and Culture: The Anthropology of Disaster.* Santa Fe, NM: School of American Research Advanced Seminar Series.

Hofmann, A.F., Peltzer, E.T., Walz, P.M. & Brewer, P.G. 2011. Hypoxia by degrees: Establishing definitions for a changing ocean. *Deep-Sea Research Part I-Oceanographic Research Papers* **58**, 1212-1226.

Hofmann, G.E., & Place, S.P. 2007. Genomics-enabled research in marine ecology: challenges, risks and pay-offs. *Marine Ecology Progress Series* **332**, 244-259.

Hofmann, G.E., & Todgham, A.E. 2010. Living in the now: physiological mechanisms to tolerate a rapidly changing environment. *Annual Review of Physiology* **72**, 172-145.

Hofmann, G.E., Barry, J.P., Edmunds, P.J., Gates, R.D., Hutchins, D.A., Klinger, T. & Sewell, M.A. 2010. The effect of ocean acidification on calcifying organisms in marine ecosystems: an organism to ecosystem perspective. *Annual Review of Ecology, Evolution, and Systematics* **41**, 127-147.

Hofstra, N.. *In press.* Quantifying the impact of climate change on enteric waterborne pathogen concentrations in surface water. *Current Opinion in Environment Sustainability.*

Holbrook, S.J., Schmitt, R.J. & Stephens, J.S. Jr. 1997. Changes in an assemblage of temperate reef fishes associated with a climate shift. *Ecological Applications* **7**, 1299-1310.

Holland, D.S., Pinto da Silva, P., & Wiersma, J. 2010. *A survey of social capital and attitudes toward management in the New England groundfish fishery.* Washington, D.C.: U.S. Department of Commerce.

Holland, M.M., Bitz, C.M., Eby, M., & Weaver, A.J., 2001. The role of ice, ocean interactions in the variability of the North Atlantic Thermohaline Circulation. *Journal of Climate* **14**, 656-675.

Holland, S. M. and R. B. Ditton. 1992. Fishing Trip Satisfaction: A Typology of Anglers. *North American Journal of Fisheries Management* 12(1):28-33.

Hollowed, A., Ito, S., Kim, S., Loeng, H. & Peck, M. 2011. Climate change effects on fish and fisheries: Forecasting impacts, assessing ecosystem responses, and evaluating management strategies. *ICES Journal of Marine Science* **68**, 983-1373.

Hollowed, A.B., Bond, N.A., Wilderbuer, T.K., Stockhausen, W.T., A'mar, Z.T., Beamish, R.J., Overland, J.E., & Schirripa, M.J. 2009. A framework for modelling fish and shellfish responses to future climate change. *ICES Journal of Marine Science* **66**, 1584-1594.

Hönisch, B., Ridgwell, A., Schmidt, D.N., Thomas, E., Gibbs, S.J., Sluijs, A., Zeebe, R., Kump, L., Martindale, R.C., Greene, S.E., Kiessling, W., Ries, J., Zachos, J.C., Royer, D.L., Barker, S., Marchitto, Jr., T.M., Moyer, R., Pelejero, C., Ziveri, P., Foster, G.L. & Williams, B. 2012. The geological record of ocean acidification. *Science* **335**, 1058-1063.

Hooff, R.C., & Peterson, W.T. 2006 Recent increases in copepod biodiversity as an indicator of changes in ocean and climate conditions in the northern California current ecosystem. *Limnology and Oceanography* **51**, 2042-2051.

Horseman, M. & Surani, S. 2011. A comprehensive review of *Vibrio* vulnificus: an important cause of severe sepsis and skin and soft-tissue infection. *International Journal of Infectious Diseases* **15**, e157-e166.

Hospital, J. and C. Beavers. 2012. Catch Disposition and Other Aspects of Fishing in the Commonwealth of the Northern Mariana Islands. Pacific Islands Fisheries Science Center. PIFSC Internal Report IR-12-XX.

Hospital, J., S. S. Bruce, and M. Pan. 2011. Economic and social characteristics of the Hawaii small boat pelagic fishery. Pacific Islands Fish. Sci. Cent., Natl. Mar. Fish. Serv., NOAA, Honolulu, HI 96822-2396. Pacific Islands Fish. Sci. Cent. Admin. Rep. H-11-01, 50 p. + Appendices.

Hospitality Advisors, LLC. 2008. *Economic Impact Analysis of the Potential Erosion of Waikiki Beach.* Waikiki Improvement Association. September 3, 2008

Hsieh, C.H., C.S. Reiss, R.P. Hewitt, and G. Sugihara. 2008. Spatial analysis shows that fishing enhances the climatic sensitivity of marine fishes. *Canadian Journal of Fisheries and Aquatic Sciences* **6**, 947-961.

Hsueh, P., C. Lin, H. Tang, H. Lee, J. Liu, Y. Liu, and Y. Chuang. 2004. *Vibrio vulnificus* in Taiwan. Emerging Infectious Diseases 10(8).

http://www.climatescience.gov/Library/sap/sap4-4/final-report/

http://www.gulfhypoxia.net/

http://www.iucn.org/about/work/programmes/marine/marine_resources marine_publications /?524/A-reef-managers-guide-to-coral-bleaching

http://www.nwf.org/News-and-Magazines/Media-Center/Reports/Archive/2009/New-Era-for-Conservation.aspx

Hu, A., Meehl, G.A., Han, W., & Yin, J. 2011. Effect of the potential melting of the Greenland ice sheet on the meridional overturning circulation and global climate in the future. *Deep Sea Research Part II: Topical Studies in Oceanography* **58**, 1914-1926.

Hughes, T.P., Graham, N.A.J., Jackson, J.B.C., Mumby, P.J. & Steneck, R.S. 2010. Rising to the challenge of sustaining coral reef resilience. *Trends in Ecology and Evolution* **25**, 33-642.

Hughes, T.P., Rodrigues, M.J., Bellwood, D.R., Ceccarelli, D., Hoegh-Guldberg, O., McCook, L. Moltschaniwskyj, N., Pratchett, M.S., Steneck, R.S. & Willis, B. 2007. Phase shifts, herbivory, and the resilience of coral reefs to climate change. *Current Biology* **17**, 360-365.

Hui, Y., Kitts, D. and Stanfield, P. 2001. *Foodborne Disease Handbook.* Second Edition. Vol 4: Seafood and Environmental Toxins. New York: Marcel Dekkar, Inc.

Huisman, J. & Weissing, F.J. 1999. Biodiversity of plankton by species oscillations and chaos. *Nature* **402**, 407-410.

Huisman, J., Pham Thi, N.N., Karl, D. M. & Sommeijer, B. 2006. Reduced mixing generates oscillations and chaos in the oceanic deep chlorophyll maximum. *Nature* **439**, 322-325.

Humston, R., Olson, D.B. & Ault, J.S. 2004. Behavioral assumptions in models of fish movement and their influence on population dynamics. *Transactions of the American Fisheries Society* **133**, 1304-1328.

Hunt, G.L. Jr., Coyle, K.O., Eisner. L.B., Farley, E.V., Heintz, R.A., Mueter, F., Napp, J.M., Overland, J.E., Ressler, P.H., Salo, S. & Stabeno, P.J. 2011. Climate impacts on eastern Bering Sea foodwebs: a synthesis of new data and an assessment of the Oscillating Control Hypothesis. *ICES Journal of Marine Science* **68**, 1230-1243.

Hunt, J., A. Rosenberger, D. Daum, and D. Solie. 2008. Changes in Water Temperature from Three Locales along the Yukon River. 2008 AAAS Arctic Science Conference. http://articaaas.org/meetings/2008/2008_aaas_abstracts_v1.4.pdf.

Hurrell, J. W. & Deser, C. 2010. North Atlantic climate variability: The role of the North Atlantic Oscillation. *Journal of Marine Systems* **79**, 231-244.

Hutchins, D.A. 2011. Oceanography: forecasting the rain ratio. *Nature* **476**, 41-42.

Hutchins, D.A., Fu, F.-X., Zhang, Y., Warner, M.E., Feng, Y., Portune, K., Bernhardt, P.W. & Mulholland, M.R. 2007. CO_2 control of *Trichodesmium* N_2 fixation, photosynthesis, growth rates, and elemental ratios: implications for past, present, and future ocean biogeochemistry. *Limnology and Oceanography* **52**, 1293-1304.

Hutchins, D.A., Mulholland, M.R. & Fu, F. 2009. Nutrient cycles and marine microbes in a CO_2-enriched ocean. *Oceanography* **22**,128-145.

Hyrenbach, K.D. & Veit, R.R. 2003. Ocean warming and seabird communities of the southern California Current System (1987-98): responses at multiple temporal scales. *Deep Sea Research II* **50**, 2537-2565.

Ibbitson, J., 2011. Canada and Denmark make headway in dispute over Hans Island. Published by the Canadian *Globe and Mail* January 27, 2011 and retrieved from http://aol.theglobeandmail.com/servlet/ArticleNews/aolstory/TGAM/20110127/NWARCTIC0127ATL.

Idjadi, J.A., & Edmunds, P.J. 2006. Scleractinian corals as facilitators for other invertebrates on a Caribbean reef. *Marine Ecology Progress Series* **319**, 117-127.

Iles, A. C., T. C. Gouhier, et al. (2012). "Climate-driven trends and ecological implications of event-scale upwelling in the California Current System." Global Change Biology **18**(2): 783-796.

Impact Assessment. 2005. Identifying and Compiling Subsistence Information for Use in Pacific Northwest Fishing Community Research. A report for NOAA's Northwest Fisheries Science Center. La Jolla, CA: Impact Assessment influences on changes in northern gannet populations and diets in the north-west Atlantic: implications for climate change. ICES Journal of Marine Science 54: 608-614.

Ingles, P. & McIlvaine-Newsad, H. 2007. Any Port in the Storm: The Effects of Hurricane Katrina on Two Fishing Communities in Louisiana. *NAPA Bulletin 28: Anthropology and Fisheries Management in the United States: Methodology for Research Issue.* **28**, 69-86.

Inoue, J., & Hori, M.E., 2011. Arctic cyclogenesis at the marginal ice zone: A contributory mechanism for the temperature amplification? Geophysical Research Letters **38**, L12502.

Institute of Medicine, 2006. Seafood Choices http://www.iom.edu/~/media/Files/Report%20Files/2006/Seafood-Choices-Balancing-Benefits-and-Risks/11762_SeafoodChoicesReportBrief.ashx

Intergovernmental Panel on Climate Change (IPCC). 2007a. *Climate Change 2007: Impacts, adaptation and vulnerability. Contribution of Working Group II to the Fourth Assessment Report of the Intergovernmental Panel on Climate Change*, Parry, M.L. et al. (eds). Cambridge, U.K: Cambridge University Press.

Intergovernmental Panel on Climate Change (IPCC). 2007b. Climate Change 2007: The Physical Science Basis. Contribution of Working Group I to the Fourth Assessment Report of the Intergovernmental Panel on Climate Change, Solomon, S., D. Qin, M. Manning, Z. Chen, M. Marquis, K.B. Averyt, M. Tignor and H.L. Miller (eds.) Cambridge University Press, Cambridge, United Kingdom and New York, NY, USA.

Intergovernmental Panel on Climatic Change (IPCC) 33rd Session Abu Dhabi, 10-13 May 2011. Activities of the Task Force on National Greenhouse Gas Inventories. http://www.ipcc.ch/meetings/session33/doc07_p33_tfi_activities.pdf

Intergovernmental Panel on Climatic Change (IPCC). (2011). IPCC Workshop on Impacts of Ocean Acidification on Marine Biology and Ecosystems. Bankoku Shinryokan, Okinawa, Japan, 17-19 January 2011. 174 pp.

Irwin, A.J., and M.J. Oliver. 2009. Are ocean deserts getting larger? *Geophysical Research Letters* **36**, L18609.

Iwasaki, S, Razafindrabe, B.H.N., & Shaw, R. 2009. Fishery livelihoods and adaptation to climate change: a case study of Chilika lagoon, India. *Mitigation and Adaptation Strategies for Global Change* **14**, 339-55.

Jaccard, M, Rivers, N. 2007. Heterogeneous capital stocks and the optimal timing for CO_2 abatement. Resource and Energy Economics 29(1) 1-16.

Jacob, S and M Jepson. Creating a community context for the Fishery Stock Sustainability Index, Fisheries; 2009 34(5):228-231.

Jacob, S. and J. Witman. 2006. Human Ecological Sources of Fishing Heritage and its Use in and Impact on Coastal Tourism. Proceedings of the 2006 Northeastern Recreation Research Symposium. GTR-NRS-P-14.

Jallow BP, Toure S, Barrow MMK, Mathieu AA. 1999. Coastal zone of The Gambia and the Abidjan region in Cˆote d'Ivoire: sea level rise vulnerability, response strategies, and adaptation options. Climate Research 12(2-3):129-36;

Jallow, B.P., Toure, S., Barrow, M.M.K., & Mathieu, A.A. 1999. Coastal zone of the Gambia and the Abidjan region in Cˆote d'Ivoire: sea level rise vulnerability, response strategies, and adaptation options. *Climate Research* **12**, 129-36;

Jansen, J.M., Pronker, A.E., Kube, S., Sokolowski, A., Sola, J.C., Marquiegui, M.A., Schiedek, D., Wendelaar Bonga, S., Wolowicz, M. & Hummel, H. 2007. Geographic and seasonal patterns and limits on the adaptive response to temperature of European *Mytilus* spp. and *Macoma balthica* populations. *Oecologia* **154**, 23-34.

Jepson, M and S Jacob. Social indicator measurements of vulnerability for Gulf Coast fishing communities, NAPA Bulletin; 2007 28(1):57-68.

Jepson, M. E. 2004. The Impact of Tourism on a natural resource community: Cultural Resistance in Cortez Florida. Ph.D. Dissertation. Department of Anthropology, University of Florida.

Jevrejeva, S., J. C. Moore, et al. (2010). "How will sea level respond to changes in natural and anthropogenic forcings by 2100?" Geophys. Res. Lett. **37**(7): L07703.

Jevrejeva, S., Moore, J.C., Grinsted, A., Woodworth, P.L. 2008. Recent global sea level acceleration started over 200 years ago? *Geophysical. Research Letters* **35**, 8-11

Jiang, L., & Pu, Z. 2009. Different effects of species diversity on temporal stability in single trophic and multitrophic communities. *The American Naturalist* **174**, 651-659.

Johansson, J.O.R. 2002. Historical and current observations on macroalgae in the Hillsborough Bay Estuary (Tampa Bay), Florida. In: *Understanding the role of macroalgae in shallow estuaries.* Linthicum, MD: Maryland Department of Natural Resources Maritime Institute, 26-28.

Johnson, J.E., & Welch, D.J. 2010. Marine fisheries management in a changing climate: a review of vulnerability and future options. *Reviews in Fisheries Science* **18**, 106-124.

Jones, K., Patel, N.G., Levy, M.A., Storeygard, A., Balk, D., Gittleman J.L. & Daszak, P. 2008. Global trends in emerging infectious diseases. *Nature* **451**(7181), 990-993.

Jones, M. K., and J. D. Oliver. 2009. *Vibrio Vulnificus:* Disease and Pathogenisis. *Infection and Immunity* 77(5):1723-1733.

Jones, S.J., Mieszkowska, N. & Wethey, D.S. 2009. Linking thermal tolerances and biogeography: *Mytilus edulis* (L.) at its southern limit on the East Coast of the United States. *Biological Bulletin* **217**, 73-85.

Joseph, J.E., Chiu, C.S., 2010. A computational assessment of the sensitivity of ambient noise level to ocean acidification. *The Journal of the Acoustical Society of America* **128**, EL144-EL149.

Justic´, D., Bierman, V.J. Jr, Scavia, D., & Hetland, R.D. 2007. Forecasting gulf's hypoxia: the next 50 years? *Estuaries and Coasts* **30**, 791-801.

Kaje, J.H. and D.D. Huppert. 2007. The Value of Short-Run Climate Forecasts in Managing the Coastal Coho Salmon (*Oncorhynchus Kisutch*) Fishery in Washington State. *Natural Resource Modeling* 20(2):321-349.

Kanzow, T., Cunningham, S.A., Johns, W.E., Hirschi, J.J.-M., Marotzke, J., Baringer, M.O., Meinen, C.S., Chidichimo, M.P. Atkinson, C., Beal, L.M., Bryden, H.L. & Collins, J. 2010. Seasonal Variability of the Atlantic Meridional Overturning Circulation at 26.5°N. *Journal of Climate* **23**, 5678-5698.

Karl, T.R., Melillo, J.M. & Peterson, T.C. (eds.) 2009. *Global Climate Change Impacts in the United States.* New York, NY: Cambridge University Press.

Karlson, K., Rosenberg, R., Bonsdorff, E. 2002. Temporal and spatial large-scale effects on Eutrophication and oxygen deficiency on the benthic fauna in Scandinavian and Baltic Waters - A Review. *Oceanography and Marine Biology: an Annual Review* **40**, 427-489.

Karnovsky, N.J., A. Harding, W. Walkusz, S. Kwasniewski, et al. 2010. Foraging distributions of little auks *Alle alle* across the Greenland Sea: implications of present and future Arctic climate change. Marine Ecology Progress Series 415: 282-293.

Kaschner, K., Tittensor, D.P., Ready, J., Gerrodette, T. & Worm, B. 2011. Current and future patterns of global marine mammal biodiversity. *PLoS ONE* **6**, e19653.

Katz, A.R., A.E. Buchholz, K. Hinson, S.Y. Park, and P.V. Effler. 2011. Leptospirosis in Hawaii, USA, 1999-2008. Emerging Infectious Diseases 17: 221-226.

Kavry, V.I., Boltunov, A.N. & Nikiforov, V.V. 2008. New coastal haulouts of walruses (*Odobenus rosmarus*) - response to the climate changes. In: *Collection of scientific papers from the Marine Mammals of the Holarctic V Conference*, Odessa, Ukraine, 248-251.

Keister, J.E., and W.T. Peterson. 2003. Zonal and seasonal variations in zooplankton community structure off the central Oregon coast, 1998-2000. Progress in Oceanography 57: 341-361.

Kelly, K.A., Singh, S. & Huang, R. X., 1999: Seasonal variations of the sea surface height in the Gulf Stream region. *Journal of Geophysical Research*. **29**, 313-327.

Kelly, M.W., Sanford, E. & Grosberg, R.K. 2012. Limited potential for adaptation to climate change in a broadly distributed marine crustacean. *Proceedings of the Royal Society B* **279**, 349-356.

Kelly, R.P, Foley, M.M., Fisher, W.S., Feely, R.A., Halpern, B.S., Waldbusser, G.G. & Caldwell, M.R. 2011. Mitigating local causes of ocean acidification with existing laws. *Science* **332**, 1036-37.

Kerr, R.A. 2002. A warmer Arctic means change for all. *Science* **297**, 1490-1492.

Khon, V.C., Mokhov, I.I., Latif, M., Semenov, V.A. & Park, W. 2010. Perspectives of Northern Sea Route and Northwest Passage in the twenty-first century. *Climate Change* **100**, 757-768.

Kik, M.J.L., Goris, M.G., Bos, J.H., Hartskeerl, R.A. & Dorrestein, G.M. 2006. An outbreak of leptospirosis in seals (*Phoca vitulina*) in captivity. *Veterinary Quarterly* **38**, 33-39.

Kilarski, S., D. Klaus, J. Lipscomb, K. Matsoukas, R. Newton, and A. Nugent. 2006. *Decision Support for Coral Reef Fisheries Management: Community Input as a Means of Informing Policy in American Samoa.* A Group Project submitted in partial satisfaction of the requirements of the degree of Master's in Environmental Science and Management for the Donald Bren School of Environmental Management. University of California, Santa Barbara.

Kildow, J.T., C.S. Colgan, J.S Scorse. 2009. State of the U.S. Ocean and Coastal Economies. National Ocean Economics Program. http://www.oceaneconomics.org/default.asp

King, J.R., Agostini, V.N., Harvey, C.J., McFarlane, G.A., Foreman, M.G.G., Overland, J.E., Di Lorenzo, E., Bond, N.A. & Aydin, K.Y. 2011. Climate forcing and the California Current ecosystem. *ICES Journal of Marine Science* **68**, 1199-1216

King, M., and L. Lambeth. 2000. Fisheries Management by Communities: A Manual on Promoting the Management of Subsistence Fisheries by Pacific Island Communities. Secretariat of the Pacific Community, Noumea, New Caledonia.

King, M., and U. Fa'asili. 1999. A new network of small, community-owned village fish reserves in Samoa. SPC Traditional marine resource management and knowledge information bulletin No. 11, September 1999. 2-6. 69

Kitagawa, H. 2008. Arctic routing: Challenges and opportunities. *WMU Journal of Maritime Affairs* **7**, 485-503.

Kitts, A, E Bing-Sawyer, J Walden, C Demarest, M McPherson, P Christman, S Steinback, J Olson, P Clay. 2010 Final Report on the Performance of the Northeast Multispecies (Groundfish) Fishery (May 2010 - April 2011). US Dept Commer, Northeast Fish Sci Cent Ref Doc. 11-19; 97 p; 2011. Available from: National Marine Fisheries Service, 166 Water Street, Woods Hole, MA 02543-1026, or on the internet at <http://www.nefsc.noaa.gov/publications/crd/crd1119/> [Accessed February 6, 2012]

Kitts, A., Pinto da Silva, P. & Rountree, B. 2007. The evolution of collaborative management in the Northeast USA Tilefish fishery. *Marine Policy* **31**, 192-200.

Klein, C.J., Chan, A., Kircher, L., Cundiff, A.J., Gardner, N., Hrovat, Y., Scholz, A., Kendall, B.E. & Airamé, S. 2008. Striking a balance between biodiversity conservation and socioeconomic viability in the design of marine protected areas. *Conservation Biology* **22**, 691-700.

Kleypas, J.A. & Langdon, C. 2006. Coral reefs and changing seawater chemistry. In: *Coral Reefs and Climate Change: Science and Management. AGU Monograph Series, Coastal and Estuarine Studies*, J. Phinney et al. (eds.) Washington, DC: American Geophysical Union, vol. 61, pp. 73-110.

Knudson, K. E. 1987. Non-commercial production and distribution in the Guam fishery. Contract WPC-0983. Micronesian Area Research Center, University of Guam. 116 p.

Knutson, T.R., McBride, J.L., Chan, J., Emanuel, K., Holland, G., Landsea, C., Held, I., Kossin, J.P., Srivastava, A.K., & Sugi, M., 2010. Tropical cyclones and climate change. *Nature Geoscience* **3**, 157-163.

Kordas, R.L., Harley, C.D.G. & O'Connor, M.I. 2011. Community ecology in a warming world: the influence of temperature on interspecific interactions. *Journal of Experimental Marine Biology and Ecology* **400**, 218-226.

Kostyack, J., & Rohlf, D. 2008. Conserving endangered species in an era of global warming. *Environmental Law Reporter News and Analysis* **38**, 10203-10213.

Kovach, R.P., Joyce, J.E., Echave, J.D., Lindberg, M.S. & Tallmon, D.A. 2013. Earlier Migration Timing, Decreasing Phenotypic Variation, and Biocomplexity in Multiple Salmonid Species. PLoS ONE 8, e53807 Kovacs, K.M., and C. Lydersen. 2008. Climate change impacts on seals and whales in the North Atlantic Arctic and adjacent shelf seas. Science Progress 92: 117-150.

Kovacs, K.M., C. Lydersen, J.E. Overland, and S.E. Moore. 2010. Impacts of changing sea-ice conditions on Arctic marine mammals. Marine Biodiversity 41(1): 181-194.

Kovacs, K.M., Lydersen, C., Overland, J.E. & Moore, S.E. 2010. Impacts of changing sea-ice conditions on Arctic marine mammals. *Marine Biodiversity* **41**(1): 181-194.

Kraak, S.B.M., Daan, N. & Pastoors, M.A. 2009. Biased stock assessment when using multiple, hardly overlapping, tuning series if fishing trends vary spatially. *ICES Journal of Marine Science* **66**, 2272-2277.

Kreiger, N. 2007. Why Epidemiologists Cannot Afford to Ignore Poverty. *Epidemiology* 18(6): 658-663.

Kreuder, C., Mazet, J.A.K., Bossart, G.D., Carpenter, T.E., Holyoak, M., Elie, M.S. & Wright, S.D. 2002. Clinicopathologic features of suspected brevetoxicosis in double-crested cormorants (*Phalacrocorax auritus*) along the Florida Gulf Coast. *Journal of Zoo and Wildlife Medicine* **33**, 8-15.

Kreuder, C., Miller, M.A., Lowenstine, L.J., Conrad, P.A., Carpenter, T.E., Jessup, D.A. & Mazet, J.A.K. 2005. Evaluation of cardiac lesions and risk factors associated with myocarditis and dilated cardiomyopathy in southern sea otters. *American Journal of Veterinary Research* **66**, 289-299.

Kristiansen, T., Drinkwater, K.F., Lough, R.G. & Sundby, S. 2011. Recruitment variability in North Atlantic cod and match-mismatch dynamics. *PLoS One* **6**, e17456.

Kroeker, K., Kordas, R.L., Crim, R.N. & Singh, G.G. 2010. Meta-analysis reveals negative yet variable effects of ocean acidification on marine organisms. *Ecology Letters* **13**, 1419-1434.

Kroeker, K.J., Micheli, F., Gambi, M.C. & Martz, T.R. 2011. Divergent ecosystem responses within a benthic marine community to ocean acidification. *Proceedings of the National Academy of Sciences USA* **108**, 14515-14520.

Krupnik, I., Aporta, C., Gearheard, S., Laidler, G., & Kielse Holm, L., (eds). 2010. *SIKU: Knowing Our Ice. Documenting Inuit Sea Ice Knowledge and Use*. New York, NY: Springer.

Kuhlbrodt, T., Rahmstorf, S., Zickfeld, K., Vikebø, F., Sundby, S., Hofmann, M., Link, P., Bondeau, A., Cramer, W., & Jaeger, C. 2009. An Integrated Assessment of changes in the thermohaline circulation. *Climatic Change* **96**, 489-537.

Kunkel, K.E., Bromirski, P.D., Brooks, H.E., Cavazos, T., Douglas, A.V., Easterling, D.R., Emanuel, K.A., Ya. P., Groisman, P.Y., Holland, G.J., Knutson, T.R., Kossin, J.P., Komar, P.D., Levinson, D.H. & Smith, R.L. 2008. Observed changes in weather and climate extremes. In: *Weather and Climate Extremes in a Changing Climate: Regions of Focus: North America, Hawaii, Caribbean, and U.S. Pacific Islands. Synthesis and Assessment Product 3.3*. T.R. Karl et al. (eds). Washington, D.C.: U.S. Climate Change Science Program, 35-80.

Kuo, E.S.L., & Sanford, E. 2009. Geographic variation in the upper thermal limits of an intertidal snail: implications for climate envelope models. *Marine Ecology Progress Series* **388**, 137-146.

Kushnir, Y., Seager, R., Ting, M., Naik, N. & Nakamura, J. 2010. Mechanisms of tropical Atlantic SST influence on North American precipitation variability. *Journal of Climate* **23**, 5610-5628.

Kutil, S. M. 2011. Scientific Certainty Thresholds in Fisheries Management: a Response to Changing Climate. Environmental Law 41: 233.

Kwok, R. & Untersteiner, N. 2011. The thinning of Arctic sea ice. *Physics Today*, **64**, 36.

Lackenbauer, P.W. (ed.) 2011. *Canadian Arctic Sovereignty and Security: Historical Perspectives*. Calgary Papers in Military and Strategic Studies. Calgary: Centre for Military and Strategic Studies and University of Calgary Press.

Lafferty, K.D. 2009. The ecology of climate change and infectious diseases. *Ecology* **90**, 888-900.

Laidre, K. L., Ian Stirling, Lloyd F. Lowry, Øystein Wiig, Mads Peter Heide-Jørgensen, and Steven H. Ferguson. 2008. "Quantifying the Sensitivity of Arctic Marine Mammals to Climate-Induced Habitat Change." Ecological Applications 18 (2) (March 1): S97-S125.

Lal, P., J. R. R. Alavalapati and E. D. Mercer. 2011. Socio-economic impacts of climate change on rural United States. *Mitig Adapt Strat Glob Change* 16(and):819-844.

Lambert, E., Hunter, C., Pierce G.J., & MacLeod, C.D. 2010. Sustainable whale-watching tourism and climate change: towards a framework of resilience. *Journal of Sustainable Tourism* **18**, 409-427.

Landsberg, J.H., Flewelling, L.J. & Naar, J. 2009. *Karenia brevis* red tides, brevetoxins in the food web, and impacts on natural resources: decadal advancements. *Harmful Algae* **8**, 598-607.

Lapointe, B.E., Barile, P.J., Littler, M.M., Littler, D.S., Bedford, B.J. & Gasque, C. 2005. Macroalgal blooms on southeast Florida coral reefs I. Nutrient stoichiometry of the invasive green alga *Codium isthmocladum* in the wider Caribbean indicates nutrient enrichment. *Harmful Algae* **4**, 1092-1105.

Large, S.I., Smee, D.L. & Trussell, G.C. 2011. Environmental conditions influence the frequency of prey responses to predation risk. *Marine Ecology Progress Series* **422**, 41-49.

Lasker, R. 1981. The role of a stable ocean in larval fish survival and subsequent recruitment. pp. 79-87 In: R. Lasker, ed. Marine fish larvae: morphology, ecology, and relation to fisheries. Washington Sea Grant Program, Seattle, WA.

Last, P.R., White, W.T., Gledhill, D.C., Hobday, A.J., Brown, R., Edgar, G.J. & Pecl, G. 2011. Long-term shifts in abundance and distribution of a temperate fish fauna: a response to climate change and fishing practices. *Global Ecology and Biogeography* **20**, 58-72.

Lau, C.L., L.D. Smythe, S.B. Craig, and P. Weinstein. 2010. Climate change, flooding, urbanisation and leptospirosis: fuelling the fire? Transactions of the Royal Society of Tropical Medicine and Hygiene 104: 631-638.

Lawler, J. J., T. H. Tear, C. Pyke, M. R. Shaw, Patrick Gonzalez, Peter Kareiva, Lara Hansen, et al. 2010. "Resource management in a changing and uncertain climate." Frontiers in Ecology and the Environment 8 (1) (February): 35-43. doi:10.1890/070146.

Laxon, S.W., Giles, K.A., Ridout, A.L., Winghm, D.J., Willatt, R., Cullen, R., Kwok, R., Schweiger, A., Zhang, J., Haas, C., Hendricks, S., Krishfield, R., Kurtz, N., Farrell, S. & Davidson, M. 2013. CryoSat-2 estimates of Arctic sea ice thickness and volume, Geophysical Research Letters DOI: 10.1002/grl.50193. Lazarus, R.J. 2009. Super wicked problems and climate change: restraining the present to liberate the future. *Cornell Law Review* **94**, 1153-1232.

Le Quere, C., Raupach, M.R., Canadell, J.G. et al. (2009) Trends in the sources and sinks of carbon dioxide. Nature Geoscience, 2, 831-6.

Le Treut, H., Somerville, R., Cubasch, U., Ding, Y., Mauritzen, C., Mokssit, A., Peterson, T., & Prather, M. (2007). Historical Overview of Climate Change. In S. Solomon, D. Qin, M. Manning, Z. Chen, M. Marquis, K.B. Averyt, M. Tignor, & H.L. Miller, Climate Change 2007: The Physical Science Basis. Contribution of Working Group I to the Fourth Assessment Report of the Intergovernmental Panel on Climate Change (pp. 95-127). Cambridge UK and New York, NY, USA: Cambridge University Press

Leatherman, T., and A. Goodman. 1998. Expanding the Biocultural Synthesis Toward a Biology of Poverty. *American Journal of Physical Anthropology* 101(1): 1-3.

Lee Y.J., Lwiza K.M.M. 2007 Characteristics of bottom dissolved oxygen in Long Island Sound, New York. Estuar. Coast. Shelf Sci, 76:187-200.

Lehodey, P, I. Senina, J. Sibert, L. Bopp, B. Calmettes, J. Hampton, and R. Murtgudde. 2010. Preliminary forecasts of Pacific bigeye tuna population trends under the A2 IPCC scenario. Progress in Oceanography 86(1-2): 302-315.

Lehodey, P, Senina, I., Sibert, J., Bopp, L., Calmettes, B., Hampton, J. & Murtgudde, R. 2010. Preliminary forecasts of Pacific bigeye tuna population trends under the A2 IPCC scenario. *Progress in Oceanography* **86**, 302-315.

Lehodey, P., Chai, F. & Hampton, J. 2003. Modelling climate-related variability of tuna popula-tions from a coupled ocean-biogeochemical populations dynamics model. *Fisheries Oceanog-raphy* **12**, 483-494.

Lesser, M.P., Bailey, M.A., Merselis, D.G. & Morrison, J.R. 2010. Physiological response of the blue mussel *Mytilus edulis* to differences in food and temperature in the Gulf of Maine. *Comparative Biochemistry and Physiology Part A* **156**, 541-551.

Levett, P.N. 2001. Leptospirosis. Clinical Microbiology Reviews 14(2): 296-326.

Levine, A., and S. Allen. 2009. American Samoa as a fishing community. U.S. Dep. Commer., NOAA Tech. Memo., NOAA-TM-NOAA Fisheries-PIFSC-19, 74 p.

Levitus, S., Antonov, J.I., Boyer, T.P., and Stephens, C., *Science.* **287**, 2225-2229 (2000).

Levitus, S., Antonov, J.I., Boyer, T.P., Locarnini, R.A., Garcia, H.E., & Mishonov, A.V. 2009. Global ocean heat content 1955-2008 in light of recently revealed instrumentation problems. *Geophysical Research Letters* **36**, L07608. doi:10.1029/2008GL037155

Levy, J. M., , Ed., 2011: Global Oceans [in "State of the Climate in 2010"]. *Bull. Amer. Meteor. Soc.,* **92** (6), S77-108.

Lewandowska, A., and U. Sommer. 2010. Climate change and the spring bloom: a mesocosm study on the influence of light and temperature on phytoplankton and mesozooplankton. Marine Ecology Progress Series 405:101-111.

Li, Y., He, R., McGillicuddy, D.J. Jr., Anderson, D.M. & Keafer, B.A. 2009. Investigation of the 2006 *Alexandrium fundyense* bloom in the Gulf of Maine: *In situ* observations and numerical modeling. *Continental Shelf Research* **29**, 2069-2082.

Lindsay, R. W., J. Zhang, A. Schweiger, M. Steele, and H. Stern 2009 Arctic Sea Ice Retreat in 2007 Follows Thinning Trend. Journal of Climate 22:165-176.

Lindsay, R.W. & Zhang, J. 2005. The thinning of Arctic sea ice, 1988-2003: have we passed a tipping point? *Journal of Climate* **18**, 4879-4894.

Ling, S. D., C. R. Johnson, S. D. Frusher, and K. R. Ridgway. 2009. Overfishing reduces resilience of kelp beds to climate-driven catastrophic phase shift. *Proceedings of the National Academy USA* **106**, 22341-22345.

Link, J.S., Fulton, E.A. & Gamble, R.J, 2010. The northeast U.S. application of ATLANTIS: A full system model exploring marine ecosystem dynamics in a living marine resource manage-ment context. *Progress in Oceanography* **87**, 214-234.

Link, J.S., Nye, J.A. & Hare, J.A. 2011. Guidelines for incorporating fish distribution shifts into a fisheries management context. *Fish and Fisheries* **12**, 461-469.

Lipp, E, and Rose, J.1997. The role of seafood in foodborne diseases in the United States of America. *Revue Scientifique et Technique* 16(2):620-640.

Lipp, E. 2011. Dust in the Wind: How Global Desertification Is Affecting Pathogenic Marine Vibrios. Presented at the American Association for the Advancement of Science Annual Meeting, 17-21 February 2011, Washington D.C. Retrieved February 6, 2012 from http://aaas.confex.com/aaas/2011/webprogram/Paper3497.html.

Lipp, E., Huq, A. & Colwell, R. 2002. Effects of global climate on infectious disease: The cholera model. *Clinical Microbiology Reviews* **15**, 757-770.

Lloyd-Smith, J.O., Greig, D.J., Hietala, S., Ghneim, G.S., Palmer, L., St. Leger, J., Grenfell, B.T. & Gulland, F.M. 2007. Cyclical changes in seroprevalence of leptospirosis in California sea lions: endemic and epidemic disease in one host species? *BMC Infectious Diseases* **7**, 125-136.

Lockwood, B.L. & Somero, G.N. 2011a. Transcriptomic responses to salinity stress in invasive and native blue mussels (genus *Mytilus*). *Molecular Ecology* **20**, 517-529.

Lockwood, B.L. & Somero, G.N. 2011b. Invasive and native blue mussels (genus *Mytilus*) on the California coast: the role of physiology in a biological invasion. *Journal of Experimental Marine Biology and Ecology* **400**, 167-174.

Lockwood, B.L., and G.N. Somero. 2011b. Invasive and native blue mussels (genus *Mytilus*) on the California coast: the role of physiology in a biological invasion. Journal of Experimental Marine Biology and Ecology 400: 167-174.

Loher, T. 2001. Recruitment variability in southeast Bering Sea red king crab (*Paralithodes camtschaticus*): The roles of early juvenile habitat requirements, spatial population structure, and physical forcing mechanisms. Ph.D. dissertation. University of Washington, Seattle, WA. 436 pp.

Lombard, F., da Rocha, R.E., Bijma, J. & Gattuso, J.P. 2010. Effect of carbonate ion concentration and irradiance on calcification in planktonic foraminifera. *Biogeosciences* **7**, 247-255.

Loomis, D.K. and R. B. Ditton. 1987. Analysis of Motive and Participation Differences between Saltwater Sport and Tournament Fishermen. *North American Journal of Fisheries Management* 7(4):482-487.

Loomis, J. and J. Crespi. 1999. Estimated Effects of Climate Change on Selected Outdoor Recreation Activities in the United States. In R. Mendelsohn and J.E. Neumann Eds. *The Impact of Climate Change on the United States Economy*, pp. 289-314. New York: Cambridge University Press.

Lopez, C.B., Dortch, Q., Jewett, E.B. & Garrison, D. 2008. *Scientific Assessment of Marine Harmful Algal Blooms*. Washington, D.C.: Interagency Working Group on Harmful Algal Blooms, Hypoxia, and Human Health of the Joint Subcommittee on Ocean Science and Technology.

Loring, P.A. & Gerlach, S.C. 2009. Food, culture, and human health in Alaska: an integrative health approach to food security. *Environmental Science and Policy* **12**, 466-478.

Loring, P.A., & Gerlach, S.C. 2010. Food security and conservation of Yukon River salmon: Are we asking too much of the Yukon River? *Sustainability* **2**, 2965-2987.

Loring, P.A., Gerlach, S.C., Atkinson, D.E. & Murray, M.S. 2011. Ways to help and ways to hinder: Governance for successful livelihoods in a changing climate. *Arctic* **64**, 73-88.

Love, D., Robman, S. Nef, R. and Nachman, K. 2011. Veterinary drug residues in seafood inspected by the European Union, United States, Canada, and Japan from 2000 to 2009. *Environmental Science and Technology* 45(17):7230-7240.

Low, L. 2008. "United States of America - Alaska Region". In R. Beamish (Ed.), "Impacts of climate and climate change on key species in the fisheries of the North Pacific" (p. 163-205). North Pacific Marine Science Organization (PICES), Sidney, B.C. PICES Sci. Rep. No. 35.

Lubchenco, J. & L.E. Petes. 2010. The interconnected biosphere: Science at the ocean's tipping points. *Oceanography* **23**, 115-129.

Luber, G. & McGeehin, M. 2008. *Climate Change and Extreme Heat Events. American Journal of Preventive Medicine* **35**, 429-435.

Luettich Jr., R.A., Carr, S.D., Reynolds-Fleming, J.V., Fulcher, C.W. & McNinch, J.E. 2002. Semi-diurnal seiching in a shallow, micro-tidal lagoonal estuary. *Continental Shelf Research* **22**, 1669-1681.

Lumpkin, R., & Speer, K., 2007. Global ocean meridional overturning. *Journal of Physical Oceanography* **37**, 2550-2562.

Luthi, D., Le Floch, M., Bereiter, B., Blunier, T., Barnola, J.-M., Siegenthaler, U., Raynaud, D., Jouzel, J., Fischer, H., Kawamura, K., Stocker, T.F., 2008. High-resolution carbon dioxide concentration record 650,000-800,000[thinsp]years before present. Nature 453, 379-382.

Lynch, A., and R. Brunner. 2007. "Context and climate change: an integrated assessment for Barrow, Alaska." Climatic Change 82: 93-111.

Macdonald, R. W., Harner, T. & Fyfe, J. 2005. Recent climate change in the Arctic and its impact on contaminant pathways and interpretation of temporal trend data. *Science of the Total Environment* **342**, 5-86.

Macey, B.M., Achilihu, I.O., Burnett, K.G. & Burnett, L.E. 2008. Effects of hypercapnic hypoxia on inactivation and elimination of *Vibrio campbellii* in the eastern oyster, *Crassostrea virginica*. *Applied and Environmental Microbiology* **74**, 6077-6084.

MacNeil, M. A., N. A. J. Graham, J. E. Cinner, N. K. Dulvy, P. A. Loring, S. Jennings, N. V. C. Polunin, A. T. Fisk, and T. R. McClanahan. 2010. Transitional states in marine fisheries: adapting to predicted global change. Philosophical Transactions of the Royal Society B: Biological Sciences, 365(1558), pp.3753 -3763.

Mahon, R. 2002. *Adaptation of fisheries and fishing communities to the impacts of climate change in the CARICOM region: issue paper-draft. Mainstreaming adaptation to climate change (MACC) of the Caribbean Center for Climate Change (CCCC)*, Washington, D.C.: Organization of American States,

Mann, K.H. 1973. Seaweeds: their productivity and strategy for growth. *Science* **182**, 975-981.

Mantua, N.J., Hare, S.R., Zhang, Y., Wallace, J.M. & Francis, R.C. 1997. A Pacific interdecadal climate oscillation with impacts on salmon production. *Bulletin of the American Meteorological Society* **78**, 1069-1079.

Markowski, M., Knapp, A., Neumann, J.E. & Gates, J. 1999. The Economic Impact of Climate Change on the U.S. Commercial Fishing Industry. In: *The Impact of Climate Change on the United States Economy*, R. Mendelsohn & J.E. Neumann (eds.). New York, N.Y.: Cambridge University Press, 237-264.

Marshall, P.A. and H. Schuttenberg. 2006. *A reef manager's guide to coral bleaching. Great Barrier Reef Marine Park Authority, Townsville, Australia.*

Martin, S., & Gattuso, J.P. 2009. Response of Mediterranean coralline algae to ocean acidification and elevated temperature. *Global Change Biology* **15**, 2089-2100.

Martin, S., M. Killorin, and S. Colt. 2008. Fuel Costs, Migration, and Community Viability. Anchorage: Institute of Social and Economic Research, University of Alaska Anchorage. http://www.denali.gov/images/announcements/fuelcost_viability_final.pdf

Martínez, E.A., Cárdenas, L. & Pinto, R. 2003. Recovery and genetic diversity of the intertidal kelp *Lessonia nigrescens* (Phaeophyceae) 20 years after El Niño 1982/83. *Journal of Phycology* **39**, 504-508.

Martinez-Urtaza, J., Bowers, J.C., Trinanes, J. & DePaola, A. 2010. Climate anomalies and the increasing risk of *Vibrio parahaemolyticus* and *Vibrio vulnificus* illnesses. *Food Research International* **43**, 1780-1790.

Maslanik, J., Fowler, C., Stroeve, J., Drobot, S., Zwally, J., Yi, D. & Emery, W., 2007. A younger, thinner Arctic ice cover: Increased potential for rapid, extensive sea-ice loss. *Geophysical Research Letters*, **34**, L24501.

Massachusetts General Court House of Representatives Committee on History of the Emblem of the Codfish. 1895. *History of the emblem of the codfish in the hall of the House of representatives.* Boston: Wright and Potter Printing Co., state printers. Available at http://www.archive.org/details/historyofemblem00mass.

Matson, P.G., & Edwards, M.S. 2007. Effects of ocean temperature on the southern range limits of two understory kelps, *Pterygophora californica* and *Eisenia arborea*, at multiple life-stages. *Marine Biology* **151**, 1941-1949.

Matsuda, F., Ishimura, S., Wagatsuma, Y., Higashi, T., Hayashi, T. and Faruque, A. 2008. Prediction of epidemic cholera due to *Vibrio cholerae* O1 in children younger than 10 years using climate data in Bangladesh. *Epidemiology and Infection* 136:73-79.

Maurstad, A. 2000. To fish or not to fish: Small-scale fishing and changing regulations of the cod fishery in northern Norway. *Human Organization* 59, 37-47.

McCarthy, J. J., Osvaldo F. Canziani, Neil A. Leary, David J. Dokken, and Kasey White, eds. 2001. Climate Change 2001: Impacts, Adaptation and Vulnerability. Intergovernmental Panel on Climate Change. Cambridge: Cambridge University Press; Turner, Bryan. 2006. *Vulnerability and Human Rights*. University Park, PA: Penn State Press.

McCay, B., K. St. Martin, J. Lamarque, B. Jones, B. Oles, and B. Stoffle. 2006. Mid-Atlantic Fishing Community Profiles: A Report to NOAA Fisheries, Northeast Fisheries Science Center. New Brunswick, N.J.: Department of Human Ecology, Rutgers the State University. Individual profiles available at http://aesop.rutgers.edu/~fisheries/mid-atlanticfishingcommunities.htm.

McCay, B.J., Brandt, S. & Creed, C.F. 2011. Human dimensions of climate change and fisheries in a coupled system: the Atlantic surfclam case. *ICES Journal of Marine Science* **68**,1354.

McClain, C.R., Signorini, S.R. & Christian, J.R. 2004. Subtropical gyre variability observed by ocean-color satellites *Deep-Sea Research II* **51**, 281-301.

McClatchie, S., R. Goericke, G. Auad and K. Hill. 2010. Re-assessment of the stock-recruit and temperature-recruit relationships for Pacific sardine (*Sardinops sagax*). Ca. J. Fish. Aquat. Sci. 67: 1782-1790.

McCook, L.J., J. Jompa, and G. Diaz-Pullido. 2001. Competition between corals and algae on coral reefs: a review of evidence and mechanisms. Coral Reefs 19: 400-417.

McGeehin, M. 2007. CDC's Role in Addressing the Health Effects of Climate Change. CDC Conference: Safe Healthier People, May 4-5, Atlanta, Georgia.

McGeehin, M., & Mirabelli M. 2001. *The potential impacts of climate variability and change on temperature-related morbidity and mortality in the United States. Environmental Health Perspectives* **109**, 191-198.

Mcilgorm, A. 2010. Economic impacts of climate change on sustainable tuna and billfish management: Insights from the Western Pacific. *Progress in Oceanography* **86**, 187-191.

McLaughlin, A., DePaola, C., Bopp, K., Martinek, N., Napolilli & Allison, C. 2005. Outbreak of *Vibrio* parahaemolyticus gastroenteritis associated with Alaskan oysters. *New England Journal of Medicine* **353**,1463-1470.

McLeod, E., Chmura, G.L., Bouillon, S., Salm, R., Björk, M., Duarte, C.M., Lovelock, C.E., Schlesinger, W.H. & Silliman, B.R. 2011. A blueprint for blue carbon: toward an improved understanding of the role of vegetated coastal habitats in sequestering CO_2. *Frontiers in Ecology and the Environment* **9**, 552-560.

McLeod, E., Salm, R., Green, A. & Almany, J. 2009. Designing marine protected area networks to address the impacts of climate change. *Frontiers in Ecology and the Environment* **7**, 362-370.

McLeod, K.E, B.S. Halpern, A.A. Rosenberg, J.F. Samhouri, C. Longo, et al. In review. A new perspective on ocean health for people and the planet. Conservation Letters.

McMichael, A., H. Campbell-Lendrum, C. Corvalan, K. Ebi, A.K. Githeko, J.D. Schwraga, and A. Woodward, eds. 2003. Climate Change and Human Health: Risks and Responses. Geneva: World Health Organization.

McNeeley, S. 2009. *Seasons out of Balance: Vulnerability and Sustainable Adaptation in Alaska.* PhD Dissertation, University of Alaska Fairbanks.

McNeil, B.I., & R.J. Matear. 2008. Southern Ocean acidification: a tipping point at 450-ppm atmospheric CO_2. *Proceedings of the National Academy of Sciences USA* **105**, 18860-18864.

Meadow, A., C. Meek, and S. McNeeley. 2009. "Towards Integrative Planning for Climate Change Impacts on Rural-Urban Migration in Interior Alaska: A Role for Anthropological and Interdisciplinary Perspectives." Alaska Journal of Anthropology 7 (1): 57-69.

Meehl, G., H. Teng, and G. Branstator, 2006: Future changes of El Niño in two global coupled climate models. Climate Dynamics, 26, 549{566, dOI:10.1007/s00382-005-0098-0.

Meehl, G.A., Stocker, T.F., Collins, W.D., Friedlingstein, P., Gaye, A.T., Gregory, J.M., Kitoh, A., Knutti, R., Murphy, J.M., Noda, A., Raper, S.P.B., Watterson, I.G., Weaver, A.J., & Zhao, Z.-C. 2007. Global Climate Projections. In *Climate Change 2007: The Physical Science Basis. Contribution of Working Group I to the Fourth Assessment Report of the Intergovernmental Panel on Climate Change* S. Solomon et al. (eds). Cambridge, U.K. and New York, NY, USA.: Cambridge University Press, 589-662.

Meier, H., Kjellstroem, E., & Graham, L. (2006) Estimating uncertainties of projected Baltic Sea salinity in the late 21st century. Geophysical Research Letters, 33, 4 pp.

Meites, E., M.T. Jay, S. Deresinski, W.J. Shieh, S.R. Zaki, L. Tompkins, and D.S. Smith. 2004. Reemerging leptospirosis, California. Emerging Infectious Diseases 10: 406-412.

Mendelsohn, R. and M. Markowski. 1999. The Impact of Climate Change on Outdoor Recreation. In R. Mendelsohn and J.E. Neumann Eds. *The Impact of Climate Change on the United States Economy*, pp. 267-288. New York: Cambridge University Press.

Menge, B.A., Chan, F., & Lubchenco, J. 2008. Response of a rocky intertidal ecosystem engineer and community dominant to climate change. *Ecology Letters* **11**, 151-162.

Merico, A, Tyrrell, T., Lessard, E.J., Oguz, T., Stabeno, P.J., Zeeman, S.I. & Whitledge, T.E. 2004. Modelling phytoplankton succession on the Bering Sea shelf: role of climate influences and trophic interactions in generating *Emiliania huxleyi* blooms 1997-2000. *Deep Sea Research Part I: Oceanographic Research Papers* **51**, 1803-1826.

Merino, G., M. Barange, and C. Mullon. 2010. Climate variability and change scenarios for a marine commodity: Modeling small pelagic fish, fisheries and fishmeal in a globalized market. *Journal of Marine Systems* 81:196-205.

Merryfield, W. J.: Changes to ENSO under CO_2 doubling in a multimodel ensemble, J. Climate, 19, 4009-4027, 2006. 2179

Merzouk, A., & Johnson, L.E. 2011. Kelp distribution in the northwest Atlantic Ocean under a changing climate. *Journal of Experimental Marine Biology and Ecology* **400**, 90-98.

Mikulski, C.M., Burnett, L.E. & Burnett, K.G. 2000. The effects of hypercapnic hypoxia on the survival of shrimp challenged with *Vibrio parahaemolyticus*. *Journal of Shellfish Research* **19**, 301-311.

Miller, G., Alley, R., Brigham-Grette, J., Fitzpatrick, J., Polyak, L., Serreze, M. & White, J., 2010. Arctic amplification: can the past constrain the future? *Quaternary Science Reviews* **29**, 1779-1790.

Miller, J., Muller, E., Rogers, C., Waara, R., Atkinson, A., Whelan, K.R.T., Patterson, M. & Witcher, B. 2009. Coral disease following massive bleaching in 2005 causes 60 percent decline in coral cover on reefs in the U.S. Virgin Islands. *Coral Reefs* **28**, 925-937.

Miller, J.R., & Russell, G.L. 1992. The impact of global warming on river runoff. *Journal of Geophysical Research* **97**, 2757-2764.

Miller, K.A. & Munro, G.R. 2003. Climate and cooperation: A new perspective on the management of shared fish stocks. *Marine Resource Economics* **19**, 367-393.

Miller, M.A., Conrad, P.A., Harris, M., Hatfield, B., Langlois, G., Jessup, D.A., Magargal, S.L., Packham, A.E., Toy-Choutka, S., Melli, A.C., Murray, M.A., Gulland, F.M. & Grigg, M.E. 2010. A protozoal-associated epizootic impacting marine wildlife: mass-mortality of southern sea otters (*Enhydra lutris nereis*) due to *Sarcocystis neurona* infection. *Veterinary Parasitology* **172**, 183-194.

Mills, J., Gage, K. & Khan, A.S. 2010. Potential influence of climate change on vector-borne and zoonotic diseases: a review and proposed research plan. *Environmental Health Perspectives* **118**, 1507-1514.

Mills, J., J. Hunt, A. Rosenberger, D. Daum, and D. Solie. 2008. EPSCoR Climate Change Research: Examination of Long-term Ice Break-up and Freeze-up Trends in the Yukon River Based on Historical Data Sources Collected for the Yukon River Temperature Archive Project. In Growing Sustainability Science in the North: Science, Policy. Education, Legacy in the International Polar Year, 41. Fairbanks, AK: Arctic Division AAAS. Retrieved February 6, 2012 from http://arcticaaas.org/meetings/2008/2008_aaas_abstracts_v1.4.pdf.

Mislan, K.A.S., & Wethey, D.S. 2011. Gridded meteorological data as a resource for mechanistic macroecology in coastal environments. *Ecological Applications* **21**, 2678-2690.

Moerlein, K. and C. Carothers. *In press.* Total environment of change: Impacts of climate change and social transitions on subsistence fisheries in Northwest Alaska.

Moeser, G.M., Leba, H. & Carrington, E. 2006. Seasonal influence of wave action on thread production in *Mytilus edulis*. *The Journal of Experimental Biology* **209**, 881-890.

Moffitt C.M., Stewart, B., Lapatra, S., Brunson, R., Bartholomew, J., Peterson J. & Amos, K.H. 1998. Pathogens and diseases of fish in aquatic ecosystems: implications in fisheries management. *Journal of Aquatic Animal Health.* **10**, 95-100.

Moffitt, C. (2004) Evaluating and Understanding Fish Health Risks and Their Consequences in Propagated and Free-Ranging Fish Populations. American Fisheries Society Symposium 44:529-537.

Moller, H., F. Berkes, P. O. Lyver, and M. Kislalioglu. 2004. Combining science and traditional ecological knowledge: monitoring populations for co-management. Ecology and Society 9: 2.

Monaco, C.J., & Helmuth, B. 2011. Tipping points, thresholds, and the keystone role of physiology in marine climate change research. *Advances in Marine Biology* **60**, 123-160.

Monahan, A.M., I.S. Miller and J.E. Nally. 2009. Leptospirosis: risks during recreational activities. 2009. Journal of Applied Microbiology 107: 707-716.

Montane, M.M. and H.M. Austin. 2005. Effects of Hurricanes on Atlantic Croaker (*Micropogonias undulatus*) Recruitment to Chesapeake Bay, pp.185-192. In K.G. Sellner (ed.) *Hurricane Isabel in Perspective*. Chesapeake Research Consortium, CRC Publication 05-160, Edgewater, MD.

Montevecchi, W.A., & Myers, R.A. 1997. Centurial and decadal oceanographic influences on changes in northern gannet populations and diets in the north-west Atlantic: implications for climate change. *ICES Journal of Marine Science* **54**(4), 608-614.

Moore, C. 2011. *Welfare impacts of ocean acidification: an integrated assessment model of the U.S. mollusk fishery, Working Paper 11-06.* Washington, DC: National Center for Environmental Economics, U.S. Environmental Protection Agency.

Moore, G.E., L.F. Guptill, N.W. Glickman, R. J. Caldanaro, D. Aucoin, and L.T. Glickman. 2006. Canine leptospirosis, United States, 2002-2004. Emerging Infectious Diseases 12: 501-503.

Moore, K.A., & Jarvis, J.C. 2008. Environmental factors affecting recent summertime eelgrass diebacks in the lower Chesapeake Bay: implications for long-term persistence. *Journal of Coastal Research* **55**, 135-147.

Moore, S., and Gill, M. 2011. Marine ecosystems summary. In: *Arctic Report Card 2011.* Richter-Menge, J., Jeffries, M.O. & Overland, J.E. (eds.). pp. 63-64.

Moore, S.E. 2008. Marine mammals as ecosystem sentinels. *Journal of Mammalogy* **89**, 534-540.

Moore, S.E., Grebmeier, J.M. & Davies, J.R. 2003. Gray whale distribution relative to forage habitat in the northern Bering Sea: current conditions and retrospective summary. *Canadian Journal of Zoology* **81**, 734-742.

Moore, S.K., Mantua, N.J., Hickey, B.M. & Trainer, V.L. 2009. Recent trends in paralytic shellfish toxins in Puget Sound, relationships to climate, and capacity for prediction of toxic events. *Harmful Algae* **8**, 463-477.

Moore, S.K., Trainer, V.L., Mantua, N.J., Parker, M.S., Laws, E.A., Backer, L.C., & Fleming, L.E. 2008 Impacts of climate variability and future climate change on harmful algal blooms and human health. *Environmental Health* **7**, S4.

Morán, X.A.G., López-Urrutia, Á., Calvo-Díaz, A. & Li, W.K.W. 2010. Increasing importance of small phytoplankton in a warmer ocean. *Global Change Biology* **16**, 1137-1144.

Moreno, A. & Becken, S. 2009. A climate change vulnerability assessment methodology for coastal tourism. *Journal of Sustainable Tourism* **17**, 473-488.

Morreale, S.J., Meylan, A.B., Sadove, S.B. & Standora, E.A. 1992. Annual occurrence and winter mortality of marine turtles in New York waters. *Journal of Herpetology* **26**, 301-308.

Moser, S. & Luers, A.L. 2008. Managing climate risks in California: the need to engage resource managers for successful adaptation to change. *Climatic Change* **87**, 309-322.

Moy, A.D., Howard, W.R., Bray, S.G. & Trull, T.W. 2009. Reduced calcification in modern Southern Ocean planktonic foraminifera. *Nature Geoscience* **2**, 276-280.

Mozaffarian, D, and E. B. Rimm. (2006) Fish intake, contaminants, and human health: evaluating the risks and the benefits. JAMA 296: 1885-1899.

MPA Monitoring Enterprise. 2012. Monitoring climate effects on temperature marine ecosystems. California Ocean Science Trust, Oakland, CA. Prepared by EcoAdapt (A. Score, R.M. Gregg, and L.J. Hansen.) Available for download at: http://monitoringenterprise.org/pdf/Monitoring_climate_change_effects_in_temperate_marine_ecosystems.pdf

Mueter, F.J., and Litzow, M.A. 2008. Sea ice retreat alters the biogeography of the Bering Sea continental shelf. Ecological Applications 18(2), 309-320.

Mueter, F.J., Bond, N.A., Lanelli, J.N. & Hollowed, A.B. 2011. Expected declines in recruitment of walleye Pollock (*Theragra chalcogramma*) in the eastern Bering Sea under future climate change. *ICES Journal of Marine Science* **68**, 1284-1296.

Mueter, F.J., Siddon, E.C., & Hunt, G.L. Jr. 2011. Climate change brings uncertain future for subarctic marine ecosystems and fishes. In *North by 2020: Perspectives on Alaska's Changing Social-Ecological Systems, A.L.* Lovecraft & H. Eicken (eds). Fairbanks, AK: University of Alaska Press, 329-357.

Mumby, P.J., Elliott, I.A., Eakin, C.M., Skirving, W., Paris, C.B., Edwards, H.J., Enríquez, S., Iglesias-Prieto, R., Cherubin L.M. & Stevens, J.R. 2011. Reserve design for uncertain responses of coral reefs to climate change. *Ecology Letters* **14**, 132-140.

Mumby, P.J., Iglesias-Prieto, R., Hooten, A.J., Sale, P.F., Hoegh-Guldberg, O., Edwards, A.J., Harvell, C.D., Gomez, E.D., Knowlton, N., Hatziolos, M.E., Kyewalyanga, M.S. & Muthiga, N. 2011. Revisiting climate thresholds and ecosystem collapse. *Frontiers in Ecology and the Environment* **9**, 94-95.

Mundy, P.R., and D. F. Evenson. 2011. "Environmental controls of phenology of high-latitude Chinook salmon populations of the Yukon River, North America, with application to fishery management." ICES Journal of Marine Science: Journal du Conseil 68 (6) (July 1): 1155 -1164. doi:10.1093/icesjms/fsr080.

Munro, G., Van Houtte, A. & Willmann, R. 2004. *The conservation and management of shared fish stocks: legal and economic aspects.* FAO Fisheries Technical Paper No. 465, Rome: Food and Agricultural Organization of the United Nations.

Murawski, S. 2011. Summing up Sendai: progress integrating climate change science and fisheries. ICES Journal of Marine Science 68(6): 1368-1372.

Murawski, S.A. 1993. Climate Change and Marine Fish Distributions: Forecasting from Historical Analogy. *Transactions of the American Fisheries Society*, 122(5):647-658.

Musick, J.A., & Limpus, C.J. 1997. Habitat utilization and migration in juvenile sea turtles. In: *The biology of sea turtles, Vol 1*, P.L. Lutz & J.A. Musick (eds.). Boca Raton, FL: CRC Press, 137-163.

Nam, S., Kim, H.-J. & Send, U. 2011. Amplification of hypoxic and acidic events by La Niña conditions on the continental shelf off California. *Geophysical Research Letters*, **38**, L22602.

Narita, D., K. Rehdanz, Tol, R.S.J. (2011). Economic Costs of Ocean Acidification: A Look into the Impacts on Global Shellfish Production, Kiel Working Papers 1710 Kiel Institute for the World Economy (forthcoming in Climatic Change).

NASA: Ocean in Motion (2012) http://oceanmotion.org/html/background/climate.htm

National Climate Assessment (NCA) Report Series. 2010a. Volume 5b. *Monitoring climate changes and its impacts: physical climate indicators.* Washington, D.C.: U.S. Global Change Research Program.

National Climate Assessment (NCA) Report Series. 2010b.Volume 5a. *Ecosystem responses to climate change: selecting indicators and integrating observation networks.* Washington, D.C.: U.S. Global Change Research Program.

National Climate Assessment (NCA) Report Series. 2011. Volume 5c. *Climate change impacts and responses: societal indicators for the National Climate Assessment.* Washington, D.C.: U.S. Global Change Research Program.

National Marine Fisheries Ser vice. 2009c. Our Living Oceans. Report on the status of U.S. living marine resources. 6th edition. U.S. Department of Commerce, NOAA Technical Memo. NMFS-F/SPO-80. 369 p.

National Marine Fisheries Service (NOAA Fisheries) 2009b. Gulf of Mexico Red Snapper Recovering: Science-based management has helped end overfishing for iconic fish. Press release published December 11, 2009 at http://www.noaanews.noaa.gov/stories2009/20091211_redsnapper.html.

National Marine Fisheries Service (NOAA Fisheries). 2009a. Fishing Communities of the United States, 2006. U.S. Dept. Commerce, NOAA Tech. Memo. NOAA Fisheries-F/SPO-98.

National Marine Fisheries Service (NOAA Fisheries). 2009c. *Our Living Oceans. Report on the status of U.S. living marine resources. 6th edition.* U.S. Department of Commerce, NOAA Technical Memo. NMFS-F/SPO-80.

National Marine Fisheries Service (NOAA Fisheries). 2010. *Fisheries Economics of the United States, 2009.* U.S. Dept. Commerce, NOAA Tech. Memo. NOAA Fisheries-F/SPO-118, 172p.

National Marine Fisheries Service (NOAA Fisheries). 2012. Gulf of Maine Cod News. Available at http://www.nero.noaa.gov/nero/hotnews/gomcod/.

National Marine Fisheries Service. 2011. Fisheries of the United States, 2010. Current Fishery Statistics No. 2010. Silver Spring, Maryland: Office of Science and Technology August 2011.

National Ocean Council. 2012. Draft National Ocean Policy Implementation Plan. Washington, D.C. http://www.whitehouse.gov/administration/eop/oceans/implementationplan

National Research Counci (NRC)l. 2000a. *Long-Term Institutional Management of U.S. Department of Energy Legacy Waste Sites.* Washington, D.C.: The National Academies Press.

National Research Council (NRC) Committee on the Science of Climate Change (2001). Climate Change Science: An Analysis of Some Key Questions, The National Academies Press.

National Research Council (NRC). 1996. Understanding Risk: Informing Decisions in a Democratic Society. Washington, D.C.: The National Academies Press.

National Research Council (NRC). 2000b. *A Risk Management Strategy for PCB-contaminated Sediments.* Washington, DC: The National Academies Press.

National Research Council (NRC). 2010. *Adapting to the Impacts of Climate Change. America's Climate Choices.* Washington, D.C.: The National Academies Press.

National Research Council (NRC). 2010. Ocean Acidification: A National Strategy to Meet the Challenges of a Changing Ocean. Washington, DC: The National Academies Press.

Naylor, R. L., Goldburg, R. J., Primavera, J.H., Kautsky, N., Beveridge, M. C. M., Clay, J., Folke, C., Lubchenco, J., Mooney, H., and M. Troell. 2000. Nature 405: 1017-1024.

Naylor, R.L., R.W. Hardy, D.P. Bureau, A. Chiu, M. Elliott, A.P. Farrell, I. Forster, D.M. Gatlin, R.J. Goldburg, K. Hua, and P.D. Nichols. 2009. Feeding aquaculture in an era of finite resources. Proceedings of the National Academy of Sciences, 106(36):15103-15110.

Nazarea, V. D. 2003. Introduction -- A View from a Point: Ethnoecology as Situated Knowledge, pp. 1-20. In V.D. Nazarea, ed. *Ethno-ecology: Situated Knowledge/Located Lives.* Tucson, U of Arizona Press. 300p.

NCDC. 2010. 2009/2010 cold season. Accessed March 12, 2012 from http://www.ncdc.noaa.gov/special-reports/2009-2010-cold-season.html.

Nelson, K. 2011. Agencies to Study Cook Inlet Tidal Energy. Article published in the Anchorage Daily News August 22, 2011. Retrieved January 9, 2012 from http://www.adn.com/2011/08/22/2026742/agencies-to-study-cook-inlet-tidal.html.

Nelson, R.K. 1969. Hunters of the Northern Ice. Chicago, IL: University of Chicago Press.

Nelson, R.K. 1986. Hunters of the Northern Forest: Designs for Survival among the Alaska Kutchin. Chicago, IL: University of Chicago Press.

Neuwald, J. L. & Valenzuala, N. 2011. The lesser known challenge of climate change: Thermal variance and sex-reversal in vertebrates with temperature-dependent sex determination. *PLoS ONE* **6**, 18117.

Newsome, S. D., M. A. Etnier, C. M. Kurle, J. R. Waldbauer, C. P. Chamberlain, and P. L. Koch. 2007. Historic decline in primary productivity in western Gulf of Alaska and eastern Bering Sea: isotopic analysis of northern fur seal teeth. Marine Ecology Progress Series 332 (March 5): 211-224. doi:10.3354/meps332211.

Nghiem, S., Rigor, I., Perovich, D., Clemente-Colon, P., Weatherly, J. & Neumann, G. 2007. Rapid reduction of Arctic perennial sea ice. *Geophysical Research Letters* **34**, L19504.

Nigro, O., Hou, A., Vithanage, G., Fujioka, R. and Steward, G. 2011. Temporal and spatial variability in culturable pathogenic *Vibrio* spp. in Lake Pontchartrain, Louisiana, following hurricanes Katrina and Rita. *Applied and Environmental Microbiology* 77(15):5384-5393.

NOAA National Ocean Service. n.d. Harmful Algal Blooms: Simple Plants With Toxic Implications. (photo credit: Kai Schumann, California Department of Public Health volunteer). Retrieved January 30, 2012 from http://oceanservice.noaa.gov/hazards/hab/.

NOAA. 1997a. Wetlands, fisheries, and economics in the New England coastal states. *Habitat Connections* **1**:3.

NOAA. 1997b. Wetlands, fisheries, and economics in the Mid-Atlantic coastal states. *Habitat Connections* **1**:5.

NOAA. 2010. Bering Climate and Ecosystem - Bering Sea status and overview. National Oceanic and Atmospheric Administration. Retrieved February 6, 2012 from http://www.beringclimate.noaa.gov/bering_status_overview.html.

NOAA: 2011 a year of climate extremes in the United States http://www.noaanews.noaa.gov/stories2012/20120119_global_stats.html (accessed January 2012).

Norris, F. 2002. Alaska Subsistence. Anchorage, AK: National Park Service.

Norris, F.H., Speier, A., Henderson, A.K., Davis, S.I., Purcell, D.W., Stratford, B.D., Baker, C.K., Reissman, D.B., & Daley, W.R. 2006. Assessment of health-related needs after Hurricanes

Katrina and Rita --- Orleans and Jefferson Parishes, New Orleans area, Louisiana, October 17--22, 2005. *Morbidity and Mortality Weekly Report* **55**, 38-41.

North Pacific Fishery Management Council (NPFMC). 2009. Fishery Management Plan for fish resources of the Arctic Management Area. http://www.fakr.noaa.gov/npfmc/PDFdocuments/fmp/Arctic/ArcticFMP.pdf

North Pacific Fishery Management Council (NPFMC). 2011. Stock Assessment and Fishery Evaluation Report for King and Tanner Crab Fisheries of the Bering Sea and Aleutian Islands Region: Crab SAFE 2011. http://www.fakr.noaa.gov/npfmc/PDFdocuments/resources/SAFE/CrabSAFE/CrabSAFE2011.pdf

Norton, J.G., and J.E. Mason. 2003. Environmental influences on species composition of the commercial harvest of finfish and invertebrates off California. Calif. Coop. Oceanic Fish. Invest. Rep. 44:123-133.

Norton, J.G., and J.E. Mason. 2004. Locally and remotely forced environmental influences on California commercial fish and invertebrate landings. Calif. Coop. Oceanic Fish. Invest. Rep. 45:136-145.

Norton, J.G., and J.E. Mason. 2005. Relationship of California sardine (Sardinops sagax) abundance to climate-scale ecological changes in the California Current system. Calif. Coop. Oceanic Fish. Invest. Rep. 46:83-92.

Nuttal, M. & T.V. Callaghan, eds. 2000. *The Arctic: Environment, people, policy.* New York, NY: Taylor and Francis.

Nuttal, M. 2001. Indigenous peoples and climate change research in the Arctic. *Indigenous Affairs* 4(1):26-33.

Nuttal, M., F. Berkes, B. Forbes, G. Kofinas, T. Vlassova, and G. Wenzel. 2004. Hunting, Herding, Fishing and Gathering: Indigenous Peoples and Renewable Resources. *Impacts of a Warming Arctic: Arctic Climate Impact Assessment.* Carolyn Symon, Lelani Arris, and Bill Heal, eds. Pp.649-690. Cambridge: Cambridge University Press.

Nuttall, M., and T. V. Callaghan, eds. 2000. *The Arctic: Environment, People, Policy.* Amsterdam: Harwood Academic Publishers; Harris, Paul G. 2001. *The Environment, International Relations, and U.S. Foreign Policy.* Washington, DC: Georgetown University Press.

Nye, J., Link, J.S., Hare, J.A. & Overholtz, W.J. 2009. Changing spatial distributions of Northwest Atlantic Fish stocks in relation to temperature and stock size. *Marine Ecology Progress Series* **393**, 111-129.

Nye, J.A., Joyce, T.M., Kwon, Y. & Link, J.A. 2011. Silver hake tracks changes in Northwest Atlantic circulation. *Nature Communications* **2**, 412.

O'Connor, M.I. 2009. Warming strengthens an herbivore-plant interaction. *Ecology* **90**, 388-398.

O'Connor, M.I., Bruno, J.F., Gaines, S.D., Halpern, B.S., Lester, S.E., Kinlan, B.P. & Weiss, J.M. 2007. Temperature control of larval dispersal and the implications for marine ecology, evolution, and conservation. *Proceedings of the National Academy of Sciences USA* **104**, 1266-1271.

O'Connor, M.I., Piehler, M.F., Leech, D.M., Anton, A. & Bruno, J.F. 2009. Warming and resource availability shift food web structure and metabolism. *PLoS Biology* **7**, e1000178.

O'Corry-Crowe, G. 2008. Climate change and the molecular ecology of Arctic Mammals. Ecological Applications 18: S56-S76.

OECD. 2010. The Economics of Adapting Fisheries to Climate Change. OECD Publishing. Retrieved January 17, 2012, from http://www.oecd-ilibrary.org/agriculture-and-food/the-economics-of-adapting-fisheries-to-climate-change_9789264090415-en.

Office of Travel and Tourism Industries (OTTI). 2011a. Fast Facts: United States Travel and Tourism Industry 2010. June 2011, U.S. Department of Commerce, ITA, Office of Travel and

Tourism Industries, Washington, D.C. Retrieved February 6, 2012, from http://tinet.ita.doc.gov/outreachpages/inbound.general_information.inbound_overview.html.

Office of Travel and Tourism Industries (OTTI). 2011b. Overseas Visitation Estimates for U.S. States, Cities, and Census Regions: 2010. May 2011, U.S. Department of Commerce, ITA, Office of Travel and Tourism Industries, Washington, D.C. Retrieved February 6, 2012, from http://tinet.ita.doc.gov/outreachpages/inbound.general_information.inbound_overview.html.

Oliver, J. & Kaper, J. 2007. *Vibrio* species, In *Food microbiology: Fundamentals and Frontiers, 3rd Edition*. M.P. Doyle & L.R. Beuchat (eds). Washington, D.C.: ASM Press, 343-379.

Oliver, J. 1989. *Vibrio vulnificus*. In *Foodborne Bacterial Pathogens,* M.P. Doyle (ed.). New York, N.Y.: Marcel Dekker, 569-599.

Oliver, J. 2005. Wound infections caused by *Vibrio vulnificus* and other marine bacteria. *Epidemiology and Infection* **122**, 383-391.

Orensanz, J., D. Armstrong, P. Stabeno, and P. Livingston. 2004. Contraction of the geographic range of distribution of Snow Crab (*Chionoecetes opilio*) in the eastern Bering Sea-An environmental ratchet? CalCOFI Report 45, pp. 67-79.

Orr, J.C., Fabry, V.J., Aumont, O., Bopp, L., Doney, S.C., Feely, R.A., Gnanadesikan, A., Gruber, N., Ishida, A., Joos, F., Key, R.M., Lindsay, K., Maier-Reimer, E., Matear, R., Monfray, P., Mouchet, A., Najjar, R.G., Plattner, G.-K., Rodgers, K.B., Sabine, C.L., Sarmiento, J.L., Schlitzer, R., Slater, R.D., Totterdell, I.J., Weirig, M.-F., Yamanaka, Y. & Yool, A. 2005. Anthropogenic ocean acidification over the twenty-first century and its impact on calcifying organisms. *Nature* **437**, 681-686.

Osgood, K.E. (ed.). 2008. *Climate Impacts on U.S. Living Marine Resources: National Marine Fisheries Service Concerns, Activities and Needs*. U.S. Department of Commerce, NOAA Technical Memo. NMFS-F/SPO-89. 118 pg.

Ostreng, W. 2010. Science without boundaries: Interdisciplinarity in research, society, and politics. New York: University Press of America.

Ottersen, G, Kim, S., Huse, G., Polovina, J.J. & Stenseth, N.C. 2010. Major pathways by which climate may force marine fish populations. *Journal of Marine Systems* **79**, 343-360.

Ottersen, G., B. Planque, A. Belgrano, E. Post, P. Reid, et al. 2001. Ecological effects of the North Atlantic Oscillation. Oecologia 128: 1-14.

Overholtz, W.J., Hare, J.A. & Keith, C.M. 2011. Impacts of interannual environmental forcing and climate change on the distribution of Atlantic mackerel in the U.S. Northeast continental shelf. *Marine and Coastal Fisheries: Dynamics, Management and Ecosystem Science* **3**, 219-252.

Overland, J., Wood, K., & Wang, M. 2011. Warm Arctic - cold continents: climate impacts of the newly open Arctic Sea. *Polar Research* **30,** 15787.

Overland, J.E., Alheit, J., Bakun, A., Hurrell, J.W., Mackas, D.L. & Miller, A.J. 2010. Climate controls on marine ecosystems and fish populations. *Journal of Marine Systems* **79**, 305-315.

Owen, D. 2012. Critical habitat and the challenge of regulating small harms. *Florida Law Review* **64**, 141-200.

Pacific Fishery Management Council (PFMC). 1998. Amendment 8 (to the northern anchovy fishery management plan) incorporating a name change to: The Coastal Pelagic Species Fishery Management Plan. Pacific Fishery Management Council, Portland, OR.

Paerl HW, Huisman J. 2008. Blooms like it hot. *Science* 320:57-58; Paerl HW, Huisman J. 2009. Climate change: a catalyst for global expansion of harmful cyanobacterial blooms. *Environmental Microbiology Reports* 1(1): 27-37.

Paine, R.T. 1992. Food web analysis through field measurements of per capita interaction strength. *Nature* **355**, 73-75.

Palacios, D.M., Bograd, S.J., Mendelsshon, R. & Schwing, F.B. 2004: Long-term and seasonal trends in stratification in the California Current, 1950-1993. *Journal of Geophysical Research* **109**, C10016.

Pallab, M. E. Flugman and T. Randhir. Adaptation behavior in the face of global climate change: Survey responses from experts and decision makers serving the Florida Keys. *Ocean & Coastal Management* 54: 37-44.

Palm, H.W. 2011. Fish parasites as biological indicators in a changing world: can we monitor environmental impact and climate change? In *Progress in Parasitology*. H. Mehlhorn (ed.). Berlin: Springer Verlag, 223-250.

Palmer, C. T. 1991. Kin-selection, reciprocal altruism, and information sharing among Maine lobstermen. *Ethology and Sociobiology* 12(3):221-235.

Palmer, C.T. 1990. Telling the truth (up to a point): Radio communication among Maine lobstermen. *Human Organization* **49**,157-163.

Pandolfi, J.M., Connolly, S.R., Marshall, D.J. & Cohen, A.L. 2011. Projecting coral reef futures under global warming and ocean acidification. *Science* **333**, 418-422.

Papiez, C. 2009. Climate change implications for the Quileute and Hoh Tribes. Retrieved January 4, 2011, from http://nativecases.evergreen.edu/collection/cases/climate-change-implications.html.

Park, K., Kim, C.K., & Schroeder, W.W. 2007. Temporal variability in summertime bottom hypoxia in shallow areas of Mobile Bay, Alabama. *Estuaries and Coasts* **30**, 54-65.

Parmenter, R.R., Yadav, E.P., Parmenter, C.A., Ettestad, P. & Gage, K.L. 1999. Incidence of plague associated with increased winter-spring precipitation in New Mexico, USA. *American Journal of Tropical Medicine and Hygiene* **61**, 814-821.

Parmesan, C., & Yohe, G. 2003. A globally coherent fingerprint of climate change impacts across natural systems. *Nature* **421**, 37-42.

Paskal, C.. 2007. How climate change is pushing the boundaries of security and national policy. Briefing Paper, June. Energy, Environment, and Development Program, CC BP 07/01. P.11.

Paul, A., J. Paul, and K. Coyle. 1989. Energy sources for first-feeding zoeae of king crab *Paralithodes camtschatica* (Tilesius). Journal of Experimental Marine Biology and Ecology, 130, 55-69.

Pauly, D. 2010. Gasping fish and panting squids: oxygen, temperature and the growth of water-breathing animals, Vol. 22 in Excellence in Ecology Series (ed. Kinne, O.) (International Ecology Institute.

Paz, S., Bisharat, N., Paz, E., Kidar, O. and Cohen, D. 2007. Climate change and the emergence of *Vibrio vulnificus* disease in Israel. *Environmental Research* 103(3):390-396.

Peeters, F., Straile, D., Lorke, A. & Livingstone, D.M. 2007. Earlier onset of the spring phytoplankton bloom in lakes of the temperate zone in a warmer climate. *Global Change Biology* **13,** 1898-1909.

Peloso, M. 2010. *Adapting to Rising Sea Levels*. PhD Dissertation, Marine Science and Conservation Division, Nicholas School of the Environment, Duke University. 418 pp.

Pendleton, L., Donato, D.C., Murray, B., Crooks, S., Jenkins, W.A., Sifleet, S., Craft, C., Fourqurean, J.W., Kauffman, J.B.,Marbà, N., Megonigal, P., Pidgeon, E., Bilbao-Bastida, V., Ullman, R., Herr, D., Gordon D., and Baldera, A. (in review). Estimating global "blue carbon" emissions from conversion and degradation of coastal ecosystems. PlosOne (in review).

Pendleton, L.H. and R. Mendelsohn. 1998. Estimating the Economic Impact of Climate Change on the Freshwater Sportsfisheries of the Northeastern U.S. *Land Economics* 74(4):483-96.

Penland, R., Boniuk, M. and Wilhelmus, K. 2000. *Vibrio* ocular infections on the U.S. Gulf Coast. *Cornea* 19 (1): 26-9.

Peperzak, L. 2003. Climate change and harmful algal blooms in the North Sea. *Acta Oecologia* **24**, 139-144.

Pereira, H.M., Leadley, P.W., Proença, V., Alkemade, R., Scharlemann, J.P.W., Fernandez-Manjarrés, J.F., Araújo, M.B., Balvanera, P., Biggs, R., Cheung, W.W.L., Chini, L., Cooper, H.D., Gilman, E.L., Guénette, S., Hurtt, G.C., Huntington, H.P., Mace, G.M., Oberdorff, T., Revenga, C., Rodrigues, P., Scholes, R.J., Sumaila, U.R. & Walpole, M. 2010. Scenarios for global biodiversity in the 21st century. *Science* **330**, 1496-1501.

Perovich, D., Meier, W., Maslanik, J., & Richter-Menge, J. 2011. *Sea ice. Marine ecosystems study. Arctic report card: update for 2011.* Retrieved December 6, 21011 from http://www.arctic.noaa.gov/reportcard/sea_ice.html.

Perry, A., Low, P., Ellis, J. & Reynolds, J. 2005. Climate change and distribution shifts in marine fishes. *Science* **308**, 1912-1915.

Perry, R.W., and Quarantelli, E.L., eds. 2005. *What is a Disaster? New Answers to Old Questions.* International Research Committee on Disasters. Philadelphia: Xlibris Corporation.

Peterson, T.C, & Baringer, M.O. (eds.). 2009. State of the climate in 2008. Special Supplement to the Bulletin of the American Meteorological Society 90(8): S1-S196.

Petes, L.E., Menge, B.A. & Harris, A.L. 2008. Intertidal mussels exhibit energetic trade-offs between reproduction and stress resistance. *Ecological Monographs* **78**, 387-402.

Petes, L.E., Menge, B.A. & Murphy, G.D. 2007. Environmental stress reduces survival, growth, and reproduction in New Zealand intertidal mussels. *Journal of Experimental Marine Biology and Ecology* **351**, 83-91.

Petit, J.R., Jouzel, J., Raynaud, D., Barkov, N.I., Barnola, J.-M., Basile, I., Bender, M., Chappellaz, J., Davis, M., Delaygue, G., Delmotte, M., Kotlyakov, V.M., Legrand, M., Lipenkov, V.Y., Lorius, C., Pépin, L., Ritz, C., Saltzman, E. & Stievenard, M. 1999. Climate and atmospheric history of the past 420,000 years from the Vostok ice core, Antarctica. *Nature* **399**, 429-436.

Pfeffer, W. T., J. T. Harper, et al. (2008). "Kinematic Constraints on Glacier Contributions to 21st-Century Sea-Level Rise." Science **321**(5894): 1340-1343.

PFEL Cliamte and Marine Fisheries, Atmosphere-Ocean Climate Interactions, (2012) http://www.pfeg.noaa.gov/research/climatemarine/cmfoceanatm/cmfoceanatm.html (accessed in February 2012)

Philip, S. and G. J. van Oldenborgh (2006). "Shifts in ENSO coupling processes under global warming." Geophys. Res. Lett. **33**(11): L11704.

Philippart, C.J.M., van Aken, H.M., Beukema, J.J., Bos, O.G., Cadée, G.C. & Dekker, R. 2003. Climate-related changes in recruitment of the bivalve *Macoma balthica. Limnology and Oceanography* **48**, 2171-2185.

Phillips, B.D. 1993. Cultural diversity in disasters: Sheltering, housing, and long-term recovery. *International Journal of Mass Emergencies and Disasters* 11(1): 99-110.

Phillips, P., & Morrow, B.H. 2007. Social science research needs: Focus on vulnerable populations, forecasting, and warnings. *Natural Hazards Review* **8**, 61-68.

Pierre Friedlingstein, Laurent Bopp, Philippe Ciais (2001) Positive feedback between future climate change and the carbon cycle. Geophysical Research Letters, VOL. 28, NO. 8, PAGES 1543-1546.

Pimm, S. 2008. Biodiversity: climate change or habitat loss — which will kill more species? *Current Biology* **18**, R117-R119.

Pimm, S., Russell, G., Gittleman, J. & Brooks, T. 1995. The future of biodiversity. *Science* **269**, 347-350.

Pincebourde, S., Sanford, E. & Helmuth, B. 2008. Body temperature during low tide alters the feeding performance of a top intertidal predator. *Limnology and Oceanography* **53**, 1562-1573.

Pindyck, RS. 2002. Optimal timing problems in environmental economics. Journal of Economic Dynamics and Control 26: 1677-97.

Pitchon A. and K. Norman. Under review. Pier Fishing in Los Angeles County: Demographics, Behaviors and Risks. Journal of Environmental Management.

Place, S.P., Menge, B.A. & Hofmann, G.E. 2012. Transcriptome profiles link environmental variation and physiological response of *Mytilus californianus* between Pacific tides. *Functional Ecology* **2**, 144-155.

Place, S.P., O'Donnell, M.J. & Hofmann, G.E. 2008. Gene expression in the intertidal mussel *Mytilus californianus*: physiological response to environmental factors on a biogeographic scale. *Marine Ecology Progress Series* **356**, 1-14.

Pollnac, R. B. and J. J. Poggie, Jr. 1988. The Structure of Job Satisfaction among New England Fishermen and Its Application to Fisheries Management Policy. *American Anthropologist* 90(4): 888-901.

Pollnac, R. B., S. Abbott-Jamieson, C. Smith, M. L. Miller, P. M. Clay, and B. Oles. 2006. A model for fisheries social impact assessment. *Mar. Fish. Rev.* 68(1-4):1-18. http://spo.nmfs.noaa.gov/mfr681-4/mfr681-41.pdf

Poloczanska, E.S., Hawkins, S.J., Southward, A.J. & Burrows, M.T. 2008. Modeling the response of populations of competing species to climate change. *Ecology* **89**, 3138-3149.

Poloczanska, E.S., Limpus, C.J. & Hays, G.C. 2009. Vulnerability of marine turtles to climate change. *Advances in Marine Biology* **56**, 151-211.

Polovina, J., Uchida, I., Balazs, G., Howell, E.A., Parker, D. & Dutton, P. The Kuroshio Extension Bifurcation Region: a pelagic hotspot for juvenile loggerhead sea turtles. *Deep Sea Research II* **53**, 326-339.

Polovina, J.J. & Haight, W.R. 1999. Climate variation, ecosystem dynamics, and fisheries management in the Northwestern Hawaiian Islands. In *Ecosystem approaches for fisheries management*, Fairbanks, AK: Alaska Sea Grant College Program AK-SG-99-01, 23-32.

Polovina, J.J. 2005. Climate variation, regime shifts, and implications for sustainable fisheries. *Bulletin of Marine Science* **76**, 233-244.

Polovina, J.J. 2007. Decadal variation in the trans-Pacific migration of northern bluefin tuna (*Thunnus thynnus*) coherent with climate-induced change in prey abundance. *Fisheries Oceanography* **5**, 114-119.

Polovina, J.J., Dunne, J.P., Woodworth, P.A. & Howell, E.A. 2011. Projected expansion of the subtropical biome and contraction of the temperate and equatorial upwelling biomes in the North Pacific under global warming. *ICES Journal of Marine Science* **68**, 986-995.

Polovina, J.J., Howell, E.A. & Abecassis, M. 2008. Ocean's least productive waters are expanding. *Geophysical Research Letters* **35**, L03618.

Pörtner, H. O. & Knust, R. 2007. Climate change affects marine fishes through the oxygen limitation of thermal tolerance. Science 315, 95-97.

Pörtner, H.O. & Farrell, A.P. 2008. Physiology and climate change. *Science* **322**, 690-692.

Pörtner, H.O. 2008. Ecosystem effects of ocean acidification in times of ocean warming: a physiologist's view. *Marine Ecology Progress Series* **373**, 203-217.

Pörtner, H.O. 2010. Oxygen- and capacity-limitation of thermal tolerance: a matrix for integrating climate-related stressor effects in marine ecosystems. *The Journal of Experimental Biology* **213**, 881-893.

Pörtner, H.O., Bock, C., Knust, R., Lannig, G., Lucassen, M., Mark, F.C. & Sartoris, F.J. 2008. Cod and climate in a latitudinal cline: physiological analyses of climate effects in marine fishes. *Climate Research* **37**, 253-270.

Porzio, L., Buia, M.C. & Hall-Spencer, J.M. 2011. Effects of ocean acidification on macroalgal communities. *Journal of Experimental Marine Biology and Ecology* **400**, 278-287.

Powers, S.P., Peterson, C.H., Grabowski, J.H. & Lenihan, H.S. 2009. Success of constructed oyster reefs in no-harvest sanctuaries: implications for restoration. *Marine Ecology Progress Series* **389**, 159-170.

Pratchett, M.S., Munday, M.S., Wilson, S.K., Graham, N.A.J., Cinner, J.E., Bellwood, D.R., Jones, G.P., Polunin, N.V.C. & McClanahan, T.R. 2008. Effects of climate-induced coral bleaching on coral reef fishes: ecological and economic consequences. *Oceanography and Marine Biology: an Annual Review* **46**, 251-296.

Pratchett, M.S., Munday, P.L., Graham, N.A.J., Kronen, M., Pinca, S., Friedman, K., Brewer, T.D., Bell, J.D., Wilson, S.K., Cinner, J.E., Kinch, J.P., Lawton, R.J., Williams, A.J., Chapman, L., Magron F. & Webb, A. 2011. Vulnerability of coastal fisheries in the tropical Pacific to climate change. In *Vulnerability of Tropical Pacific Fisheries and Aquaculture to Climate Change*, J.D. Bell et al. (eds) Noumea, New Caledonia: Secretariat of the Pacific Community.

Rabalais, N.N., Turner, R.E., Diaz, R.J. & Justić, D. 2009. Global change and eutrophication of coastal waters. *ICES Journal of Marine Science* **66**, 1528-1537.

Rabalais, N.N., Turner, R.E., Sen Gupta, B.K., Boesch, D.F., Chapman, P., & Murrell, M.C. 2007. Hypoxia in the northern Gulf of Mexico: Does the science support the plan to reduce, mitigate and control hypoxia? *Estuaries and Coasts* **30**, 753-772.

Rahel, F.J., & Olden, J.D. 2008. Assessing the effects of climate change on aquatic invasive species. *Conservation Biology* **22**, 521-533.

Randall, D.A., Wood, R.A., Bony, S., Colman, R., Fichefet, T., Fyfe, J., Kattsov, V., Pitman, A., Shukla, J., Srinivasan, J., Stouffer, R.J., Sumi, A., & Taylor, K.E. 2007. Cilmate Models and Their Evaluation. In *Climate Change 2007: The Physical Science Basis. Contribution of Working Group I to the Fourth Assessment Report of the Intergovernmental Panel on Climate Change* S. Solomon et al. (eds). Cambridge UK and New York, NY: Cambridge University Press, 589-662.

Rasher, D.B., E.P. Stout, S. Engel, J. Kubanek, and M.E. Hay. 2011. Macroalgal terpenes function as allelopathic agents against reef corals. Proceedings of the National Academy of Science USA 108: 17726-17731.

Rattenbury, K., Kielland, K., Finstad, G. & Schneider, W. 2009. A reindeer herder's perspective on caribou, weather and socio-economic change on the Seward Peninsula, Alaska. *Polar Research* **28**, 71-88.

Rayner, N.A., Brohan, P., Parker, D.E., Folland, C.K., Kennedy, J.J., Vanicek, M., Ansell, T.J. & Tett, S.F.B. 2006. Improved analyses of changes and uncertainties in sea surface temperature measured in situ since the mid-nineteenth century: the HadSST2 dataset. *Journal of Climate* **19**, 446-469.

Reed, D.C., Rassweiler, A. & Arkema, K.K. 2008. Biomass rather than growth rate determines variation in net primary production by giant kelp. *Ecology* **89**, 2493-2505.

Reeder, D.B. & Chiu, C.S. 2010. Ocean acidification and its impact on ocean noise: Phenomenology and analysis. *The Journal of the Acoustical Society of America* **128**, EL137-EL143.

Regehr, E.V., Amstrup, S.C. & Stirling, I. 2006. *Polar bear population status in the southern Beaufort Sea*. Open-File Report 2006-1337, Reston, VA: U.S. Geological Survey.

Regehr, E.V., Hunter, C.M., Caswell, H., Armstrup, S.C. & Stirling, I. 2010. Survival and breeding of polar bears in the southern Beaufort Sea in relation to sea ice. *Journal of Animal Ecology* **79**, 117-127.

Reid, P.C., Johns, D.G., Edwards, M., Starr, M., Poulin, M. & Snoeijs, P. 2007. A biological consequence of reducing Arctic ice cover: arrival of the Pacific diatom *Neodenticula seminae* in the
North Atlantic for the first time in 800,000 years. *Global Change Biology* **13**, 1910-1921.

Reusch, T.B.H., A. Ehlers, A. Hämmerli, and B. Worm. 2005. Ecosystem recovery after climatic
extremes enhanced by genotypic diversity. Proceedings of the National Academy of Sciences
USA 102: 2826-2831.

Revelle, R. & Seuss, H.E. 1957. Carbon dioxide exchange between atmosphere and ocean and the
question of an increase of atmospheric CO_2 during past decades. *Tellus* **9**, 18-27.

Reynolds-Fleming, J.V. & Luettich, R.A. Jr 2004. Wind-driven lateral variability in a partially
mixed estuary. *Estuarine, Coastal and Shelf Science* **60**, 395-407.

Riebesell, U, Zondervan, I., Rost, B. Tortell, P.D., Zeebe, R.E. & Morel, F.M.M. 2000. Reduced
calcification of marine plankton in response to increased atmospheric CO_2. *Nature* **407**,
364-367

Riebesell, U., Schulz, K.G., Bellerby, R.G.J., Botros, M., Fritsche, P., Meyerhöfer, M., Neill,
C., Nondal, G., Oschlies, A., Wohlers, J. & Zöllner, E. 2007. Enhanced biological carbon
consumption in a high CO_2 ocean. *Nature* **450**, 545-548.

Ries, J., Cohen, A., & McCorkle, D. 2009. Marine calcifiers exhibit mixed responses to
CO_2-induced ocean acidification, *Geology* **37**(12), 1131-1134.

Ries, J.B. 2010. Review: geological and experimental evidence for secular variation in seawater
Mg/Ca (calcite-aragonite seas) and its effects on marine biological calcification. *Biogeosciences*
7, 2795-849

Rignot, E., I. Velicogna, et al. (2011). "Acceleration of the contribution of the Greenland and
Antarctic ice sheets to sea level rise." Geophys. Res. Lett. **38**(5): L05503.

Rind, D., Perlwitz, J., Lonergan, P. & Lerner, J. 2005. AO/NAO response to climate change: 2.
Relative importance of low- and high-latitude temperature changes, *Journal of Geophysical
Research* **110**, D12108.

Robinson, R.A., Learmouth, J.A., Hutson, A.M., Macleod, C.D., Sparks, T.H., Leech, D.I., Pierce,
G.J., Rehfisch, M.M. & Crick, H.Q.P. 2005. Climate change and migratory species. BTO
Research Report 414

Robinson, S. and the Gloucester Community Panel. 2003. A study of Gloucester's commercial
fishing infrastructure. Community Panels Project. Retrieved January 17, 2012, from http://
seagrant.mit.edu/cmss/comm_mtgs/commmtgs.html

Robinson, S. and the Gloucester Community Panel. 2005. Commercial fishing industry needs on
Gloucester Harbor, now and in the future. Community Panels Project. Retrieved January 17,
2012, from http://seagrant.mit.edu/cmss/comm_mtgs/commmtgs.html

Rodó, X., Pascual, M., Fuchs, G. & Faruque, A.S.G. 2002. ENSO and cholera: A nonstationary
link related to climate change? *Proceedings of the National Academy of Sciences* USA **99**,
12901-12906.

Rodriguez-Sanchez, R., D. Lluch-Belda, Villalobos, H., and Ortega-Garcia, S. 2002. Dynamic
geography of small pelagic fish populations in the California current ecosystem on the
regime time scale. Can. J. Fish.Aquat. Sci. 59(12):1980-1988.

Roemmich, D., McGowan, J., 1995. Climatic Warming and the Decline of Zooplankton in the
California Current. Science 267, 1324-1326.

Roff, D.A. 1992. *Evolution of life histories: theory and analysis.* New York, NY, USA: Chapman and
Hall,

Roleda, M.Y., Morris, J.N., McGraw, C.M. & Hurd, C.L. 2012. Ocean acidification and seaweed
reproduction: increased CO_2 ameliorates the negative effect of lowered pH on meiospore

germination in the giant kelp *Macrocystis pyrifera* (Laminariales, Phaeophyceae). *Global Change Biology* **18**, 854-864.

Rosenkranz, G.E., A. Tyler, G. Kruse, and J. Niebauer. 1998. Relationship between winds and year-class strength of Tanner crabs in the Southeastern Bering Sea. Alaska Fisheries Research Bulletin, 5, 18-24.

Rosenkranz, G.E., A. Tyler, G. Kruse. 2001. Effects of water temperature and wind on recruitment of Tanner crabs in Bristol Bay, Alaska. Fisheries Oceanography, 10, 1-12.

Rotstein, D.S., Burdett, LG W McLellan, L Schwacke, T Rowles, K A Terio, S Schultz, and A Pabst. 2009. Lobomycosis in Offshore Bottlenose Dolphins (Tursiops truncatus), North Carolina. *Emerging Infectious Diseases* **15**, 588-590.

Rubinstein, D. 2001. A Sociocultural Study of Pelagic Fishing Activities in Guam. Final progress report available from University of Hawaii Joint Institute for Marine and Atmospheric Research, Pelagic Fisheries Research Program. Retrieved January 6, 2012 from http://www.soest.hawaii.edu/PFRP/pdf/rubinstein01.pdf.

Ruhl, J.B. 2010. Climate change adaptation and the structural transformation of environmental law. *Environmental Law* **40**, 363-431.

Ruiz, G.M., Fofonoff, P., Hines, A.H. & Grosholz, E.D. 1999. Non-indigenous species as stressors in estuarine and marine communities: assessing invasion impacts and interactions. *Limnology and Oceanography* **44**, 950-972.

Ruiz, G.M., Fofonoff, P.W., Carlton, J.T., Wonham, M.J. & Hines, A.H. 2000. Invasion of coastal marine communities in North America: apparent patterns, processes, and biases. *Annual Review of Ecology and Systematics* **31**, 481-531.

Runge, J.A., Kovach, A.I., Churchill, J.H., Kerr, L.A., Morrison, J.R., Beardsley, R.C., Berlinsky, D.L., Chen, C., Cadrin, S.X., Davis, C.S., Ford, K.H., Grabowski, J.H., Howell, W.H., Ji, R., Jones, R.J., Pershing, A.J., Record, N.R., Thomas, A.C., Sherwood, G.D., Tallack, S.M.L. & Townsend, D.W. 2010. Understanding climate impacts on recruitment and spatial dynamics of Atlantic cod in the Gulf of Maine: integration of observations and modeling. *Progress in Oceanography* **87**, 251-263.

Russell, B.D., Harley, C.D.G., Wernberg, T., Mieszkowska, N., Widdicombe, S., Hall-Spencer, J.M. & Connell, S.D. 2011. Predicting ecosystem shifts requires new approaches that integrate the effects of climate change across entire systems. *Biology Letters* doi: 10.1098/rsbl.2011.0779.

Rust, M.B. 2002. Chapter 7: Nutritional Physiology. In *Fish Nutrition, 3rd ed*. R.W. Hardy & J. H. Halver (eds.). San Diego, CA: Elsevier Science, 367-452.

Rust, M.B., Barrows, F.T., Hardy, R.W., Lazur, A., Naughten, K. & Silverstein, J. 2010. The Future of Aquafeeds. Draft report to the NOAA/USDA Alternative Feeds Initiative. 103 pp.

Rykaczewski, R.R., & Dunne, J.P. (2010) Enhanced nutrient supply to the California Current Ecosystem with global warming and increased stratification in an earth system model. Geophysical Research Letters, 37

Sabater, M. 2007. Prehistoric, historic, and current reef fish utilization and the consequence of human population grown on fish density, biomass, and community composition in American Samoa. Biological Report Series No: 2007-01. Department of Marine and Wildlife Resources, American Samoa.

Sabine, C.L., Feely, R.A., Gruber, N., Key, R.M., Lee, K., Bullister, J.L., Wanninkhof, R., Wong, C.S., Wallace, D.W.R., Tilbrook, B., Millero, F.J., Peng, T.-S., Kozyr, A., Ono, T., & Rios, A.F. 2004. The oceanic sink for anthropogenic CO_2. *Science* **305**, 367-371.

Sagarin, R.D. & Gaines, S.D. 2002. The "abundant centre" distribution: to what extent is it a biogeographical rule? *Ecology Letters* **5**, 137-147.

OCEANS AND MARINE RESOURCES IN A CHANGING CLIMATE

Sagarin, R.D., J.P. Barry, S.E. Gilman, and C.H. Baxter. 1999. Climate-related change in an intertidal community over short and long time scales. Ecological Monographs 69(4): 465-490.

Sala, E., & Knowlton, N. 2006. Global marine biodiversity trends. *Annual Review of Environment
and Resources* **31**, 93-122.

Salibury J, Green M, Hunt C, Campbell J. (2008). Coastal acidification by rivers: a new threat to
shellfish? EOS Transactions, AGU, 89, 513.

Sandin, S.A., J.E. Smith, E.E. DeMartini, E.A. Dinsdale, S.D. Donner, et al. 2008. Baselines and
degradation of coral reefs in the Northern Line Islands. PLoS ONE 3: e1548.

Sanford, E. & Kelly, M.W. 2011. Local adaptation in marine invertebrates. *Annual Review of
Marine Science* **3**, 509-535.

Sanford, E. 1999. Regulation of keystone predation by small changes in ocean temperature.
Science **283**, 2095-2097.

Sarch, M.T., & Allison, E.H. 2000. Fluctuating fisheries in Africa's inland waters: well adapted
livelihoods, maladapted management. In *Proceedings of the 10th international conference of the
Institute of Fisheries Economics and Trade, Corvallis, Oregon, July 9-14, 2000*

Sarmiento, J. L. et al. 2004. Response of ocean ecosystems to climate warming. Glob. Biogeochem.
Cycles 18, GB3003.

Sarmiento, J.L., Gruber, N., Brzezinski, M.A., & Dunne, J.P., 2004. High-latitude controls of thermocline nutrients and low latitude biological productivity*Nature.* **427**, 56-60.

Sartoris, F.J., Bock, C., Serendero, I., Lannig, G. & Pörtner, H.O. 2003. Temperature-dependent
changes in energy metabolism, intracellular pH and blood oxygen tension in the Atlantic
cod. *Journal of Fish Biology* **62**, 1239-1253.

Scallan, E., Hoekstra, R., Angulo, F., Tauxe, R., Widdowson, M-A., Roy, S., Jones, J. & Griffin, P.
2011. Foodborne Illness Acquired in the United States—Major Pathogens. *Emerging Infectious
Diseases* **17**, 7-15.

Scheffer, M. & Carpenter, S.R. 2003. Catastrophic regime shifts in ecosystems: linking theory to
observation. *Trends in Ecology and Evolution* **18**, 648-656.

Scheffer, M., Bascompte, J., Brock, W.A., Brovkin, V., Carpenter, S.R., Dakos, V., Held, H., van
Nes, E.H., Rietkerk, M. & Sugihara, G. 2009. Early-warning signals for critical transitions.
Nature **461**, 53-59.

Scheffer, M., Carpenter, S., Foley, J., Folke, C. & Walker, B. 2001. Catastrophic shifts in ecosystems. *Nature* **413**, 591-596.

Schewe, J., Levermann, A., & Meinshausen, M. 2010 Climate change under a scenario near 1.5C
of global warming: monsoon intensification, ocean warming and steric sea level rise. *Earth
System Dynamics Discussions,* **1**, 297-324.

Schiedek, D., Sundelin, B., Readman, J.W. & Macdonald, R.W. 2007. Interactions between climate
change and contaminants. *Marine Pollution Bulletin* **54**, 1845-1856.

Schiel, D.R., Steinbeck, J.R. & Foster, M.S. 2004. Ten years of induced ocean warming causes
comprehensive changes in marine benthic communities. *Ecology* **85**, 1833-1839.

Schindler, D.E., Hilborn, R., Chasco, B., Boatright, C.P., Quinn, T.P., Rogers, L.A. & Webster, M.S.
2010. Population diversity and the portfolio effect in an exploited species. *Nature* **465**, 609-613.

Schmidt, P.S., Serrao, E.A., Pearson, G.A., Riginos, C., Rawson, P.D., Hilbish, T.J., Brawley, S.H.,
Trussell, G.C., Carrington, E., Wethey, D.S., Grahame, J.W., Bonhomme, F. & Rand, D.M.
2008. Ecological genetics in the North Atlantic: Environmental gradients and adaptation at
specific loci. *Ecology* **89**, S91-S107.

Schmittner, A. 2005. Decline of the marine ecosystem caused by a reduction in the Atlantic overturning circulation. *Nature* **434**, 628-633.

Schneider, K.R., Van Thiel, L.E. & Helmuth, B. 2010. Interactive effects of food availability and aerial body temperature on the survival of two intertidal *Mytilus* species. *Journal of Thermal Biology* **35**, 161-166.

Scholin, C.A., F. Gulland, G.J. Doucette, S. Benson, M. Busman, et al. 2000. Mortality of sea lions along the central California coast linked to toxic diatom bloom. Nature 403: 80-84.

Scott, D., McBoyle, G. & Schwartzentruber, M. 2004. Climate change and the distribution of climatic resources for tourism in North America. *Climate Research* **27**, 105-117.

Screen, J.A. & Simmonds, I., 2010. The central role of diminishing sea ice in recent Arctic temperature amplification. *Nature* **464**, 1334-1337.

Sechena, R., S.Liao, R.Lorenzana, C.Nakano, N.Polissar and R.Fenske, 2003. Asian and Pacific Islander Seafood Consumption – a community based study in King County Washington. *Journal of Exposure Analysis and Environmental Epidemiology* 13, 256-266.

Segner, H. 2011. Moving beyond a descriptive aquatic toxicology: the value of biological process and trait information. *Aquatic Toxicology* **105**, 50-55.

Selkoe, K.A., B.S. Halpern, and R.J. Toonen. 2008. Evaluating anthropogenic threats to the Northwestern Hawaiian Islands. Aquatic Conservation 18: 1149-1165.

Selkoe, K.A., B.S. Halpern, C.M. Ebert, E.C. Franklin, E.R. Selig, K.S. Casey, J. Bruno, and R.J. Toonen. 2009. A map of human impacts to a 'pristine' coral reef ecosystem, the Papahānaumokuākea Marine National Monument. Coral Reefs 28: 635-650.

Semenza, J., McCullough, J., Flanders, W., & McGeehin, M. 1999. *Excess hospital admissions during the July 1995 heat wave in Chicago. American Journal of Preventive Medicine* **16**, 269-277.

Sepez, J. 2001.Political and Social Ecology of Contemporary Makah Subsistence Hunting, Fishing, and Shellfish Collecting Practices. University of Washington.

Serreze, M. & Barry, R., 2011. Processes and impacts of Arctic Amplification. *Global and Planetary Change* **77**, 85-96.

Serreze, M. & Francis, J. 2006. The arctic amplification debate. *Climatic Change* **76**, 241-264.

Shepard, C.J., I.H. Pike, and S.M. Barlow. 2005. Sustainable feed resources of marine origin. European Aquaculture Society Special Publication No 35, pp59-66.

Shillinger G., Palacios, D.L., Bailey, H. & Bograd, S. 2008. Persistent leatherback turtle migrations present opportunities for conservation. *PLoS Biology* **6**, e171.

Shivlani, M. 2009. Examination of non-fishery factors on the welfare of fishing communities in the Florida Keys: A focus on the cumulative effects of trade, economic, energy, and aid policies, macroeconomic (county and regional) conditions and coastal development on the Monroe County commercial fishing industry. MARFIN Grant NA05NMF4331079.

Shuntov, V.P., and O.S. Temnykh. 2009. Current status and tendencies in the dynamics of biota of the Bering Sea macroecosystem. N. Pac. Anadr. Fish Comm. Bull. 5: 332-331.

Siegel, D.A., Doney, S.C., and Yoder, J.A., *Science.* **296,** 730-733 (2002).

Sigler, M.F., Renner, M., Danielson, S.L., Eisner, L.B., Lauth, R.R., Kuletz, K.J., Logerwell, E.A. & Hunt, G.L. Jr. 2011. Fluxes, fins, and feathers: relationships among the Bering, Chukchi, and Beaufort Seas in a time of climate change. *Oceanography* **24**, 250-265.

Silverman, J., Lazar, B., Cao, L., Caldiera, K. & Erez, J. 2009. Coral reefs may start dissolving when atmospheric CO_2 doubles. *Geophysical Research Letters* **36**, L05606.

Simmonds, M. & Eliott, W.J. 2009. Climate change and cetaceans: Concerns and recent developments. *Journal of the Marine Biological Association of the UK* **89**, 203-210.

Simmonds, M.P., & Isaac, S.J. 2007 The impacts of climate change on marine mammals: early signs of significant problems. *Oryx* **41**, 19-26.

Singer, M.1994. AIDS and the Health Crisis of the U.S. Urban Poor: The Perspective of Critical Medical Anthropology. *Social Science and Medicine* 39(7): 931-948.

Skirrow, G. & Whitfield, M. 1975. The effect of increases in the atmospheric carbon dioxide content on the carbonate ion concentration of surface ocean water at 25°C. *Limnology and Oceanography* **20**, 103-108.

Slenning, B.D. 2010. Global climate change and implications for disease emergence. *Veterinary Pathology Online* **47**, 28-33.

Smith Travel Research (STR). 2008. *Hotel lodging and financial survey data.*

Smith, C. L. and P. M. Clay. 2010. Measuring Subjective and Objective Well-being: Analyses from Five Marine Commercial Fisheries. *Human Organization* 69(2): 158-168.

Smith, C.R., Grange, L.J., Honig, D.L., Naudts, L., Huber, B., Guidi, L. & Domack, E. 2011. A large population of king crabs in Palmer Deep on the west Antarctic Peninsula shelf and potential invasive impacts. *Proceedings of the Royal Society B* **279**, 1017-1026.

Smith, J.E., Shaw, M., Edwards, R.A., Obura, D., Pantos, O. Sala, E., Sandin, S.A., Smriga, S., Hatay, M. & Rohwer, F.L. 2006. Indirect effects of algae on coral: algae-mediated, microbe-induced coral mortality. *Ecology Letters* **9**, 835-845.

Smith, J.R., Fong, P. & Ambrose, R.F. 2006. Dramatic declines in mussel bed community diversity. *Ecology* **87**, 1153-1161.

Snyder, M. A., L. C. Sloan, et al. (2003). "Future climate change and upwelling in the California Current." Geophys. Res. Lett. **30**(15): 1823.

Soden, B.J. & Held, I.M. 2006. An assessment of climate feedbacks in coupled ocean atmosphere models. *Journal of Climate* **19**, 3354-3360.

Sokolova, I.M. & Lannig, G. 2008. Interactive effects of metal pollution and temperature on metabolism in aquatic ectotherms: implications of global climate change. *Climate Research* **37**, 181-201.

Solomon S., Qin D., Manning M., Chen Z., Marquis M., Averyt K.B., Tignor M., Miller H.L. *2007, (eds.), Climate Change 2007: The Physical Science Basis: Contribution of Working Group I to the Fourth Assessment Report of the Intergovernmental Panel on Climate Change: Cambridge, UK Cambridge University Press 996 p.*

Somero, G.N. 2011. Comparative physiology: a "crystal ball" for predicting consequences of global change. *American Journal of Physiology – Regulatory, Integrative, and Comparative Physiology* **301**, R1-R14.

Sorte, C.J.B., Jones, S.J. & Miller, L.P. 2011. Geographic variation in temperature tolerance as an indicator of potential population responses to climate change. *Journal of Experimental Marine Biology and Ecology* **400**, 209-217.

Sorte, C.J.B., Williams, S.L. & Carlton, J.T. 2010b. Marine range shifts and species introductions: comparative spread rates and community impacts. *Global Ecology and Biogeography* **19**, 303-316.

Sorte, C.J.B., Williams, S.L. & Zerebecki, R.A.. 2010a. Ocean warming increases threat of invasive species in a marine fouling community. *Ecology* **91**, 2198-2204.

Southward, A.J., Langmead, O., Hardman-Mountford, N.J., Aitken, J., Boalch, G.T., Dando, P.R., Genner, M.J., Joint, I., Kendall, M.A., Halliday, N.C., Harris, R.P., Leaper, R., Mieszkowska, N., Pingree, R.D., Richardson, A.J., Sims, D.W., Smith, T., Waine, A.W., Hawkins, S.J.2005. Long-term oceanographic and ecological research in the western English Channel. *Advances in Marine Biology* **47**, 1-105.

Springer, A.M., Byrd, G.V. & Iverson, S.J. 2007. Hot oceanography: Planktivorous seabirds reveal ecosystem responses to warming of the Bering Sea. *Marine Ecology Progress Series* **352**, 289-297.

St. Martin, K. & Hall-Arber, M. 2008. The missing layer: Geo-technologies, communities, and implications for marine spatial planning, *Marine Policy* **32**, 779-786.

Stachowicz, J., Bruno, J. & Duffy, J. 2007. Understanding the effects of marine biodiversity on communities and ecosystems. *Annual Review of Ecology, Evolution, and Systematics* **38**, 739-766.

Stachowicz, J.J., Fried, H., Osman, R.W. & Whitlatch, R.B. 2002a. Biodiversity, invasion resistance, and marine ecosystem function: reconciling pattern and process. *Ecology* **83**, 2575-2590.

Stachowicz, J.J., Terwin, J.R., Whitlatch, R.B. & Osman, R.W. 2002b. Linking climate change and biological invasions: ocean warming facilitates nonindigenous species invasions. *Proceedings of the National Academy of Sciences USA* **99**, 15497-15500.

Stafford, K.M., Moore, S.E. & Spillane, M. 2007. Gray whale calls recorded near Barrow, Alaska, throughout the winter of 2003-04. *Arctic* **60**, 167-172.

State of Hawai'i Department of Business, Economic Development and Tourism (DBEDT). 2008. *Visitor Statistics.*

Stearns, S.C. 1992. *The evolution of life histories.* Oxford, U.K.: Oxford University Press.

Steinacher, M., Joos, F., Frölicher, T. L., Plattner, G.-K. & Doney, S. C. 2009. Imminent ocean acidification in the Arctic projected with the NCAR global coupled carbon cycle-climate model, *Biogeosciences* 6, 515-533.

Steinacher, M., Joos, F., Frölicher, T.L., Bopp, L., Cadule, P., Cocco, V., Doney, S.C., Gehlen, M., Lindsay, K., Moore, J.K., Schneider, B. & Segschneider, J. 2010. Projected 21st century decrease in marine productivity: a multi-model analysis. *Biogeosciences* **7**, 979-1005.

Steinback, S., K. Wallmo and P. M. Clay.2007. Assessing Subsistence On A Regional Scale For The Northeast U.S. Presented at the American Anthropological Association annual meeting. November. Washington, DC.

Steinback, S., Wallmo, K. & Clay, P. M. 2009. Saltwater sport fishing for food or income in the Northeastern U.S.: Statistical estimates and policy implications. *Marine Policy* **33**, 49-57.

Stenseth, N.C., Mysterud, A., Ottersen, G., Hurrell, J.W., Chan, K.-S. & Lima, M. 2002. Ecological effects of climate fluctuations. *Science* **297**, 1292-1296.

Stephens, G. L., et al. (2008), CloudSat mission: Performance and early science after the first year of operation, *Journal of Geophysical Research*, 113, D00A18.

Stephens, G. L., et al. (2008), CloudSat mission: Performance and early science after the first year of operation, J. Geophys. Res., 113, D00A18, doi:10.1029/2008JD009982.

Stevens, B. 1990. Temperature-dependent growth of juvenile red king crab (*Paralithodes camtschatica*), and its effects on size-at-age and subsequent recruitment in the eastern Bering Sea. Canadian Journal of Fisheries and Aquatic Sciences, 47, 1307-1317.

Stewart, E.J., Howell, S.E.L., Draper, D., Yackel, J. & Tivy, A. 2007. Sea ice in Canada's Arctic: Implications for cruise tourism. *Arctic* **60**, 370-380.

Stirling, I. 2005. Reproductive rates of ringed seals and survival of pups in northwestern Hudson Bay, Canada, 1991-2000. Polar Biology 28: 381-387.

Stirling, I., & Parkinson, C.L. 2006. Possible effects of climate warming on selected populations of polar bears (*Ursus maritimus*) in the Canadian Arctic. *Arctic* **59**, 261-275.

Stirling, I., Lunn, N.J. & Iacozza, J. 1999. Long-term trends in the population ecology of polar bears in western Hudson Bay in relation to climate change. *Arctic* **52**, 294-306.

Stirling, I., N.J. Lunn, and J. Iacozza. 1999. Long-term trends in the population ecology of polar bears in western Hudson Bay in relation to climate change. Arctic 52: 294-306.

Stock, C.A., Alexander, M.A., Bond, N.A., Brander, K.M., Cheung, W.W.L., Curchitser, E.N., Delworth, T.L., Dunne, J.P., Griffies, S.M., Haltuch, M.A., Hare, J.A., Hollowed, A.B., Lehodey, P., Levin, S.A., Link, J.S., Rose, K.A., Rykaczewski, R.R., Sarmiento, J.L., Stouffer,

R.J., Schwing, F.B., Vecchi, G.A., & Werner, F.E. 2011. On the use of IPCC-class models to assess the impact of climate on Living Marine Resources. *Progress in Oceanography*, **88**, 1-27.

Stockholm Convention 2005. Ridding the World of Pops: A Guide to the Stockholm Convention on Persistent Organic Pollutants. United Nations Environment Programme (UNEP), Geneva, Switzerland.

Storck, C.H., D. Postic, I. Lamaury, and J.M. Perez. 2008. Changes in epidemiology of Lepto-spirosis in 2003-2004, a two El Niño Southern Oscillation period, Guadeloupe archipelago, French West Indies. Epidemiology and Infection 136: 1407-1415.

Stouffer, R.J., Yin, J., Gregory, J.M., Dixon, K.W., Spelman, M.J., Hurlin, W., Weaver, A.J., Eby, M., Flato, G.M., Hasumi, H., Hu, A., Jungclaus, J.H., Kamenkovich, I.V., Levermann, A., Motoya, M., Murakami, S., Nawrath, S., Oka, A., Peltier, W.R., Robitaille, D.Y., Sokolov, A., Vettoretta, G. & Weber, S.L. 2006. Investigating the causes of the response of the thermohaline circula-tion to past and future climate changes. *Journal of Climate* **19**, 1365-1387.

Stram, D.L., and Evans, D.C.K. 2009. Fishery management responses to climate change in the North Pacific. *ICES Journal of Marine Science* **66**, 1633-1639.

Stroeve, J., Serreze, M., Drobot, S., Gearheard, S., Holland, M., Maslanik, J., Meier, W. & Scambos, T. 2008. Arctic Sea Ice Extent Plummets in 2007. *Eos Transactions of the American Geophysical Union* **89**, 13..

Stroeve, J.C., Serreze, M.C., Holland, M.M., Kay, J.E., Meier, W., & Barrett, A.P. 2011. The Arctic's rapidly shrinking sea ice cover: A research synthesis. *Climatic Change* doi: 10.1007/s10584-011-0101-1.

Stumm, W. and Morgan, J.J. (1970). *Aquatic chemistry*, New York, NY: Wiley..

Sturges, W. 1974. Sea level slope along continental boundaries. *Journal of Geophysical Research* **79**, 825-830.

Su, N.-J., Sun, C.-L., Punt, A.E., Yeh, S.-Z. & DiNardo, G. 2011. Modeling the impacts of environ-mental variation on the distribution of blue marlin, *Makaira nigricans*, in the Pacific Ocean. *ICES Journal of Marine Science* **68**, 1072-1080.

Sumaila, U.R., Cheung, W., Lam, V.W.Y., Pauly, D. & Herrick, S. 2011. Climate change impacts on the biophysics and economics of world fisheries. *Nature Climate Change* **1**, 449-456.

Sun, J., Hutchins, D.A., Feng, Y., Seubert, E.L., Caron, D.A. & Fu, F.-X. 2011. Effects of changing pCO_2 and phosphate availability on domoic acid production and physiology of the marine harmful bloom diatom *Pseudo-nitzschia multiseries*. *Limnology and Oceanography* **56**, 829-840.

Sussman, E., Major, D.C., Deming, R., Esterman, P.R., Fadil, A., Fisher, A., Fucci, F., Gordon, R., Harris, C., Healy, J.K., Howe, C., Robb, K. & Smith, J. 2010. Climate change adapta-tion: fostering progress through law and regulation. *New York University Environmental Law Journal* **18**, 55-155.

Sweet, W. V., C. Zervas, and S. Gill, 2009: Elevated East Coast sea level anomaly: July-July 2009. Tech. Rep. NOS CO-OPS 051, NOAA, 30 pp.

Sydeman, W., & Thompson, S.A. 2010. The California Current integrated ecosystem assessment (IEA) Module II: Trends and variability in climate-ecosystem state. Final Report to NOAA/NMFS/Environmental Research Division

Sydeman, W.J., Mills, K., Santora, J. & Thompson, S.A. 2009. *Seabirds and climate in the California Current – a synthesis of change.* CalCOFI Report Vol. 50.

Tacon, A.G.J. and Metian, M. 2009. Fishing for feed or fishing for food: Increasing global competi-tion for small pelagic forage fish. Ambio, 38(6):294-302.

Tacon, A.G.J., Hasan, M.R., & Metian, M. 2011. *Demand and supply of feed ingredients for farmed fish and crustaceans: trends and prospects.* FAO Fisheries and Aquaculture Technical Paper No. 564.

Takasuka, A., Oozeki, Y. & Aoki, I. 2007. Optimal growth temperature hypothesis: why do anchovy flourish and sardine collapse or vice versa under the same ocean regime? *Canadian Journal of Fisheries and Aquatic Sciences* **64**, 768-776.

Talmage, S.C., & Gobler, C.J. 2011. Effects of elevated temperature and carbon dioxide on the growth and survival of larvae and juveniles of three species of Northwest Atlantic bivalves. *PLoS One* **6**, e26941.

Tao, Z., Bullard, S. and Arias C. 2011. High Numbers of *Vibrio* vulnificus in Tar Balls Collected from Oiled Areas of the North-Central Gulf of Mexico Following the 2010 BP Deepwater Horizon Oil Spill. *Ecohealth* Epub ahead of print.

Tatters, A.O., Fu, F.-X. & Hutchins, D.A. In press. High CO_2 and silicate limitation synergistically increase the toxicity of a harmful bloom diatom. PLoS ONE.

Taylor, D.I., Nixon, S.W., Granger, S.L., Buckley, B.A., McMahon, J.P. & Lin, H.J. 1995. Responses of coastal lagoon plant communities to different forms of nutrient enrichment: A mesocosm experiment. *Aquatic Botany* 52: 19-34.

Taylor, S.J. 2008. Climate warming causes phenological shift in pink salmon, Oncorhynchus gorbuscha, behavior at Auke Creek, Alaska. Global Change in Biology 14, 229-235. Teichberg, M., Fox, S.E., Aguila, C., Olsen, Y.S. & Valiela, I. 2008. Macroalgal responses to experimental nutrient enrichment in shallow coastal waters: growth, internal nutrient pools, and isotopic signatures. *Marine Ecology Progress Series* **368**, 117-126.

Thienkrua, W., Cardozo BL, Chakkraband ML, Guadamuz TE, Pengjuntr W, Tantipiwatanaskul P, Sakornsatian S, Ekassawin S, Panyayong B, Varangrat A, Tappero JW, Schreiber M, van Griensven F. 2006. Thailand post-tsunami mental health study group. *JAMA* 296(5):549-559.

Thomas, K., J.T. Harvey, T. Goldstein, J. Barakos, and F. Gulland. 2010. Movement, dive behavior, and survival of California sea lions (*Zalophus californianus*) posttreatment for domoic acid toxicosis. Marine Mammal Science 26: 36-52.

Thomas, P. O. and K. L. Laidre. 2011. Biodiversity- Cetaceans and Pinnipeds (Whales and Seals). http://www.arctic.noaa.gov/reportcard/biodiv_whales_walrus.html

Thornber, C.S., DiMilla, P., Nixon, S.W. & McKinney, R.A. 2008. Natural and anthropogenic nitrogen uptake by bloom-forming macroalgae. *Marine Pollution Bulletin* **56**, 261-269.

Tierney, K.J., Lindell, M.K., and Perry, R.W., eds. 2001. *Facing the Unexpected: Disaster Preparedness and Response in the United States.* Washington, DC: Joseph Henry Press

Tomanek, L. & Zuzow, M.J. 2010. The proteomic response of the mussel congeners *Mytilus galloprovincialis* and *M. trossulus* to acute heat stress: implications for thermal tolerance limits and metabolic costs of thermal stress. *The Journal of Experimental Biology* **213**, 3559-3574.

Tomanek, L. 2011. Environmental proteomics: changes in the proteome of marine organisms in response to environmental stress, pollutants, infection, symbiosis, and development. *Annual Review of Marine Sciences* **3**, 373-399.

Torres de la Riva, G., C. Kreuder Johnson, F.M.D. Gulland, G.W. Langlois, J.E. Heyning, T.K. Rowles, and J.A.K. Mazet. 2009. Association of an unusual marine mammal mortality event with *Pseudo-nitzschia spp.* blooms along the southern California coastline. Journal of Wildlife Diseases 45: 109-121.

Torrisson, O., Olsen, R.E., Toresen, R., Hemre, G.I., Tacon, A.G.J., Asche, F., Hardy, R.W. & Lall, S. 2011. Atlantic Salmon (Salmo salar): The "Super-Chicken" of the Sea? *Reviews in Fisheries Science* **19**, 257-278.

Toth, J. F. Jr. and R. B. Brown. 1997. Racial and gender meanings of why people participate in recreational fishing. *Leisure Sciences* 19(2):129-146.

Townsend, D.W., Keller, M.D., Sieracki, M.E., and Ackleson, S.G., *Nature.* **360**, 59-62 (1992).

Trenberth K.E. & Fasullo, J. 2007: Water and energy bud- gets of hurricanes and implications for cli- mate change. *Journal of Geophysical Research* **112**, D23107.

Trenberth KE, et al. Intergovernmental Panel on Climate Change (IPCC). 2007. Climate Change 2007: The Physical Science Basis. Solomon, S., D. Qin, M. Manning, Z. Chen, M. Marquis, K.B. Averyt, M. Tignor and H.L. Miller (eds.) Cambridge, United Kingdom: Cambridge University Press.

Trenberth, K. E., Fasullo, J. & Smith, L. Trends and variability in column-integrated atmospheric water vapor. *Clim. Dyn.*, **24:** 741-758 (2005).

Trenberth, K.E. Changes in precipitation with climate change. *Climate Research* **47**, 123-138 (2011).

Trenberth, K.E., Dai, A., Rasmussen, R.M., & Parsons, D.B. 2003 The changing character of precipitation. *Bulletin of the American Meteorological Society* **84,**1205-1217.

Trenberth, K.E., Fasullo, J., & Kiehl, J. 2009. Earth's global energy budget. *Bulletin of the American Meteorological Society* **90**, 311-323

Trenberth, K.E., Jones, P.D., Ambenje, P., Bojariu, R., Easterling, D., Klein Tank, A. Parker, D, Rahimzadeh, F., Renwick, J.A., Rusticucci, M., Soden B. & Zhai, P. . 2007: Observations: Surface and Atmospheric Climate Change. In *Climate Change 2007: The Physical Science Basis. Contribution of Working Group I to the Fourth Assessment Report of the Intergovernmental Panel on Climate Change* S. Solomon, et al. (eds.). Cambridge, United Kingdom and New York, NY, USA: Cambridge University Press.

Trussell, G.C. & Etter, R.J. 2001. Integrating genetic and environmental forces that shape the evolution of geographic variation in a marine snail. *Genetica* **112-113**, 321-337.

Tuler, S, J Agyeman, P Pinto da Silva, K Roth LoRusso and R Kay. Improving the social sustain- ability of fisheries management by assessing stakeholder vulnerability, Special section on Vulnerability and resilience in the fisheries, P Pinto da Silva and M Hall-Arber (Guest Editors), Human Ecology Review; 2008 15(2):171-184. [Accessed August, 20, 2009]

Turchini, G.M., D.S. Francis, R.S.J. Keast, and A.J. Sinclair. 2011. Transforming salmonid aqua- culture from a consumer to a producer of long chain omega-3 fatty acids. Food Chemistry 124:609-614.

U.S. Census Bur., Washington, DC

U.S. Climate Change Science Program (CCSP). 2008. Preliminary review of adaptation options for climate-sensitive ecosystems and resources. In *A report by the U.S. Climate Change Science Program and the Subcommittee on Global Change Research*, S.H. Julius, S.H. et al. (eds.) Wash- ington, D.C.: U.S. Environmental Protection Agency .

U.S. Climate Change Science Program (CCSP). 2008: Weather and Climate Extremes in a Changing Climate. Regions of Focus: North America, Hawaii, Caribbean, and U.S. Pacific Islands. A Report by the U.S. Climate Change Science Program and the Subcom- mittee on Global Change Research. Thomas R. Karl, Gerald A. Meehl, Christopher D. Miller, Susan J. Hassol, Anne M. Waple, and William L. Murray (eds.). Department of Commerce, NOAA's National Climatic Data Center, Washington, D.C., USA, 164 pp.

U.S. Commission on Ocean Policy. 2004. *An Ocean Blueprint for the 21ˢᵗ Century. Final Report.* Washington, DC. ISBN#0–9759462–0–X

U.S. Environmental Protection Agency (EPA). 2008. *Effects of Climate Change for Aquatic Invasive Species and Implications for Management and Research.* Washington, D.C.: National Center for Environmental Assessment .

U.S. Environmental Protection Agency (EPA). 2010. Memorandum on integrated reporting and listing decisions related to ocean acidification, from Denise Keehner, Director, to Water Divi- sion Directors, Regions 1-10. Nov. 15, 2010.

U.S. Food and Drug Administration. 2008. *Enhanced Aquaculture and Seafood Inspection - Report to Congress.* Washington, D.C.: U.S. Food and Drug Administration.

U.S. Government Accountability Office (GAO). 1998. Food Safety: Federal Effort to Ensure the Safety of Imported Food Are Inconsistent and Unreliable. Washington, DC: Government Accounting Office. pp. 2-3.

U.S. Government Accountability Office (GAO). 2004. *Food Safety: FDA's Imported Seafood Safety Program Shows Some Progress, but Further Improvements are Needed. Report GAO 246.* Washington, DC: GAO.

U.S. Government Accountability Office (GAO). 2007. *Climate change: agencies should develop guidance for addressing the effects on Federal land and water resources.* Report to Congressional requesters. GAO-07-863.

U.S. Travel Association. 2011. U.S. Travel Answer Sheet. July 2011. U.S. Travel Association, Washington, D.C. Retrieved January 20, 2012 from http://www.ustravel.org.

U.S. Travel Association. 2012. Travel Industry Facts: In Advance of the President's Speech at Walt Disney World. January 18, 2012. U.S. Travel Association, Washington, D.C. Retrieved January 19, 2012, from http://www.ustravel.org/news/press-releases/travel-industry-facts-advance-president percentE2 percent80 percent99s-speech-walt-disney-world.

Udovydchenkov, I.A., Duda, T.F., Doney, S.C., & Lima, I.D., 2010. Modeling deep ocean shipping noise in varying acidity conditions. *The Journal of the Acoustical Society of America* **128**, EL130-EL136.

Ulbrich, U., Pinto, J., Kupfer, H., Leckebusch, G., Spangehl, T. & Reyers, M. 2008. Changing northern hemisphere storm tracks in an ensemble of IPCC climate change simulations. *Journal of Climate* **21**, 1669-1679.

Umana, E. 2003. Erysipelothrix rhusiopathiae: an unusual pathogen of infective endocarditis. International *Journal of Cardiology* 88: 297-299.

UNFCCC SBSTA report on Research and observation (2011) - FCCC/SBSTA/2011/L.27

United Nations World Tourism Organization (UNWTO). 2012. UN World Tourism Barometer. January 2012 - Statistical Annex, Volume 10, Publications Unit, World Tourism Organization, Madrid, Spain – First Printing 2012 (version 20/01/12)

United Nations. 2009. Report and recommendations from a workshop on "Climate change and Arctic sustainable development: Scientific, social, cultural and educational challenges," Monte Carlo, Monaco, March 3-6, 2009.

United States Fisheries and Wildlife Service (USFWS). 2011. Marking, Tagging and Reporting Program data bases for northern sea otter, Pacific Walrus and polar bear. Office of Marine Mammals Management. Anchorage, Alaska. Data compiled by Alaska Fisheries Information Network for Alaska Fisheries Science Center, Seattle.

Urian, A.G., Hatle, J.D. & Gilg, M.R.. 2011. Thermal constraints for range expansion of the invasive green mussel, *Perna viridis*, in the Southeastern United States. *Journal of Experimental Zoology* **315**, 12-21.

Valiela, I., McClelland, J., Hauxwell, J., Behr, P.J., Hersch, D. & Foreman, K. 1997. Macroalgal blooms in shallow estuaries: controls and ecophysiological and ecosystem consequences. *Limnology and Oceanography* **42**, 1105-1118.

Van den Hove, S., Menestrel, M.L. & de Bettignies, H-C. 2002. The oil industry and climate change: strategies and ethical dilemmas. *Climate Policy* **2**, 3-18.

van Griensven, F., Chakkraband, M.L., Thienkrua, W., Pengjuntr, W., Lopes Cardozo, B., Tantipiwatanaskul, P., Mock, P.A., Ekassawin, S., Varangrat, A., Gotway, C., Sabin, M., & Tappero,

J.W. 2006. Thailand post-tsunami mental health study group. *Journal of the American Medical Association* **296**, 537-548.

Van Houtan, K.S., & Halley, J.M.. 2011. Long-term climate forcing in loggerhead sea turtle nesting. *PLoS ONE* **6**, e19043.

Vandenbosch, R. Effects of ENSO and PDO events on seabird populations as revealed by Christmas bird count data. 2000. Waterbirds 23: 416-422.

Vaquer-Sunyer, R. & Duarte, C.M. 2008. Thresholds of hypoxia for marine biodiversity. *Proceedings of the National Academy of Sciences USA* **105**, 15452-15457.

Vecchi, G. A. and Soden, B. J. Increased tropical Atlantic wind shear in model projections of global warming. Geophys. Res. Lett. 34. L08702 (2007)

Vedros, N.A., A.W. Smith, J. Schonewald, G. Migaki, and R.C. Hubbard. 1971. Leptospirosis epizootic among California sea lions. Science 172: 1250-1251.

Vellinga, M. & Wu, P. 2004. Low-latitude freshwater influence on centennial variability of the Atlantic thermohaline circulation. *Journal of Climate* **17**, 4498-4511.

Veltre, D. W. and M. J. Veltre. 1983. *Resource Utilization in Atka, Aleutian Islands, Alaska.* Alaska Department of Fish and Game, Division of Subsistence, Technical Paper No. 88.

Verdant Power. 2012. The Roosevelt Island Tidal Energy Project. Retrieved January 9, 2012 from http://verdantpower.com/what-initiative/.

Vermeij, G.J. & Roopnarine, P.D. 2008. The coming Arctic invasion. *Science* **321**, 780-781.

Vining, I. and J. Zheng. 2004. Status of king crab stocks in the eastern Bering Sea in 2003. Alaska Department of Fish and Game, Division of Commercial Fisheries, Regional Information Report 4K03-03, Kodiak, AK. 22 pp.

Wagner, J.D.M., Cole, J.E. Beck, J.W., Patchett, P.J., Henderson, G.M. & Barnett, H.R.2010. Moisture variability in the southwestern United States linked to abrupt glacial climate change. Nature Geoscience **3**, 110-113.

Wake, C., Burakowski, L., Lines, G., McKenzie, K. & Huntington, T. 2006. Cross border indicators of climate change over the past century: Northeastern United States and Canadian Maritime Region. The Climate Change Task Force of the Gulf of Maine Council on the Marine Environment in cooperation with Environment Canada and Clean Air-Cool Planet. http://www.gulfofmaine.org/council/publications/cross-border-indicators-of-climate-change.pdf

Walker, N., Leben, R.R. & Balasubramanian, S. 2005. Hurricane forced upwelling and chlorophyll a enhancement within cold core cyclones in the Gulf of Mexico. *Geophysical Research Letters* **32,** L18610.

Walsh, J.E., & Chapman, W.L. 2001. 20th-century sea-ice variations from observational data. *Annals of Glaciology* **33**, 444-448.

Walther, K., Anger, K. & Pörtner, H.O. 2010. Effects of ocean acidification and warming on larval development of the spider crab *Hyas araneus* from different latitudes (54° vs. 79°N). *Marine Ecology Progress Series* **417**, 159-170.

Walther, K., Sartoris, F.J., Bock, C. & Pörtner, H.O. 2009. Impact of anthropogenic ocean acidification on thermal tolerance of the spider crab *Hyas araneus.* Biogeosciences **6**, 2207-2215.

Wang, F., Jiang, L., Yang, Q., Han, F., Chen, S., Pu, S., Vance, A. and Ge, B. 2011. Prevalence and antimicrobial susceptibility of major foodborne pathogens in imported seafood. *Journal of Food Protection* 74(9):1451-1461.

Wang, M., Overland, J.E., & Bond, N.A., 2010. Climate projections for selected large marine ecosystems. *Journal of Marine Systems* **79**, 258-266.

Ward, J.R., & Lafferty, K.D. 2004. The elusive baseline of marine disease: are diseases in ocean ecosystems increasing? *Plos Biology* **2**, 542-547.

Warzybok, P.M., and R.W. Bradley. 2010. Status of seabirds on Southeast Farallon Island during the 2010 breeding season. Unpublished report to the U.S. Fish and Wildlife Service. PRBO Conservation Science, Petaluma, C.A. PRBO Contribution Number 1769.

Wassmann, P. 2011. Arctic marine ecosystems an era of rapid change. *Progress in Oceanography* **90**, 1-17.

Wassmann, P., Duarte, C.M., Agustí, S. & Sejr, M.K. 2011. Footprints of climate change in the Arctic marine ecosystem. *Global Change Biology* **17**, 1235–1249.

Waycott, M., Duarte, C.M., Carruthers, T.J.B., Orth, R.J., Dennison, W.C., Olyarnik, S., Calladine, A., Fourqurean, J.W., Heck, K.L. Jr, Hughes, A.R., Kendrick, G.A., Kenworthy, W.J., Short, F.T. & Williams, S.L. 2009. Accelerating loss of seagrasses across the globe threatens coastal ecosystems. *Proceedings of the National Academy of Sciences USA* **106**,12377-12381.

Webster, P., Holland, G., Curry, J. and Chang, H-R (2005) Changes in Tropical Cyclone Number, Duration, and Intensity in a Warming Environment. Science 309:1844-1846.

Webster, P., Holland, G., Curry, J., and Chang, H. 2005. Changes in tropical cyclone number, duration, and intensity in a warming environment. *Science* 309(5742):1844-1846.

Weis, K., Hammond, R., Hutchinson, R., & Blackmore, C. 2011. *Vibrio* illness in Florida, 1998-2007. *Epidemiology and Infection* **139**, 591-598.

Wendler, G, and M Shulski. 2009. "A century of climate change for Fairbanks, AK." Arctic 62 (3): 295-300.

Wernberg, T., Russell, B.D., Moore, P.J., Ling, S.D., Smale, D.A., Campbell, A., Coleman, M.A., Steinberg, P.D., Kendrick, G.A. & Connell, S.D. 2011a. Impacts of climate change in a global hotspot for temperate marine biodiversity and ocean warming. *Journal of Experimental Marine Biology and Ecology* **400**, 7-16.

Wernberg, T., Thomsen, M.S., Tuya, F., & Kendrick, G.A. 2011b. Biogenic habitat structure of seaweeds change along a latitudinal gradient in ocean temperature. *Journal of Experimental Marine Biology and Ecology* **400**, 264-271.

Wernberg, T., Thomsen, M.S., Tuya, F., Kendrick, G.A., Staehr, P.A. & Toohey, B.D. 2010. Decreasing resilience of kelp beds along a latitudinal temperature gradient: potential implications for a warmer future. *Ecology Letters* **13**, 685-694.

Westlund, L., Poulain, F., Bage, H., & van Anrooy, R. 2007. *Disaster response and risk management in the fisheries sector.* Rome: FAO.

Wethey, D.S. 2002. Biogeography, competition, and microclimate: the barnacle *Chthamalus fragilis* in New England. *Integrative and Comparative Biology* **42**, 872-880.

Wethey, D.S., & Woodin, S.A. 2008. Ecological hindcasting of biogeographic responses to climate change in the European intertidal zone. *Hydrobiologia* **606**, 139-151.

Wethey, D.S., Woodin, S.A., Hilbish, T.J., Jones, S.J., Lima, F.P. & Brannock, P.M. 2011. Response of intertidal populations to climate: effects of extreme events versus long term change. *Journal of Experimental Marine Biology and Ecology* **400**, 132-144.

White House Council on Environmental Quality, Final Report of the Ocean Policy Task Force (July 19, 2010), available at http://www.whitehouse.gov/files/documents/ OPTF_FinalRecs.pdf

White, C., Halpern, B.S. & Kappel, C.V. 2012. Ecosystem service tradeoff analysis reveals the value of marine spatial planning for multiple ocean uses. *Proceedings of the National Academy of Sciences USA.* **109**, 4696-4701.

White, D.M., Gerlach, S.C., Loring, P.A. & Tidwell, A. 2007. Food and water security in a changing arctic climate. *Environmental Research Letters* **2**, 4.

Widdicombe, S., & Spicer, J.I.. 2008. Predicting the impact of ocean acidification on benthic biodiversity: What can animal physiology tell us? *Journal of Experimental Marine Biology and Ecology* **366**, 187-197.

Wijffels, S. and P. Durack (2009). "Global and regional 50 year ocean salinity trends." IOP Conference Series: Earth and Environmental Science **6**(3): 032004.

Wilcox, B.A., & Gubler, D.J. 2005. Disease Ecology and the Global Emergence of Zoonotic Pathogen. *Environmental Health and Preventive Medicine* **10**, 263-272,

Williams, J., & Jackson, S. 2007. Novel climates, no-analog communities, and ecological surprises. *Frontiers in Ecology and the Environment* **5**, 475-482.

Wilson, S., & Fischetti, T. 2010. Coastline population trends in the United States: 1960 to 2008. In *Current Population Reports*,Washington, D.C.: US Census Bureau.

Wilson, W.J. & Ormseth, O.A. 2009. A new management plan for the Arctic waters of the United States. *Fisheries* **34**, 555-558

Winder, M., & Schindler, D.E., 2004. Climate change uncouples trophic interactions in an aquatic ecosystem. *Ecology* **85**, 2100-2106.

Wingfield, J.C., & Sapolsky, R.M. 2003. Reproduction and resistance to stress: when and how. *Journal of Neuroendocrinology* **15**, 711-724.

Wirl F. 2006. Consequences of irreversibilities on optimal intertemporal CO_2 emission policies under uncertainty. Resource and Energy Economics 28(2): 105-123.

Wisner, B., Blaikie, P., Cannon, T., and Davis, I. 2003. *At Risk: Natural Hazards, People's Vulnerability, and Disasters*. Second Edition. London: Routledge.

Witherington, B.E. & Ehrhart, L.M. 1989. Hypothermic stunning and mortality of marine turtles in the Indian River Lagoon System, Florida. *Copeia* **1989** 696-703.

Wolf, S.G., Snyder, M.A., Sydeman, W.J., Doak, D.F. & Croll, D.A. 2010. Predicting population consequences of ocean climate change for an ecosystem sentinel, the seabird Cassin's auklet. *Global Change Biology* **16**, 1923-1935.

Wolf, S.G., Sydeman, *W.J.*, Hipfner, J.M., Abraham, C.L. Tershy, *B.R.* & Croll, D.A. 2009. Range-wide reproductive consequences of ocean climate variability for the seabird Cassin's auklet. *Ecology* **90**, 742-753.

Wolfe, R.J. 2004. Local Traditions and Subsistence: A Synopsis of Twenty Five Years of Research by the State of Alaska.

Wood R.A. Collins, M., Gregory, J., Harris, G. & Vellinga, M. 2006. Towards a risk assessment for shutdown of the atlantic thermohaline circulation. In *Avoiding Dangerous Climate Change* , H.-J. Schellnhuber et al. (eds.). Cambridge, U.K.: Cambridge University Press

Wood R.A., Vellinga, M, & Thorpe, R.B. 2003. Global warming and thermohaline circulation stability. *Philosophical Transactions of the Royal Society A* **361**, 1961-1975

Wootton, J.T., Pfister, C.A. & Forester, J.D. 2008. Dynamic patterns and ecological impacts of declining ocean pH in a high-resolution multi-year dataset. *Proceedings of the National Academy of Sciences USA* **105**, 18848-18853.

Worm B., Barbier, E.B., Beaumont, N., Duffy, J.E., Folke, C., Halpern, B.S., Jackson, J.B.C., Lotze, H.K., Micheli, F., Palumbi, S.R., Sala, E., Selkoe, K.A., Stachowicz, J.J. & Watson, R. 2006. Impacts of biodiversity loss on ocean ecosystem services. *Science* **314**, 787-790.

www.aoml.noaa.gov/phod/floridacurrent

Yamane, L., & Gilman, S.E. 2009. Opposite responses by an intertidal predator to increasing aquatic and aerial temperatures. *Marine Ecology Progress Series* **393**, 27-36.

Year of the Oceans 1998. *Year of the Oceans – Coastal Tourism*. National Oceanic & Atmospheric Administration (NOAA), U.S. Department of Commerce. http://www.yoto98.noaa.gov/yoto/meeting/tour_rec_316.html (Accessed January 19, 2012).

Yin, J., S. M. Griffies, et al. (2010). "Spatial Variability of Sea Level Rise in Twenty-First Century Projections." Journal of Climate **23**(17): 4585-4607.

Young, I.R., S. Zeiger, and A.V. Babanin (2011) Global trends in wind speed and wave height. Retrieved March 24, 2011, from www.scienceexpress.org.

Yu, G., Schwartz, Z. & Walsh, J.E. 2009. A weather-resolving index for assessing the impact of climate change on tourism related climate resources. *Climatic Change* **95**, 551-573.

Zacherl, D., Gaines, S.D. & Lonhart, S.I. 2003. The limits to biogeographical distributions: insights from the northward range extension of the marine snail, *Kelletia kelletii* (Forbes, 1852). *Journal of Biogeography* **30**, 913-924.

Zachos, J.C., Röhl, U., Schellenberg, S.A., Sluijs, A., Hodell, D.A., Kelly, D.C., Thomas, E., Nicolo, M., Raffi, I., Lourens, L.J., McCarren, H. & Krron, D. 2005. Rapid acidification of the ocean during the Paleocene-Eocene thermal maximum. *Science* **308**, 1611-1615.

Zardi, G., Nicastro, K., McQuaid, C.D., Hancke, J. & Helmuth, B. 2011. The combination of selection and dispersal helps explain genetic structure in intertidal mussels. *Oecologia* **165**, 947-958.

Zhang J, Lindsay R, Steele M, Schweiger A. (2008). What drove the dramatic retreat of Arctic sea ice during summer 2007? Geophys Res Lett 35:L11505.

Zheng, J. and G. Kruse. 2000. Recruitment patterns of Alaskan crabs and relationships to decadal shifts in climate and physical oceanography. ICES Journal of Marine Science, 57, 438-451.

Zheng, J. and G. Kruse. 2003. Stock-recruitment relationships for three major Alaskan crab stocks. Fisheries Research, 65, 103-121.

Zheng, J. and G. Kruse. 2006. Recruitment variation of eastern Bering Sea crabs: climate-forcing or topdown effects? Progress in Oceanography, 68, 184-204.

Ziemann, D., L. Conquest, M. Olaizola, and P. Bienfang. 1991. Interannual variability in the spring phytoplankton bloom in Auke Bay, Alaska. Marine Biology, 109, 321-334.

Zippay, M.L., & Hofmann, G.E. 2010. Effect of pH on gene expression and thermal tolerance of early life history stages of red abalone (*Haliotis rufescens*). *Journal of Shellfish Research* **29**, 429-439.

Zuerner, R.L, C.E. Cameron, S. Raverty, J. Robinson, K.M. Colegrove, et al. 2009. Geographical dissemination of *Leptospira interrogans* serovar Pomona during seasonal migration of California sea lions. Veterinary Microbiology 137: 105-110.